Seed Germination and Micropropagation of Ornamental Plants

Seed Germination and Micropropagation of Ornamental Plants

Editors

Konstantinos Bertsouklis
Epaminondas Kartsonas
Angela Carra

MDPI

Basel • Beijing • Wuhan • Barcelona • Belgrade • Novi Sad • Cluj • Manchester

Editors
Konstantinos Bertsouklis
Agricultural University of Athens
Athens
Greece

Epaminondas Kartsonas
University of Peloponnese
Kalamata
Greece

Angela Carra
National Research Council of Italy
Palermo
Italy

Editorial Office
MDPI AG
Grosspeteranlage 5
4052 Basel, Switzerland

This is a reprint of articles from the Special Issue published online in the open access journal *Horticulturae* (ISSN 2311-7524) (available at: https://www.mdpi.com/journal/horticulturae/special_issues/Seed_Germination_Micropropagation).

For citation purposes, cite each article independently as indicated on the article page online and as indicated below:

Lastname, A.A.; Lastname, B.B. Article Title. *Journal Name* **Year**, *Volume Number*, Page Range.

ISBN 978-3-7258-1525-8 (Hbk)
ISBN 978-3-7258-1526-5 (PDF)
doi.org/10.3390/books978-3-7258-1526-5

Contents

About the Editors

Konstantinos Bertsouklis

Dr Konstantinos Bertsouklis is a full-time assistant professor at the Laboratory of Floriculture and Landscape Architecture AUA (LFLA). He graduated from the Department of Crop Production Science of AUA in 1993; he completed (in 2000) his postgraduate studies, specializing in Floriculture (LFLA). His doctoral thesis was presented at the LFLA (in 2012)—specializing in micropropagation, seed ecophysiology, and phylogenetic studies with morphological and molecular markers. From 1993 to 2005, he worked as a manager of Urban Green in municipal districts (Department of Urban Green, Municipality of Piraeus). His teaching experience relates to lector and laboratory courses at undergraduate and postgraduate levels. His research interests focus on native species and their use as ornamental plants; on species whose populations are endangered by human activity or other factors contributing to their conservation; tissue proliferation culture; the study of soil mixtures with native species; vegetation management; the use of native species in archaeological sites; the ecophysiology of seeds; the control of the development of floricultural species; and phylogenetic studies . He has published more than 45 scientific articles since 2003. He is particularly interested in endangered mediterranean species.

Epaminondas Kartsonas

Dr Epaminondas Kartsonas is a full associate professor at the Department of Agriculture at University of Peloponnese. He graduated from the Department of Crop Production Science of AUA in 1996; he completed (in 2000) his postgraduate studies, specializing in Floriculture (LFLA). His doctoral thesis was presented at the LFLA (in 2007)—specializing in micropropagation of endemic and endangered plant species that can be used as ornamentals. From 1997 to 2003, he worked as a Greenworks manufacturer and landscape architect. Since 2003, he has worked as a lecturer, assistant professor, and associate professor at the Department of Agriculture at University of Peloponnese. His teaching experience relates to lector and laboratory courses at undergraduate and postgraduate levels. He teaches the following topics: cognitive objects of ornamental plants, plant propagation techniques, and plant anatomy and morphology. His research interests focus on native species of Greece and how to introduce them as ornamental plants, and the study of endangered species contributing to their conservation, with a focus on how to propagate them with biofriendly techniques. He has published many scientific articles (total: 23) since 1999.

Angela Carra

Dr Angela Carra is a permanent researcher at the Institute of Bioscience and BioResources of Italian National Council, CNR Palermo (IBBR-PA). She graduated from the Department of Evolutionary and Functional Biology, Plant Section, University of Parma in 1997. Her doctoral thesis was discussed in 2003 and on the topic of cytokinin-like activity of diphenylurea derivatives. Since 2003, she has worked as a researcher at CNR, working on plant physiology and biotechnology, with an emphasis on germplasm conservation using new technologies, in vitro culture, genetic and sanitary improvements of Mediterranean crops, and the development of sanitation protocols through somatic embryogenesis. Furthermore, she developed protocols for the in vitro conservation of endangered native species of Sicily. She has published more than 30 scientific articles since 2001.

horticulturae

MDPI

Editorial

Seed Germination and Micropropagation of Ornamental Plants

Konstantinos Bertsouklis [1,*], Epaminondas Kartsonas [2] and Angela Carra [3,*]

1 Laboratory of Floriculture and Landscape Architecture, Department of Crop Science, School of Plant Sciences, Agricultural University of Athens, Iera Odos 75, 11855 Athens, Greece

2 Department of Agriculture, University of Peloponnese, 24100 Kalamata, Greece; e.kartsonas@uop.gr

3 National Research Council of Italy (CNR), Institute of Biosciences and BioResources (IBBR), Via Ugo La Malfa 153, I-90146 Palermo, Italy

* Correspondence: kber@aua.gr (K.B.); angela.carra@cnr.it (A.C.)

1. Introduction

Ornamental species can improve the environmental conditions in cities, serving the European Union's biodiversity strategy to bring nature back into our lives [1,2]. They are used both in interior and exterior landscaping, as trees, shrubs, and annual and perennial species. Furthermore, they can be cultivated as pot and foliage plants, while the cut flower industry is an important agriculture sector [3]. Alongside widespread ornamental species, native species can also be introduced for use as ornamentals in the floriculture industry and landscape architecture [4,5]. In recent decades, the reproduction and use of native plants have been increased; these species have reduced water and cultivation requirements, being adapted to drought conditions caused by climate change, greatly supporting biodiversity and local pollinators [6]. Moreover, several native species populations are under anthropogenic pressure; therefore, it is crucial to develop conservation strategies using biotechnological tools. There is a high demand for ornamental plants; the global ornamental plants market is expected to reach USD 45.07 billion by the end of 2029 [7]. New ornamental plants can be produced by conventional methods or by improved micropropagation techniques. Micropropagation is an efficient tool used widely for in vitro conservation and research, concerning the production of plant clones with desirable characteristics for crosses to improve wild species for potential use as ornamental species [8]. Moreover, micropropagation is a tool that enhances the availability of necessary propagating material through the year. It requires a small laboratory working space, and facilitates the transfer of disease-free propagating material worldwide. The employment of proper nutrient media, plant growth hormones, and other physical or chemical derivatives is necessary for the establishment of efficient micropropagation protocols. The present Special Issue focuses on seed-plant material and plant tissue culture techniques. The aim was to provide a comprehensive overview of the latest achievements in new ornamental plants that could be introduced in landscaping and floriculture.

2. Overview of Published Articles

Ahmad et al. (contribution 1) performed research on *Lagerstroemia speciosa*, a valuable ornamental tree blooming in beautiful purple flowers, native to China and cultivated across other tropical regions, e.g., in Bangladesh, India, Malaysia, Thailand, the Philippines, Indonesia, and Japan. The species is widely used due to its ethnomedicinal properties, although it is unexploited. Moreover, *L. speciosa* could be extinct at national and international levels without the implementation of suitable conservation strategies. The study aimed to establish an efficient in vitro propagation method starting from the nodal explants of sprouting shoots. The study revealed that a reproducible two-stage in vitro protocol could include both the use of thidiazuron and a secondary Murashige and Skoog medium (MS) supplemented with different cytokinins. An ex vitro rooting method of in vitro

Citation: Bertsouklis, K.; Kartsonas, E.; Carra, A. Seed Germination and Micropropagation of Ornamental Plants. *Horticulturae* **2024**, *10*, 541. https://doi.org/10.3390/horticulturae10060541

Received: 13 May 2024
Accepted: 13 May 2024
Published: 22 May 2024

micro shoots was employed. Furthermore, the study examined possible somaclonal variation using molecular methods, and the genetic uniformity of the micropropagated plants was recorded.

Light quality is significant for many physiological and biochemical processes of plants; light-emitting diodes (LEDs) are widely used, with the aim of improving the regeneration process. Aslami et al. (contribution 2) examined, for the first time, the possibility of establishing an improved African violet regeneration method using leaf cuttings. The use of red and blue LEDs was effective on shoot growth, root growth, and physiological traits. Optimized LED treatments could enhance the regeneration rate of African violets.

Dyckia brevifolia (contribution 3) is an endangered Bromeliad of high ornamental value. It has a remarkable efficiency in both its tolerance to full sunlight and to large quantities of water floods. The plant is under high human pressure; the authors suggested suitable in vitro regeneration methods using solid and liquid MS media, starting from seedlings grown in vitro. The study defined the cardinal temperatures for in vitro germination in the range from 15 °C to 30 °C. Two different types of explants, lateral shoot explants and leaf explants, were used, with shoot explants achieving high multiplication rates in liquid media supplemented with cytokinins. The plantlets regenerated in vitro were successfully acclimatized.

In the next study (contribution 4), the authors investigated a suitable exploitation protocol of the endangered species *Sideritis raeseri* subsp. *Attica*. The species can be introduced as an ornamental plant into the floriculture industry, demonstrating suitability for rocky places. Furthermore, the genus *Sideritis* is well known for its medicinal properties. Hence, an effective micropropagation method could serve both conservation and exploitation strategies of *S. raeseri* subsp. *Attica*. The authors investigated the in vitro seed germination of the species and used seedlings grown in vitro as the starting plant material. The germination rate did not exceed 37% at 25 °C; on the other hand, a producible in vitro protocol was established, using MS medium and low concentrations of cytokinins. Half-strength MS was a suitable medium for in vitro rooting with median efficiency. Hyperhydrated shoots were formed during the multiplication cycles, but the use of hormone-free MS eliminated the phenomenon of hyperhydricity. The acclimatization was successful for all the rooted plantlets, indicating practical aspects for future endeavors.

The subsequent paper (contribution 5) aimed to develop, for the first time, an in vitro regeneration method via the somatic embryogenesis of different *Sanseveria* genotypes, using flower explants. *Sanseveria* presents limited sexual propagation, and the development of new variants is a challenge for modern floriculture. The authors reported that a remarkable regeneration method was obtained on MS media with various plant growth regulator combinations, reaching up to 73.3%. Furthermore, the study highlighted ploidy variations amongst the different *Sansevieria* spp. clones.

Nevertheless, effective rooting systems of stem-cuttings are of high importance for floriculture. Hence, Darras et al. (contribution 6) determined the effect of UV-C irradiation on the growth and development of plant propagation material of *Pelargonium* × *hortorum*, a widespread ornamental species. Both physiological and morphological changes were recorded during the experimentation in a greenhouse. The use of UV irradiation could promote root biomass accumulation depending on the species. In this contribution, UV-C irradiation enhanced the rooting percentage of *Pelargonium* × *hortorum*, inducing endogenous ethylene production. Furthermore, the flower size and the total flower number were increased.

Another species used in interior landscapes is *Philodendron bipinnatifidum*; the plant faces acclimatization difficulties in commercial micropropagation systems, resulting in a 10–40% plant loss. Contribution 7 examined the effect of compatible arbuscular mycorrhiza fungi (AMFs), recording morphological physiological responses. AMFs provide protection against biotic and abiotic stresses, adsorbing nutrients to the plants and improving plant growth and development. It is suggested that the AMF colonization of micropropagated plants following ex vitro transplantation is efficient for plant development. The study revealed that there was a positive AMF effect during acclimatization.

Ioannidis and Koropouli (contribution 8) employed different nutrient media with varying strengths in their study to examine the in vitro growth and organogenesis of three different individuals of *C. creticus* plants. Additionally, they used simple sequence repeats (SSRs) as molecular markers for the genetic assessment of the micropropagated plantlets. Driver and Kuniyaki Walnut medium (DKW), MS, and woody plant medium (WPM) were proved efficient in different biometric characteristics. SSR markers revealed that there was a narrow genetic base amongst the micropropagated *C. sativa* clones.

The study by Kim et al. (contribution 9), on the dormancy types and germination characteristics of *Berberis koreana* Palibin, concluded that seeds exhibit an intermediate physiological seed dormancy. Seeds germinated both in light and dark conditions; cold stratification treatment at 5 °C for 12 weeks was extremely effective in breaking their dormancy. These data are essential for improving the use of seed plant material, which has the effect of reducing the need for labor.

The next contribution (10) utilized chitosan soaking to improve the seed germination of *Platycodon grandiflorus*. The species serves as ornamental horticulture in Eastern countries (China, North Korea, South Korea, Japan, and Russia). Its seeds have low germination rates, and they show dormancy. Chitosan is a co-polymer which has recently been introduced in agriculture. It can affect plant growth and seed germination, increasing yield. The authors of the present contribution studied the effect of this compound on seed germination and growth. They suggested that 0.15–0.20% chitosan could improve various growth parameters, e.g., leaf growth, stem diameter, and plant height, while several physiological parameters were increased under these concentrations. Chitosan is a low-cost compound that could promote the exploitation of *P. grandiflorus*.

Miler and Kulus (contribution 11) examined the possibility of establishing a new crossing on a widely used ornamental species, i.e., *Chrysanthemum* × *morifolium*/Ramat./Hemsl. They investigated the in vivo seed efficiency increasing the pollination rate of suitable parental genotypes. Seven cultivars with medium-size, semi-full inflorescences were used as the initial plant material. The study outlined eight new phenotypes, varying in shape, size, and color of inflorescence. Six phenotypes received plant breeder's rights, while the two cultivars were evaluated for their distinctness, uniformity, and stability (DUS). The variety "Wda" proved the most suitable as a mother plant, producing the highest seed number.

Contribution 12 was focused on making selections from native Greek sage populations, identifying the most appropriate for ornamental use. The genus *Salvia* (known as the sage genus) is found throughout the Old World and the Americas, and high species of this genus are suitable for use as ornamental or medicinal plants. *Salvia officinalis* is of high medicinal and ornamental value. In this contribution, a collection of native Greek sage populations was evaluated for their ornamental and morphological properties. At the second stage of the study, populations with appropriate morphological characteristics were tested for their asexual propagation ability through shoot cuttings and in vitro techniques. Their vegetative propagation was studied, aiming to reinforce their introduction in the floriculture market as new varieties.

Finally, contribution 13 presents a review article on seed germination within the genus Rosa, one of the most significant plant genera. The study underlines the complexity of germination and the variety of influencing factors (seed dormancy, environmental conditions, and seed treatments). The review identifies the problems that have arisen in the germination of Rosa seeds and proposes the most appropriate methods for stimulating germination. The study outlines the use of different methods to improve the germination of this valuable ornamental species, considering the practical aspect for future endeavors.

3. Conclusions

In vitro culture techniques are effective in terms of multiplication, producing disease-free plants suitable for conservation strategies, plant breeding, and the introduction of new ornamental plants. The present Special Issue reveals that micropropagation can serve as a suitable method for the conservation and exploitation of endangered species, e.g., *Dyckia brevifolia* and *Sideritis raeseri* subsp. *Attica*. Efficient protocols for the seed germination of native species, e.g., *Berberis koreana* Palibin and *Platycodon Grandiflorus*, could promote their exploitation. Modern micropropagation and conventional cross methods have proven suitable for producing new cultivars and varieties of *Sansevieria, Chrysanthemum* and Greek *Salvia officinalis*. Furthermore, studies on *Lagerstroemia speciosa, Cistus creticus* L., *Philodendron bipinnatifidum* Schott ex Endl. have revealed suitable micropropagation methods starting from adult mother plants. However, propagation by cuttings or seeds indicated that, combined with proper techniques, this can lead to the production of favorable ornamental plants, as occurred with *Pelargonium* × *hortorum* and *Platycodon Grandiflorus*. Finally, a review manuscript on seed germination within the genus *Rosa* outlined in vitro germination and molecular methods, highlighting the *Rosa* seed germination biology with the aim of establishing new rose production systems and supporting conservation strategies as well.

Funding: This research received no external funding.

Acknowledgments: As Guest Editors of the current Special Issue, entitled "Seed Germination and Micropropagation of Ornamental Plants", we would like to express our appreciation to all authors who submitted their valuable studies and contributed to this SI.

Conflicts of Interest: The authors declare no conflicts of interest.

List of Contributions

1. Ahmad, N.; Faisal, M.; Ahmad, A.; Alatar, A.A.; Qahtan, A.A.; Alok, A. Thidiazuron Induced In Vitro Clonal Propagation of *Lagerstroemia speciosa* (L.) Pers.—An Important Avenue Tree. *Horticulturae* **2022**, *8*, 359. https://doi.org/10.3390/horticulturae8050359.
2. Aslami, Z.; Ghehsareh, M.G.; Mahdavi, S.M.E.; Nicola, S. Regeneration of African Violet in Response to Light Quality. *Horticulturae* **2024**, *10*, 78. https://doi.org/10.3390/horticulturae10010078.
3. Bertsouklis, K.; Panagaki, K.-P. In Vitro Germination and Propagation of *Dyckia brevifolia*, An Ornamental and Endangered Bromeliad. *Horticulturae* **2022**, *8*, 390. https://doi.org/10.3390/horticulturae8050390.
4. Bertsouklis, K.; Theodorou, P.; Aretaki, P.-E. In Vitro Propagation of the Mount Parnitha Endangered Species *Sideritis raeseri* subsp. *Attica*. *Horticulturae* **2022**, *8*, 1114. https://doi.org/10.3390/horticulturae8121114.
5. Catalano, C.; Carra, A.; Carimi, F.; Motisi, A.; Sajeva, M.; Butler, A.; Lucretti, S.; Giorgi, D.; Farina, A.; Abbate, L. Somatic Embryogenesis and Flow Cytometric Assessment of Nuclear Genetic Stability for *Sansevieria* spp.: An Approach for In Vitro Regeneration of Ornamental Plants. *Horticulturae* **2023**, *9*, 138. https://doi.org/10.3390/horticulturae9020138.
6. Darras, A.I.; Grigoropoulou, K.; Dimiza, K.; Zulfiqar, F. Effects of Brief UV-C Irradiation Treatments on Rooting Performance of *Pelargonium* × *hortorum* (L.H. Bailey) Stem Cuttings. *Horticulturae* **2022**, *8*, 897. https://doi.org/10.3390/horticulturae8100897.
7. Dewir, Y.H.; Habib, M.M.; AlQarawi, A.A.; Alshahrani, T.S.; Alaizari, A.A.; Malik, J.A.; Alwahibi, M.S.; Murthy, H.N. Mycorrhization Enhances Vegetative Growth, Leaf Gas Exchange, and Root Development of Micropropagated *Philodendron bipinnatifidum* Schott ex Endl. Plantlets during Acclimatization. *Horticulturae* **2023**, *9*, 276. https://doi.org/10.3390/horticulturae9020276.
8. Ioannidis, K.; Koropouli, P. Effects of Different Media and Their Strengths in In Vitro Culture of Three Different Cistus creticus L. Populations and Their Genetic Assessment Using Simple Sequence Repeat Molecular Markers. *Horticulturae* **2024**, *10*, 104. https://doi.org/10.3390/horticulturae10010104.

9. Kim, D.-H.; Kim, S.-G.; Lee, H.; Na, C.-S.; Lee, D.-H. The Dormancy Types and Germination Characteristics of the Seeds of *Berberis koreana* Palibin, an Endemic Species of Korea. *Horticulturae* **2023**, *9*, 547. https://doi.org/10.3390/horticulturae9050547.
10. Liu, H.; Zheng, Z.; Han, X.; Zhang, C.; Li, H.; Wu, M. Chitosan Soaking Improves Seed Germination of *Platycodon Grandiflorus* and Enhances Its Growth, Photosynthesis, Resistance, Yield, and Quality. *Horticulturae* **2022**, *8*, 943. https://doi.org/10.3390/horticulturae8100943.
11. Miler, N.; Kulus, D. Effect of Parental Components and Pollination Frequency on the Setting and Germination of *Chrysanthemum* Seeds. *Horticulturae* **2022**, *8*, 827. https://doi.org/10.3390/horticulturae8090827.
12. Nanos, C.; Tsoulpha, P.; Kostas, S.; Hatzilazarou, S.; Michail, I.; Anastasiadi, V.; Pipinis, E.; Gklavakis, E.; Kanellis, A.K.; Nianiou-Obeidat, I. Asexual Propagation of Greek *Salvia officinalis* L. Populations Selected for Ornamental Use. *Horticulturae* **2023**, *9*, 847. https://doi.org/10.3390/horticulturae9070847.
13. Stoian-Dod, R.L.; Dan, C.; Morar, I.M.; Sestras, A.F.; Truta, A.M.; Roman, G.; Sestras, R.E. Seed Germination within Genus Rosa: The Complexity of the Process and Influencing Factors. *Horticulturae* **2023**, *9*, 914. https://doi.org/10.3390/horticulturae9080914.

References

1. Hermoso, V.; Carvalho, S.B.; Giakoumi, S.; Goldsborough, D.; Katsanevakis, S.; Leontiou, S.; Markantonatou, V.; Rumes, B.; Vogiatzakis, I.N.; Yates, K.L. The EU Biodiversity Strategy for 2030: Opportunities and Challenges on the Path towards Biodiversity Recovery. *Environ. Sci. Policy* **2022**, *127*, 263–271. [CrossRef]
2. Di Martino, L.; Di Cecco, V.; Di Cecco, M.; Di Santo, M.; Ciaschetti, G.; Marcantonio, G. Use of native plants for ornamental purposes to conserve plant biodiversity: Case of study of Majella National Park. *J. Nat. Conserv.* **2020**, *56*, 125839. [CrossRef]
3. Darras, A. Overview of the Dynamic Role of Specialty Cut Flowers in the International Cut Flower Market. *Horticulturae* **2021**, *7*, 51. [CrossRef]
4. Manikas, I.; Malindretos, G.; Abeliotis, K. Sustainable cities through alternative urban farming: The case of floriculture. *J. Int. Food Agribus. Mark.* **2019**, *32*, 295–311. [CrossRef]
5. Krigas, N.; Tsoktouridis, G.; Anestis, I.; Khabbach, A.; Libiad, M.; Megdiche-Ksouri, W.; Ghrabi-Gammar, Z.; Lamchouri, F.; Tsiripidis, I.; Tsiafouli, M.A.; et al. Exploring the Potential of Neglected Local Endemic Plants of Three Mediterranean Regions in the Ornamental Sector: Value Chain Feasibility and Readiness Timescale for Their Sustainable Exploitation. *Sustainability* **2021**, *13*, 2539. [CrossRef]
6. Calheiros, C.S.; Pereira, S.I. Resilience of green roofs to climate change. In *Adapting the Built Environment for Climate Change*; Pacheco-Torgal, F., Granqvist, C., Eds.; Woodhead Publishing: Sawston, UK, 2023; pp. 273–296.
7. DBMR. Hybrid Cloud Market Players, Size, Share, Segmentation, & Forecast Trends by 2029. Market Research Business Consulting and Strategy Planning Firm. Data Bridge Market Research Private Ltd. 2022. Available online: https://www.databridgemarketresearch.com/reports/global-hybrid-cloud-market (accessed on 24 December 2023).
8. Bhattacharjee, S.; Washmin, N.; Borah, T.; Sarkar, A.; Mudoi, K.D.; Saikia, S.P.; Verma, J.S.; Banik, D. Conspectus on endangered carnivorous pitcher plant *Nepenthes khasiana* Hook. f. emphasizing in-vitro regeneration, pitcher development, and stability in genetic makeup. *S. Afr. J. Bot.* **2024**, *167*, 270–284. [CrossRef]

horticulturae

MDPI

Article

Thidiazuron Induced In Vitro Clonal Propagation of *Lagerstroemia speciosa* (L.) Pers.—An Important Avenue Tree

Naseem Ahmad [1], Mohammad Faisal [2,*], Anees Ahmad [1], Abdulrahman A. Alatar [2], Ahmed A. Qahtan [2] and Anshu Alok [3]

1 Plant Biotechnology Laboratory, Department of Botany, Aligarh Muslim University, Aligarh 202 002, India; naseembot@gmail.com (N.A.); aneesqadriamu@gmail.com (A.A.)
2 Department of Botany & Microbiology, College of Science, King Saud University, P.O. Box 2455, Riyadh 11451, Saudi Arabia; aalatar@ksu.edu.sa (A.A.A.); aqahtan@ksu.edu.sa (A.A.Q.)
3 Department of Plant Pathology, University of Minnesota, Saint Paul, MN 55108, USA; anshualok2@gmail.com
* Correspondence: faisalm15@yahoo.com

Abstract: A high throughput regeneration protocol has been developed for *Lagerstroemia speciosa* through node explants under the regime of various plant growth regulators (PGRs). This protocol can provide an alternative mode to seed-grown plants and minimize the cost–time of regeneration, significantly. Murashige and Skoog (MS) medium containing various combinations of PGRs exhibited a marked stimulatory effect on morphogenesis. Of the various combinations tried, node explant pretreated with thidiazuron (TDZ; 5.0 μM) for 4 weeks and followed with transfer into MS medium containing 1.0 μM 6-benzyladenine (BA) and 0.25 μM α-naphthalene acetic acid (NAA) was reported to be the best treatment as it resulted in a maximum number of 24.5 shoots with an average shoot length of 7.1 cm per explant in 90% of cultures after 12 weeks of incubation. The in vitro-generated shoots rooted satisfactorily in the adopted ex vitro method of rooting, which saves time and cost. Among the different treatments, the greatest rooting percentage (85%) was observed in the 200 μM IBA-treated shoots, with the highest root number (8.7) and length (3.4 cm) occurring after 4 weeks. Four months after being transferred to ex vitro, some of the physiological attributes of the in vitro-propagated plants were examined and compared to the ex vitro plants. Further, analysis of the genetic integrity in tissue culture-raised plantlets along with the parental tree was accomplished through DNA-based RAPD technique. The monomorphic banding pattern obtained by the RAPD primers resulted in a high level of genetic uniformity in regenerated plants.

Keywords: shoot bud induction; chlorophyll; carotenoids; genetic fidelity; micropropagation; plant growth regulator; synergistic effect; thidiazuron

Citation: Ahmad, N.; Faisal, M.; Ahmad, A.; Alatar, A.A.; Qahtan, A.A.; Alok, A. Thidiazuron Induced In Vitro Clonal Propagation of *Lagerstroemia speciosa* (L.) Pers.—An Important Avenue Tree. *Horticulturae* **2022**, *8*, 359. https://doi.org/10.3390/horticulturae8050359

Academic Editors: Konstantinos Bertsouklis, Epaminondas Kartsonas, Angela Carra and Zhiyong Guan

Received: 28 February 2022
Accepted: 18 April 2022
Published: 20 April 2022

1. Introduction

Lagerstroemia speciosa (L.) Pers., commonly known as 'Pride of India' or 'Banaba', belongs to the family Lythraceae and is a tropical deciduous tree widely distributed in the Philippines, Malaysia, India, Vietnam, and China [1,2]. It is commonly planted in gardens for its aesthetic and ornamental value, with long-lasting gorgeous blossoms of various colors, and has been explored in current research for vital medicinal properties such as anti-diabetic [1,3], hyperuricemia [4], anti-obesity [5,6], anti-septic [7], and anti-cancer [8,9].

Due to its numerous applications in the pharmaceutical, paper, pulp, small-scale building, and wood industries [10,11], *L. speciosa* is in great demand and is continually over-exploited. Although the multiplication of this species is accomplished through seeds, they have a short lifespan of viability (approximately 1–2 months); after this period germination is too difficult [12]. Due to its significant ethnomedicinal applications and unsustainable exploitation, the tree is on the verge of becoming extinct at both the national and international levels if appropriate conservation measures are not implemented immediately. Thus, it is important to provide a standardized micropropagation methodology for this multipurpose tree. This technique is currently used for multiplication of the planting stock of such types of tree species by several conversationalists [13,14]. In vitro clonal multiplication via axillary bud differentiation offers immense potential for the regeneration of healthy plantlets in a short period of time and space.

TDZ is an artificially modified phenyl urea (*N*-phenyl-*N*′-1, 2, 3-thidiazol-5-yl-urea) or non-purine cytokinin that has been commonly utilized in the in vitro regeneration of many tree species [13,15,16]. In general, the morphogenic response to TDZ is varied and depends on the concentration, exposure period, explant type, and genotype of the cells. It possesses the unique property of high efficacy at low concentrations with short exposure times, and it is less susceptible to enzymatic degradation in vivo than other naturally occurring aromatic cytokinins, and has been found to be most responsive at very low concentrations for the micropropagation of woody plants [17]. In plant tissues, it may exist in a variety of forms, including TDZ-free molecules, sequestered TDZ molecules, conjugated forms coupled to proteins or cell wall components, and TDZ-free molecules connected to cell wall components. Shoot regeneration was noted in other tree species on culture media with TDZ [18,19]. However, in many cases, a major problem is allied with micropropagated plantlets is the occurrence of somaclonal variation among the regenerants [20,21]. According to Fatima and Anis [22], somaclonal variation is a main disadvantage during the in vitro cultivation practice of any elite parental line. A number of scientific studies have also shown that it inhibited shoot elongation, created fasciated shoots (causing hyperhydricity), and induced other physiological abnormalities [23–26]. Therefore, it is important to evaluate the genetic fidelity of tissue culture-raised plantlets. Polymerase chain reaction (PCR)-based techniques, such as random amplified polymorphic DNA (RAPD), have been successfully used to investigate genetic diversity and clonal fidelity in several woody or perennial species such as *Morus alba* [21], *Arbutus* spp. [27], *Canna indica* [28], *Calycophyllum spruceanum* [29], and *Paulownia* [30]. Such study will provide a guarantee towards the in vitro propagation of normal, healthy, and true-to-type plantlets. Up to now, a few reports on in vitro regeneration in *L. speciosa* have been described using different explants [12,15] with limited success in terms of number of shoots per explant, and further efforts are needed to establish a practicable approach. To date there is no report available on in vitro regeneration using the highly active plant growth regulator (PGR) thidiazuron (TDZ). The major aim of the present study was to develop an efficient and reproducible in vitro regeneration and shoot multiplication protocol from nodal explant using thidiazuron with ex vitro rooting, acclimatization, assessment of physiological attributes like photosynthetic pigments, net photosynthetic rate, stomatal conductance, water use efficiency, transpiration rate, and genetic fidelity of regenerated plants using DNA-based RAPD markers.

2. Materials and Methods

2.1. Explants Source, Media and Culture Conditions

Newly sprouting shoots of *L. speciosa* were collected from an approximately 20-year-old mature tree growing in the Department of Botany, AMU, Aligarh, India (27°54′52.0″ N 78°04′21.8″ E). The harvested sprouting shoots were washed under running tap water for up to 10 min and then treated with fungicide (1% *w*/*v* Bavistin™, Mumbai, India) and again rinsed with running tap water for 15 min. Thereafter, they were treated with laboratory detergent (5% *v*/*v* Labolene™, Qualigens, Mumbai, India) for 8 min followed by being rinsed 5 to 6 times with autoclaved distilled water. Surface disinfection took place using 0.1% $HgCl_2$ (*w*/*v*) for 5 min in aseptic conditions followed by thorough washing with sterile distilled water to remove the disinfectant. Sterile node explants (0.5–1.0 cm) excised under laminar flow hood were used as the starting materials in all the experiments. Murashige and Skoog [31] (MS) medium (Himedia®, Mumbai, India) was used as a nutrient–substrate in all the culture experimental setups with exogenously applied sucrose (3%, *w*/*v*) (Qualigens, Mumbai, India) and agar (0.8%, *w*/*v*) (Qualigens, Mumbai, India). The pH was adjusted (5.75 to 5.85) before autoclaving at 121 °C for 20 min. The cultures were maintained at 24 ± 2 °C during a 16 h light photoperiod with a light intensity of 50–150 µmol m^{-2} s^{-1}

photon flux density provided by cool white fluorescent lamps. Relative humidity was adjusted up to 55 ± 5% and regularly monitored by thermo-hygrometer (Testo, India Pvt. Ltd., Pune, India). The sub-culturing was carried out by transfer of the cultures onto fresh medium after every 3–4 week of incubation. Data were collected on the percent response, shoot numbers, and shoot lengths per explant, after 4, 8, and 12 weeks of culture incubation.

2.2. Shoot Induction

The induction of multiple shoots from node explants was carried out in two different sets of experiments. In the first phase, node explants were grown for four weeks in a broad range of TDZ concentrations (1.0, 5.0, 10.0, 15.0, 20.0 μM) supplemented MS medium to determine the optimal concentration at which maximal bud induction could be achieved. In order to determine the optimal TDZ dosage, the shoot formation was carefully monitored in each tube. In the second step, based on the results of the experiment, the best TDZ dose among those listed in Table 1 was refined at a narrow level (3.5, 4.0, 4.5, 5.0, 5.5, 6.0, and 6.5 μM) for 4 weeks in a separate experiment to determine the optimal concentration of TDZ for further shoot bud induction from node explants.

Table 1. Primary (or wide) screening of TDZ on multiple shoot induction from node explants of *L. speciosa* after 4 weeks of incubation.

TDZ (μM)	Percent (%) Response	No. of Shoots per Explant Mean ± SE	Shoot Length (cm) Mean ± SE
0.0 (C)	25	1.00 ± 0.00^{e}	1.58 ± 0.05^{c}
1.0	45	3.48 ± 0.08^{c}	2.63 ± 0.06^{a}
5.0	75	14.52 ± 0.38^{a}	2.17 ± 0.13^{b}
10.0	55	6.05 ± 0.14^{b}	0.86 ± 0.08^{d}
15.0	30	2.08 ± 0.12^{d}	0.40 ± 0.06^{e}
20.0	00	0.00 ± 0.00^{f}	0.00 ± 0.00^{f}

Mean values (n = 30) followed by the same letter within a column are not significantly different (p = 0.05) using Duncan's multiple range test. (C, control; TDZ, thidiazuron).

2.3. Shoot Multiplication

After four weeks of bud induction period on culture medium with TDZ, responsive explants were transferred to MS culture medium (in glass tubes of 25 × 150 mm) supplemented with various cytokinins, such as 6-benzyladenine (BA), kinetin (Kin), and 2-isopentenyl adenine (2iP), each at a concentration of 0.25, 0.5, 1.0, 2.0, or 4.0 μM either singly or in combination with auxins, which included indole-3-acetic acid (IAA), indole-3-butyric acid (IBA), and α-naphthalene acetic acid (NAA) at three different concentrations (0.10, 0.25, and 0.50 μM) for shoot multiplication.

2.4. Ex Vitro Rooting and Acclimatization

Healthy in vitro-regenerated shoots of about 4–5 cm in length with 3–5 fully expended leaves were selected for the rooting experiment. The basal cut end of the shoots was dipped in different concentrations (200, 400, 800, or 1000 μM) of various auxins (IBA, IAA, or NAA) for about 30 min. The treated shoots were washed thrice with sterile distilled water and transplanted into plastic/thermocol cups full of a sterile potting mixture called Soilrite™ (Keltech Pvt. Ltd., Bangalore, India), and these cups were covered with transparent polybags to ensure high humidity up to a couple of weeks. The plantlets were irrigated with quarter-strength MS liquid medium without organic supplements at every alternate day for 2 weeks. The polybags were removed after each second day with increasing time period for the purpose of hardening. The percentage of rooting response, root numbers per shoot, and root length was noted after 4 weeks of rooting culture. Furthermore, these semi-acclimatized plantlets were again transplanted into a mixture of vermiculated garden soil with Soilrite (1:1) for the next 2 weeks. Thereafter, the healthy, acclimatized plants were transplanted on garden soil for the next 2 weeks under the same conditions. The acclimatized plants were transferred to greenhouse conditions for another couple of weeks, and, finally, they were successfully established in natural conditions.

2.5. Growth and Photosynthetic Traits

Plant height (in centimeters) and the number of branches on each plant were measured for five randomly selected micropropagated plants and ex vitro plants that were maintained in a greenhouse. At the same time, leaves from each plant were taken and analyzed spectrophotometrically to estimate the chlorophyll *a* and *b* concentrations, as well as the carotenoids content, using 80% acetone following the procedure described by Lichtenthaler and Wellburn [32]. The absorbance of the samples was measured at 645, 663, and 440 nm using a UV spectrophotometer for the determination of chlorophyll *a*, *b*, and carotenoid contents. Photosynthetic traits, i.e., net photosynthetic rate (Pn), stomatal conductance (g_s), internal CO_2 (Ci), and transpiration rate (E), were measured on the fully expanded leaves of the in vitro propagated plants and ex vitro plants with a portable photosynthetic system (LI-COR 6200, Lincoln, NE, USA) on a sunny day between 10 a.m. and 12 p.m. The measurement was repeated thrice. Water use efficiency (WUE) was calculated using the data of photosynthetic rate and stomatal conductance [33]. Five samples were used for each measurement, which was repeated twice.

2.6. Genomic DNA Isolation and Molecular Screening

Screening of the genetic fidelity in micropropagated plants was achieved using a PCR-based RAPD technique. Genomic DNA was extracted from young leaf tissues of the mother plant and 10 randomly selected 3-month-old tissue culture-raised plants using a slightly modified cetyl-trimethyl ammonium bromide (CTAB) technique established by Doyle and Doyle [34]. The quantification of genomic DNA was accomplished by Nanodrop Spectrophotometer (Implen, München, Germany). The preliminary screenings of the DNA samples were done by 20 RAPD primers (Operon Kit B primer). Of these, 11 primers were chosen based on their unambiguous, reproducible, and clear banding pattern and were employed for final screening. Polymerase chain reactions (PCR) for RAPD primers were completed in a thermocycler machine (Biometra, Göttingen, Germany). About 25 μL of DNA amplification reaction mixture consisted of various major components including 10× $(NH_4)_2SO_4$ buffer, $MgCl_2$ (25 mM), dNTPs (10 mM), primers (10 μM), *Taq* pol (2 Unit), and 50 ng/μL DNA template. The program of DNA amplification initially started with a denaturation step at 94 °C for 5 min followed by 38 repeated cycles of denaturation at 94 °C (1 min), annealing at 45–60 °C (1 min), and elongation at 72 °C (2 min), and final extension of DNA was carried out at 72 °C (8 min). The PCR assays were performed twice in order to eliminate the possibility of false positive results and to confirm the reproducibility of the RAPD markers. These amplified DNA fragments were separated by electrophoresis technique in agarose (1.2%, *v/w*) gel containing 4 μL ethidium bromide in TAE buffer run at 65 V (110 min) and imaged on a UV transilluminator (Bio Rad, Hercules, CA, USA). Thereafter, clear and well-distinct and reproducible bands were scored as present (1) or absent (0) for RAPD primers in each DNA sample. The relative size of each amplified DNA fragment was compared with 1 kb DNA ladder (Gene Ruler™ 1 kb DNA ladder, Thermo Scientific, Waltham, MA, USA). The amplification of each RAPD primer was repeated thrice to confirm the reproducibility pattern.

2.7. Statistical Analysis

All research experiments were followed on a completely randomized block design. Each treatment was repeated three times with 10 replicates. The data on different parameters were subjected one-way analysis of variance (ANOVA) using SPSS ver. 16 (SPSS Inc., Chicago, IL, USA). The significance of variances among mean values was calculated using Duncan's multiple range tests at $p = 0.05$. The results of each treatment are represented as the mean ± SE (standard error) in 3 repeated experiments.

3. Results and Discussion

3.1. Shoot Induction

A node explant procured from a mature tree produced a single shoot when cultured on a PGR-free medium even after 4 weeks of incubation. However, a response in bud induction was recorded within 4 weeks when the node explants were grown in TDZ-fortified culture medium. The TDZ treatment on node explants significantly improved the rate of shoot bud induction. Among different concentrations, the optimal medium for the highest shoot bud induction was recorded as MS medium containing 5.0 μM TDZ (Figure 1A). At this level, a maximum of 75% shoot induction, a highest number of 14.5 shoots per node explant and a mean shoot length of 2.1 cm were recorded within 4 weeks of incubation (Tables 1 and 2). However, it was noted that an increased dose of TDZ inhibited the elongation of shoots when compared to a lower dose of TDZ individually after a period of 4 weeks of culture. The MS medium enriched with 3.5 μM TDZ exhibited 6.2 shoots per explant with a mean shoot length of 2.8 cm after 4 weeks of culture (Table 2). The bud induction response declined with increased concentrations of TDZ beyond the optimal dose resulted in small, condensed shoots and most of them showed bunching of shoots with no further shoot growth even after 4 weeks of culture. According to Liu et al. [35], shoot multiplication and growth are varying in response to TDZ exposure because TDZ is a more biologically active PGR than others. However, Matand and Prakash [36] suggested that TDZ-pretreated explants subsequently transferred to secondary medium or TDZ-free medium exhibited better response in stimulations of shoot multiplication across a large number of plants species. Murch and Saxena [37] suggested that TDZ may exist in different forms such as free molecules of TDZ, sequestrated molecules of TDZ, conjugated molecules with proteins, or cell wall in TDZ-pretreated cultures. Overall, these studies indicated that TDZ molecules promotes a high rate of shoot bud induction by stimulating cell division and multiplication in apical meristem while also reprograming cells to the appropriate developmental stage for shoot bud differentiation. In many plant species, long duration exposure and high dose TDZ pretreatment exhibited morphological abnormalities such as hyper-hydricity and suppressed shoot growth [26,38,39]. Similarly, in the present study, the shoot numbers per node explant were quite high in TDZ-pretreated samples, but their shoot elongation was not obvious even after 4 weeks of incubation (Table 1). Hence, it was clear that TDZ exhibited inhibitory results on the elongation and growth of shoots. However, it shows a promoting effect on shoot bud induction from node explants. Similar kinds of TDZ effects have also already been documented in a large number of trees [13,40,41]. Although their studies suggested such problems could be overcome via a two-fold culture approach. Similarly, in the present study, when TDZ-pretreated node explants were subsequently transferred to a secondary medium (or one without TDZ) containing another cytokinin alone or in combination with auxins, it successfully improved *L. speciosa* cultures' protocol. This type of strategy using primary medium (for shoot bud induction) and secondary medium (for shoot elongation) was successfully applied in several tree species [26,42,43].

Table 2. Secondary (or narrow) screening of TDZ for multiple shoot induction from node explants of *L. speciosa* after 4 weeks of incubation.

TDZ (μM)	Percent (%) Response	No. of Shoots per Explant Mean ± SE	Shoot Length (cm) Mean ± SE
0.0 (C)	25	1.00 ± 0.00 ^{g}	1.58 ± 0.06 ^{d}
3.5	50	6.25 ± 0.18 ^{f}	2.86 ± 0.09 ^{a}
4.0	60	9.10 ± 0.33 ^{e}	2.51 ± 0.09 ^{b}
4.5	70	12.83 ± 0.15 ^{b}	2.24 ± 0.05 ^{bc}
5.0	75	14.52 ± 0.38 ^{a}	2.17 ± 0.13 ^{c}
5.5	65	11.58 ± 0.32 ^{c}	2.05 ± 0.08 ^{c}
6.0	60	10.14 ± 0.11 ^{d}	1.49 ± 0.06 ^{d}
6.5	55	8.69 ± 0.21 ^{e}	1.06 ± 0.17 ^{e}

Mean values ($n = 30$) followed by the same letter within a column are not significantly different ($p = 0.05$) using Duncan's multiple range test. (C, control; TDZ, thidiazuron).

Figure 1. (**A**) Multiple shoot bud induction from node explants of *L. speciosa* on MS medium supplemented with 5.0 μM TDZ after 4 weeks of culture (Bar = 0.8 cm); (**B**) proliferation of shoots from TDZ-pretreated node explants on MS medium supplemented with 1.0 μM BA + 0.25 μM NAA after 12 weeks of culture (Bar = 1.3 cm).

3.2. Shoot Multiplication

The TDZ-pretreated node explants were sub-cultured onto secondary MS medium containing different concentrations of cytokinins (Table 3) for 8 weeks. A maximum of 18.6 shoots per node explant and a subsequent shoot length of 5.8 cm was recorded in 80% of cultures containing 1.0 μM BA, whereas, pretreated node explants sub-cultured on hormone-free MS medium (or control medium) exhibited a remarkable declination in shoot multiplication (12.0) and elongation (2.4 cm) per node explant. Similarly, the positive response of BA over other cytokinins (Kin and 2iP) on shoot elongation and multiplication has been already documented in various plants [22,44]. Meanwhile, the subsequent transfer of TDZ-pretreated node explants onto a secondary medium having an optimal dose of BA exhibited a better effect on the rate of shoot multiplication and their subsequent elongation. This means that TDZ exposure at appropriate concentration is sufficient to induce better shoot multiplication rates in pretreated node explants. The possible reason for better shoot growth in TDZ-pretreated node cultures on secondary medium augmented with an optimum dose of BA may be its capability of changing to an active form of cytokinin from non-active storage forms at a later stage of culture through the modulation of the endogenous PGRs level, either directly or as a result of induced stress [15]. Similar results have been reported by different researchers in several woody plant species [16,45]. Furthermore, we examined the combined effect of an optimum dose of BA (1.0 μM) with auxins (IAA, IBA and NAA) at different combinations for improving the rate of shoot multiplication as well as proliferation. Among different cytokinin–auxin combinations, the highest shoot numbers (24.5) with an average shoot length of 7.1 cm per node explant were noted on 1.0 μM BA and 0.25 μM NAA supplemented MS medium in 90% node cultures after 12 weeks (Table 4; Figure 1B). Whereas 1.0 μM BA and 0.50 μM IBA yielded lesser shoot numbers (16.9) per node explant with a mean shoot length of 4.7 cm in 75% cultures after 12 weeks. Interestingly, in the present study, the rates of shoot multiplication and proliferation were positively correlated with a combination of cytokinin–auxin concentrations.

Table 3. Effect of different concentrations of cytokinins on 5.0 μM TDZ-exposed node explants of *L. speciosa* for shoot multiplication and elongation after 8 weeks of culture.

Cytokinins (μM)			Percent (%) Response	No. of Shoots per Explant Mean ± SE	Shoot Length (cm) Mean ± SE
BA	Kin	2iP			
0.0 (C)	0.0	0.0	60	12.06 ± 0.47 i	2.43 ± 0.18 h
0.25			70	15.16 ± 0.28 fg	4.32 ± 0.24 cd
0.5			70	15.85 ± 0.24 cde	4.94 ± 0.31 b
1.0			80	18.62 ± 0.75 a	5.87 ± 0.23 a
2.0			75	16.26 ± 0.31 bc	5.12 ± 0.28 b
4.0			70	15.06 ± 0.37 fgh	4.56 ± 0.19 c
	0.25		55	14.68 ± 0.16 gh	3.85 ± 0.21 e
	0.5		60	15.82 ± 0.41 cde	4.08 ± 0.21 de
	1.0		70	16.56 ± 0.44 b	5.26 ± 0.20 b
	2.0		65	16.04 ± 0.26 bcd	4.52 ± 0.22 c
	4.0		60	15.48 ± 0.21 def	4.36 ± 0.21 cd
		0.25	50	14.62 ± 0.11 gh	3.04 ± 0.11 g
		0.5	55	15.22 ± 0.32 efg	3.43 ± 0.17 f
		1.0	60	15.95 ± 0.38 bcd	3.74 ± 0.20 ef
		2.0	55	14.98 ± 0.41 fgh	3.02 ± 0.23 g
		4.0	50	14.43 ± 0.12 h	2.90 ± 0.21 g

Mean values (n = 30) followed by the same letter within a column are not significantly different (p = 0.05) using Duncan's multiple range test. (C, control; TDZ, thidiazuron).

Table 4. Combined effect of different concentrations of auxins in combination with optimal cytokinin (1.0 μM BA) on multiple shoot proliferation from node explants of *L. speciosa* after 12 weeks of culture.

Auxin (μM)			Percent (%) Response	No. of Shoots per Explant Mean ± SE	Shoot Length (cm) Mean ± SE
IAA	IBA	NAA			
0.10			75	19.45 ± 0.40 de	5.68 ± 0.09 c
0.25			80	20.38 ± 0.48 cd	6.04 ± 0.08 b
0.50			85	18.62 ± 0.42 ef	5.27 ± 0.15 de
	0.10		80	17.36 ± 0.28 g	5.05 ± 0.06 ef
	0.25		80	17.55 ± 0.22 fg	5.34 ± 0.07 de
	0.50		75	16.93 ± 0.34 g	4.76 ± 0.15 f
		0.10	85	22.08 ± 0.22 b	6.25 ± 0.10 b
		0.25	90	24.53 ± 0.61 a	7.16 ± 0.12 a
		0.50	80	20.85 ± 0.32 c	5.50 ± 0.09 cd

Mean values (n = 30) followed by the same letter within a column are not significantly different (p = 0.05) using Duncan's multiple range test. (C, Control).

However, the growth of shoots was inhibited when the levels of auxin were increased beyond the 0.25 μM in the culture medium. Besides this, after 2 weeks of culture, the new shoots emerged directly from the axillary node of node explants and were further subcultured onto the same fresh medium for further shoot growth. In the present study, these positive results (in vitro shoot proliferation from nodal explants of *L. speciosa* may be due to a suitable combination of cytokinin–auxin in a synergistic, additive, or antagonistic manner to get an effective proliferation response on post-TDZ pretreated cultures) are in agreement with several earlier reports such as Husain and Anis [41] in *Melia azedarach*, Siddique and Anis [46] in *Balanites aegyptiaca*, Javed, et al. [47] in *Acacia ehrenbergiana*, Ahmad et al. (2018) in *Pterocarpus marsupium*, and in *Tecoma stans* Hussain, Ahmad, Anis, and Hakeem [45].

3.3. Ex Vitro Rooting and Acclimatization

The efficacy of the established micropropagation protocols is determined by the proportion of rooting and acclimation success with the transfer of regenerants under natural environmental conditions. The ex vitro rooting approach has been proven to be the most efficient among the numerous methods of rooting in isolated shootlets from culture and may be utilized as an alternative method [25,48]. Among the several treatments used, 200 µM (IBA) was shown to be the most successful in promoting rooting, with 85% of shoots rooted effectively with a mean number of 8.7 and length of 3.4 cm per shootlet after 4 weeks (Table 5, Figure 2A). The applications of exogenous auxins, namely IBA, IAA, and NAA, in culture medium are commonly utilized for in vitro rooting in micropropagated shootlets.

Table 5. Effect of different auxins on ex vitro root induction in tissue culture-raised microshoots of *L. speciosa* after 4 weeks of culture.

PGRs (µM)			Percent (%) Rooting	No. of Roots per Microshoot Mean ± SE	Root Length (cm) Mean ± SE
IAA	IBA	NAA			
0.0 (C)			0	0.00 ± 0.00 [i]	0.00 ± 0.00 [h]
200			60	3.62 ± 0.07 [f]	2.10 ± 0.10 [de]
400			75	5.84 ± 0.12 [c]	2.65 ± 0.09 [b]
600			65	4.78 ± 0.13 [e]	2.37 ± 0.07 [cd]
800			55	2.93 ± 0.09 [g]	1.76 ± 0.09 [f]
	200		70	5.35 ± 0.07 [d]	2.50 ± 0.08 [bc]
	400		85	8.72 ± 0.20 [a]	3.48 ± 0.10 [a]
	600		75	6.45 ± 0.11 [b]	2.63 ± 0.11 [b]
	800		60	4.84 ± 0.12 [e]	2.35 ± 0.06 [cd]
		200	35	2.92 ± 0.14 [g]	1.85 ± 0.11 [ef]
		400	50	4.65 ± 0.05 [e]	2.32 ± 0.12 [cd]
		600	40	3.47 ± 0.08 [f]	2.05 ± 0.05 [e]
		800	30	2.25 ± 0.4 [h]	1.50 ± 0.06 [g]

Mean values ($n = 30$) followed by the same letter within a column are not significantly different ($p = 0.05$) using Duncan's multiple range test. (C, control).

It has also been reported that IBA showed superior effect over other auxins on adventitious roots formation in a large number of plants. According to Elmongy et al. [49], IBA is an important plant hormone for in vitro rhizogenesis, because it shows more stability towards photo-degradation, adherence to shootlets, and, most importantly, activates the meristematic region of pericycle. Furthermore, IBA shows an active role in the formation of adventitious roots and acts as a signaling molecule that controls gene expression and the root development process. Similar effects of IBA treatment on in vitro root formation have been established in many scientific reports [41,47,50]. By combining rooting and hardening, the ex vitro approach not only aids in the creation of appropriate root systems without basal callusing, but also aids in the removal of one extra stage of in vitro rooting. The efficiency of IBA in comparison to all other auxins has previously been documented in *Vitex negundo* [51]. After 4 weeks, well-developed rooted juvenile plants were transplanted to sterilized Soilrite for a couple of weeks and artificial light irradiated with 50 µmol m^{-2} s^{-1} PPFD. Thereafter, these juvenile plants were transplanted into a mixture of Soilrite–vermiculated garden soil (1:1) for the next couple of weeks under 100 µmol m^{-2} s^{-1} PPFD light irradiance in order to harden. Subsequently, the acclimatized plants were shifted to normal garden soil for the next couple of weeks under 150 µmol m^{-2} s^{-1} PPFD light irradiance (Figure 2B). The constructive effect of Soilrite on the acclimatization process during the early stages is because it is a spongy material, meaning it holds more nutrients along with water content, and thus it supports the development of adventitious tender roots. Also, the gradual increment of light intensity helps with the hardening process of juvenile plants during acclimatization. Later, these juvenile plantlets were further nourished under greenhouse conditions for the next couple of weeks. Finally, the 8-week-old acclimatized plants were transplanted under field conditions where they grew normally with a 93% of survival rate.

Figure 2. (**A**) Ex vitro rooted shoots of *L. speciosa* after pulse treatment with 200 μM IBA (Bar = 0.6 cm); (**B**) 1-month-old, acclimatized plantlets of *L. speciosa* (Bar = 0.6 cm).

3.4. Growth and Photosynthetic Traits

Four months after being transferred to ex vitro growth, the physiological traits of the in vitro-propagated plants were examined and compared to ex vitro plants maintained in the greenhouse. Table 6 shows the comparative data on plant height, branch number, chlorophylls, carotenoids, photosynthetic rate, stomatal conductance, water use efficiency, and transpiration rate. When comparing ex vitro plants to micropropagated plants, the height of the plants and the number of branches on each plant were both slightly higher in ex vitro plants. In contrast, the net photosynthetic rate, chlorophyll *a*, and chlorophyll *b* concentration were found to be very similar in both plants, while the carotenoids content was found to be higher in in vitro plants. The results are in agreement with the findings in *Tylophora indica* [52] and *Clitoria ternatea* [53]. The stomatal conductance of in vitro-regenerated plants was found to be lower than that of ex vitro plants. In vitro plants, on the other hand, showed a substantial improvement in water use efficiency compared to ex vitro plants. Meanwhile, ex vitro plants transpired at a higher rate than micropropagated plants.

Table 6. Comparison of plant height, number of branches per plant, chlorophylls, carotenoids, net photosynthetic rate, stomatal conductance, water use efficiency, and transpiration rate between in vitro plants and ex vitro grown plants of *L. speciosa*.

Plants	Plant Height	Branches per Plant	Chl *a* (mg g^{-1} Fresh Mass)	Chl *b* (mg g^{-1} Fresh Mass)	Car (mg g^{-1} Fresh Mass)	Pn (μmol m^{-2} s^{-1})	g_s (mol m^{-2} s^{-1})	WUE (mol m^{-2} s^{-1})	E (mmol m^{-2} s^{-1})
In vitro	8.5 ± 0.52	4.10 ± 0.20	0.83 ± 0.05	0.64 ± 0.07	0.49 ± 0.03	5.3 ± 0.21	0.43 ± 0.03	30.1 ± 1.15	0.43 ± 0.03
Ex vitro	9.5 ± 0.65	4.52 ± 0.17	0.90 ± 0.03	0.57 ± 0.02	0.41 ± 0.02	5.7 ± 0.17	0.47 ± 0.07	28.3 ± 0.88	0.57 ± 0.05

Chl *a* = chlorophyll *a*; Chl *b* = chlorophyll *b*; Car = carotenoids; Pn = photosynthetic rate; g_s = stomatal conductance; WUE = water use efficiency; E = transpiration rate.

3.5. Molecular Screening

The plants raised from node explants of *L. speciosa* through the micropropagation technique were assessed for their clonal fidelity by DNA-based RAPD primers. Out of 20 IRAPD primers of Kit-B, 11 were selected because of their unambiguous, reproducible, and clear banding pattern results to evaluate the genetic fidelity in nine randomly selected micropropagated plantlets along with the mother plant. Monomorphic RAPD banding pattern exhibited a total of 70 DNA bands with an average of 6.3 DNA fragments per primer, indicating homogeneity among regenerants with the donor tree (Table 7). Meanwhile, no polymorphism was detected, revealing the complete genetic fidelity among the regenerated clones. The monomorphic banding patterns among regenerated plants were amplified by OPB1 and OPB10 and are depicted in Figure 3A,B.

Table 7. Randomly amplified polymorphic DNA primers (Kit B) used to assessed in vitro-propagated *L. speciosa* plantlets.

S. No.	Primers	Sequence (5′-3′)	No. of Bands
1	OPB01	GTTTCGCTCG	10
2	OPB02	TGATCCCTGG	0
3	OPB03	CATCCCCCTG	6
4	OPB04	GGACTGGAGT	0
5	OPB05	TGCGCCCTTC	6
6	OPB06	TGCTCTGCCC	9
7	OPB07	GGTGACGCAG	3
8	OPB08	GTCCACACGG	0
9	OPB09	TGGGGGACTC	0
10	OPB10	CTGCTGGGAC	10
11	OPB11	GTAGACCCGT	0
12	OPB12	CCTTGACGCA	7
13	OPB13	TTCCCCCGCT	2
14	OPB14	TCCGCTCTGG	5
15	OPB15	GGAGGGTGTT	0
16	OPB16	TTTGCCCGGA	10
17	OPB17	AGGGAACGAG	0
18	OPB18	CCACAGCAGT	0
19	OPB19	ACCCCCGAAG	2
20	OPB20	GGACCCTTAC	0

Larkin and Scowcroft [54] reported that culture conditions during in vitro propagation practice are stressful and may sometimes exhibit genetic variations in the tissue cultured plants, known as somaclonal variations. The occurrence of somaclonal variations is more frequent when the in vitro cultures are exposed to stress. Thus, it is important to assess genetic integrity among micropropagated plantlets through DNA-based molecular markers. Such kinds of study provide a guarantee towards the regeneration of true-to-type plantlets. Generally, DNA-based molecular markers are not affected by any kind of environmental factors [55]. Although RAPD markers have become a more popular DNA-based molecular technique as they do not require any prior information on DNA sequences. Therefore, in the present study, RAPD primers exhibited a monomorphic DNA-banding pattern suggesting high-level genetic integrity among micropropagated plants. Consequently, the mode of clonal propagation using TDZ was appropriate for gaining true-to-type plantlets. There are a number of studies confirming the genetic fidelity of regenerated plants using DNA-based RAPD makers when cytokinin and auxin is optimized for micropropagation via axillary shoot proliferation in several plant species, such as *Terminalia bellerica* [20], *Morus alba* [21], *Withania somnifera* [25], *Rauvolfia tetraphylla* [56], *Ruta chalepensis* [57], *Vitex negundo* [58], *Zanthoxylum armatum* [59] *Carnation* [60], and *Humulus lupulus* [61].

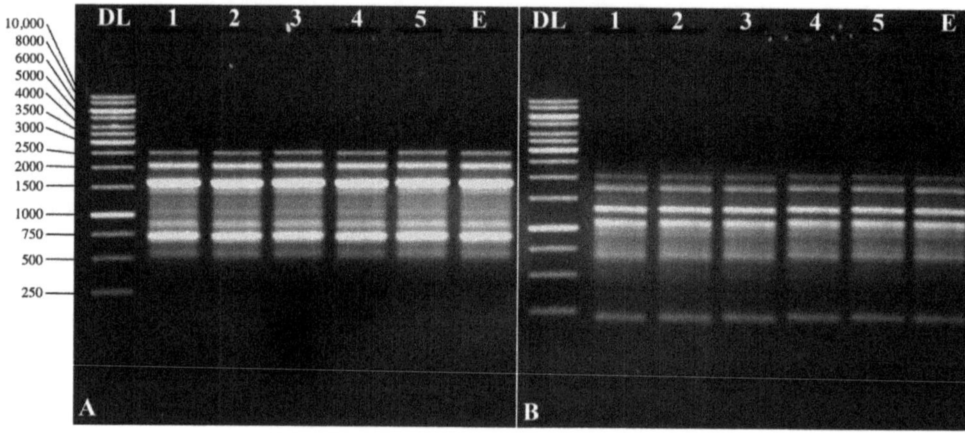

Figure 3. Showing genetic stability between randomly selected in vitro plants and donor plants using RAPD markers in *L. speciosa*. (**A**) Amplified profile from primer OPB-01; (**B**) amplified profile from primer OPB10. DL—DNA ladder; lanes 1–5: randomly selected in vitro plants; lane E: donor plant.

4. Conclusions

The present work describes a reproducible and efficient in vitro protocol useful for scaling up the propagation of *L. speciosa* using mature nodal explants. Different concentrations of TDZ were optimized, and it was found that 5.0 µM TDZ is the most effective concentration for the induction of shoot buds from node explants. However, these buds showed stunted shoots and did not grow further. To overcome this problem, 4 weeks of TDZ exposure is optimum followed by their transfer to a secondary medium for maximum shoot elongation and multiplication. It is concluded that a two-step procedure is necessary for clonal propagation. Ex vitro rooting helps in the reduction of time period as it combines both the additional steps of in vitro rooting and acclimatization. From the physiological and molecular data, it can be postulated that the regenerated progeny is true to type. Thus, the developed clonal propagation system through node explant of *L. speciosa* will not only help in low-cost multiplication but also their conservation and commercial propagation.

Author Contributions: Conceptualization, N.A. and M.F.; methodology, N.A., M.F., A.A. (Anees Ahmad), A.A.A., A.A.Q. and A.A. (Anshu Alok); validation, N.A., M.F. and A.A. (Anees Ahmad); formal analysis, N.A., M.F., A.A. (Anees Ahmad) and A.A.Q.; investigation, N.A., M.F., A.A. (Anees Ahmad), A.A.A., A.A.Q. and A.A. (Anshu Alok); resources, N.A. and M.F.; writing—original draft preparation N.A. and M.F.; writing—review and editing, N.A., M.F., A.A. (Anees Ahmad), A.A.A., A.A.Q. and A.A. (Anshu Alok); supervision, M.F. and A.A.A.; project administration, A.A.A.; funding acquisition, M.F. and A.A.A. All authors have read and agreed to the published version of the manuscript.

Funding: This study was supported by the Researchers Supporting Project (RSP-2021/86), King Saud University, Riyadh, Saudi Arabia.

Institutional Review Board Statement: Not applicable.

Informed Consent Statement: Not applicable.

Data Availability Statement: Not applicable.

Acknowledgments: The authors are thankful to the Researchers Supporting Project (RSP-2021/86), King Saud University, Riyadh, Saudi Arabia for financial support.

Conflicts of Interest: There are no conflict of interest.

References

1. Klein, G.; Kim, J.; Himmeldirk, K.; Cao, Y.; Chen, X. Antidiabetes and Anti-obesity Activity of *Lagerstroemia speciosa*. *Evid.-Based Complement. Altern. Med.* **2007**, *4*, 401–407. [CrossRef] [PubMed]
2. Rahman, F.B.; Ahmed, S.; Noor, P.; Rahman, M.M.M.; Huq, S.M.A.; Akib, M.T.E.; Shohael, A.M. A comprehensive multi-directional exploration of phytochemicals and bioactivities of flower extracts from *Delonix regia* (Bojer ex Hook.) Raf., *Cassia fistula* L. and *Lagerstroemia speciosa* L. *Biochem. Biophys. Rep.* **2020**, *24*, 100805. [CrossRef] [PubMed]
3. Kim, K.-H.; Roh, S.-G.; Li, C.-R.; Jin, C.-F.; Kim, A.; Choi, W.-C. Anti-diabetic effects of banaba leaf extracts (*Lagerstroemia speciosa* Pers.) through solvents. *J. Life Sci.* **2008**, *18*, 1305–1311. [CrossRef]
4. Unno, T.; Sugimoto, A.; Kakuda, T. Xanthine oxidase inhibitors from the leaves of *Lagerstroemia speciosa* (L.) Pers. *J. Ethnopharmacol.* **2004**, *93*, 391–395. [CrossRef] [PubMed]
5. Suzuki, Y.; Unno, T.; Ushitani, M.; Hayashi, K.; Kakuda, T. Antiobesity Activity of Extracts from *Lagerstroemia speciosa* L. Leaves on Female KK-Ay Mice. *J. Nutrf. Sci. Vitaminol.* **1999**, *45*, 791–795. [CrossRef]
6. Karsono, A.H.; Tandrasasmita, O.M.; Tjandrawinata, R.R. Bioactive fraction from *Lagerstroemia speciosa* leaves (DLBS3733) reduces fat droplet by inhibiting adipogenesis and lipogenesis. *J. Exp. Pharm.* **2019**, *11*, 39–51. [CrossRef]
7. Garo, E.; Eldridge, G.R.; Goering, M.G.; DeLancey Pulcini, E.; Hamilton, M.A.; Costerton, J.W.; James, G.A. Asiatic acid and corosolic acid enhance the susceptibility of *Pseudomonas aeruginosa* biofilms to tobramycin. *Antimicrob. Agents Chemother.* **2007**, *51*, 1813–1817. [CrossRef]
8. Thakur, R.S.; Devaraj, E. *Lagerstroemia speciosa* (L.) Pers. triggers oxidative stress mediated apoptosis via intrinsic mitochondrial pathway in HepG2 cells. *Environ. Toxicol.* **2020**, *35*, 1225–1233. [CrossRef]
9. Rohit Singh, T.; Ezhilarasan, D. Ethanolic Extract of *Lagerstroemia Speciosa* (L.) Pers., Induces Apoptosis and Cell Cycle Arrest in HepG2 Cells. *Nutr. Cancer* **2020**, *72*, 146–156. [CrossRef]
10. Zobayed, S.M.A. In vitro propagation of *Lagerstroemia* spp. from nodal explants and gaseous composition in the culture headspace. *Environ. Control Biol.* **2000**, *38*, 1–11. [CrossRef]
11. Jayakumar, K.S.; Sajan, J.S.; Aswati Nair, R.; Padmesh Pillai, P.; Deepu, S.; Padmaja, R.; Agarwal, A.; Pandurangan, A.G. Corosolic acid content and SSR markers in *Lagerstroemia speciosa* (L.) Pers.: A comparative analysis among populations across the Southern Western Ghats of India. *Phytochemistry* **2014**, *106*, 94–103. [CrossRef] [PubMed]
12. Hadiuzzaman, S.; Zobayed, S.; Roy, S.; Miah, M. In vitro cloning from seedling explants of *Lagerstroemia speciosa* Pers. and *L. thorellii* Gagnep. *Bangladesh J. Bot.* **1992**, *21*, 59–64. (In Bengali)
13. Ahmad, A.; Ahmad, N.; Anis, M.; Alatar, A.A.; Abdel-Salam, E.M.; Qahtan, A.A.; Faisal, M. Gibberellic acid and thidiazuron promote micropropagation of an endangered woody tree (*Pterocarpus marsupium* Roxb.) using in vitro seedlings. *Plant Cell Tissue Organ Cult. (PCTOC)* **2021**, *144*, 449–462. [CrossRef]
14. Faisal, M.; Alatar, A.A.; Abdel-Salam, E.M.; Qahtan, A.A. Effects of 4-CPPU on in vitro multiplication and sustainable generation of *Hibiscus rosa-sinensis* L. 'White Butterfly'. *Saudi J. Biol. Sci.* **2020**, *27*, 412–416. [CrossRef]
15. Huetteman, C.A.; Preece, J.E. Thidiazuron: A potent cytokinin for woody plant tissue culture. *Plant Cell Tissue Organ Cult.* **1993**, *33*, 105–119. [CrossRef]
16. Javed, S.B.; Alatar, A.A.; Anis, M.; El-Sheikh, M.A. In vitro regeneration of coral tree from three different explants using thidiazuron. *HortTechnology* **2019**, *29*, 946–951. [CrossRef]
17. Lu, C.-Y. The use of thidiazuron in tissue culture. *Vitr. Cell. Dev. Biol.-Plant* **1993**, *29*, 92–96. [CrossRef]
18. Corredoira, E.; Ballester, A.; Vieitez, A.M. Thidiazuron-induced high-frequency plant regeneration from leaf explants of *Paulownia tomentosa* mature trees. *Plant Cell Tissue Organ Cult.* **2008**, *95*, 197–208. [CrossRef]
19. Feng, J.-C.; Yu, X.M.; Shang, X.L.; Li, J.D.; Wu, Y.X. Factors influencing efficiency of shoot regeneration in *Ziziphus jujuba* Mill. 'Huizao'. *Plant Cell Tissue Organ Cult. (PCTOC)* **2010**, *101*, 111–117. [CrossRef]
20. Dangi, B.; Khurana-Kaul, V.; Kothari, S.L.; Kachhwaha, S. Micropropagtion of *Terminalia bellerica* from nodal explants of mature tree and assessment of genetic fidelity using ISSR and RAPD markers. *Physiol. Mol. Biol. Plants* **2014**, *20*, 509–516. [CrossRef]
21. Saha, S.; Adhikari, S.; Dey, T.; Ghosh, P. RAPD and ISSR based evaluation of genetic stability of micropropagated plantlets of *Morus alba* L. variety S-1. *Meta Gene* **2015**, *7*, 7–15. [CrossRef] [PubMed]
22. Fatima, N.; Anis, M. Role of growth regulators on in vitro regeneration and histological analysis in Indian ginseng (*Withania somnifera* L.) Dunal. *Physiol. Mol. Biol. Plants* **2012**, *18*, 59–67. [CrossRef] [PubMed]
23. Papafotiou, M.; Bertsouklis, K.F.; Trigka, M. Micropropagation of *Arbutus unedo, A. andrachne*, and their natural hybrid, *A.* x *andrachnoides* from seedling explants. *J. Hortic. Sci. Biotechnol.* **2013**, *88*, 768–775. [CrossRef]

24. Husain, M.K.; Anis, M.; Shahzad, A. In Vitro Propagation of Indian Kino (*Pterocarpus marsupium* Roxb.) Using Thidiazuron. *Vitr. Cell. Dev. Biol.-Plant* **2007**, *43*, 59–64. [CrossRef]
25. Fatima, N.; Ahmad, N.; Ahmad, I.; Anis, M. Interactive effects of growth regulators, carbon sources, pH on plant regeneration and assessment of genetic fidelity using single primer amplification reaction (SPARS) techniques in *Withania somnifera* L. *Appl. Biochem. Biotechnol.* **2015**, *177*, 118–136. [CrossRef]
26. Ahmad, A.; Ahmad, N.; Anis, M. Preconditioning of Nodal Explants in Thidiazuron-Supplemented Liquid Media Improves Shoot Multiplication in *Pterocarpus marsupium* (Roxb.). In *Thidiazuron: From Urea Derivative to Plant Growth Regulator*; Ahmad, N., Faisal, M., Eds.; Springer: Singapore, 2018; pp. 175–187.
27. Bertsouklis, K.F.; Papafotiou, M. Morphometric and Molecular Analysis of the Three Arbutus Species of Greece. *Not. Bot. Horti Agrobot.* **2016**, *44*, 423–430. [CrossRef]
28. Mishra, T.; Goyal, A.K.; Sen, A. Somatic Embryogenesis and Genetic Fidelity Study of the Micropropagated Medicinal Species, *Canna indica*. *Horticulturae* **2015**, *1*, 3–13. [CrossRef]
29. Saldaña, C.L.; Cancan, J.D.; Cruz, W.; Correa, M.Y.; Ramos, M.; Cuellar, E.; Arbizu, C.I. Genetic Diversity and Population Structure of Capirona (*Calycophyllum spruceanum* Benth.) from the Peruvian Amazon Revealed by RAPD Markers. *Forests* **2021**, *12*, 1125. [CrossRef]
30. Salem, J.; Hassanein, A.; El-Wakil, D.A.; Loutfy, N. Interaction between Growth Regulators Controls In Vitro Shoot Multiplication in Paulownia and Selection of NaCl-Tolerant Variants. *Plants* **2022**, *11*, 498. [CrossRef]
31. Murashige, T.; Skoog, F. A Revised Medium for Rapid Growth and Bio Assays with Tobacco Tissue Cultures. *Physiol. Plant.* **1962**, *15*, 473–497. [CrossRef]
32. Lichtenthaler, h.k.; Wellburn, A.R. Determinations of total carotenoids and chlorophylls a and b of leaf extracts in different solvents. *Biochem. Soc. Trans.* **1983**, *11*, 591–592. [CrossRef]
33. Dudley, S.A. Differing selection on plant physiological traits in response to environmental water availability: A test of adaptive hypotheses. *Evolution* **1996**, *50*, 92–102. [CrossRef] [PubMed]
34. Doyle, J.J.; Doyle, J.L. A rapid DNA isolation procedure for small quantities of fresh leaf tissue. *Phytochem. Bull.* **1987**, *19*, 11–15.
35. Liu, Q.; Salih, S.; Hammerschlag, F. Etiolation of 'Royal Gala' apple (*Malus* × *domestica* Borkh.) shoots promotes high-frequency shoot organogenesis and enhanced, -glucuronidase expression from stem internodes. *Plant Cell Rep.* **1998**, *18*, 32–36. [CrossRef]
36. Matand, K.; Prakash, C.S. Evaluation of peanut genotypes for in vitro plant regeneration using thidiazuron. *J. Biotechnol.* **2007**, *130*, 202–207. [CrossRef] [PubMed]
37. Murch, S.J.; Saxena, P.K. Molecular fate of thidiazuron and its effects on auxin transport in hypocotyls tissues of *Pelargonium* × *hortorum* Bailey. *Plant Growth Regul.* **2001**, *35*, 269–275. [CrossRef]
38. Murthy, B.N.S.; Murch, S.J.; Saxena, P.K. Thidiazuron: A potent regulator of in vitro plant morphogenesis. *Vitr. Cell. Dev. Biol.-Plant* **1998**, *34*, 267. [CrossRef]
39. Hussain, S.A.; Ahmad, N.; Anis, M. Synergetic effect of TDZ and BA on minimizing the post-exposure effects on axillary shoot proliferation and assessment of genetic fidelity in *Rauvolfia tetraphylla* (L.). *Rend. Lincei Sci. Fis. Nat.* **2018**, *29*, 109–115. [CrossRef]
40. Faisal, M.; Ahmad, N.; Anis, M. Shoot multiplication in *Rauvolfia tetraphylla* L. using thidiazuron. *Plant Cell Tissue Organ Cult.* **2005**, *80*, 187–190. [CrossRef]
41. Husain, M.K.; Anis, M. Rapid in vitro multiplication of *Melia azedarach* L. (a multipurpose woody tree). *Acta Physiol. Plant.* **2009**, *31*, 765–772. [CrossRef]
42. Vengadesan, G.; Ganapathi, A.; Anand, R.P.; Selvaraj, N. In vitro propagation of *Acacia sinuata* (Lour.) Merr. from nodal segments of a 10-year-old tree. *Vitr. Cell. Dev. Biol.-Plant* **2003**, *39*, 409–414. [CrossRef]
43. Thomas, T.D. Thidiazuron induced multiple shoot induction and plant regeneration from cotyledonary explants of mulberry. *Biol. Plant.* **2003**, *46*, 529–533. [CrossRef]
44. Perveen, S.; Javed, S.B.; Anis, M.; Aref, I.M. Rapid in vitro multiplication and ex vitro establishment of Caribbean copper plant (*Euphorbia cotinifolia* L.): An important medicinal shrub. *Acta Physiol. Plant.* **2013**, *35*, 3391–3400. [CrossRef]
45. Hussain, S.A.; Ahmad, N.; Anis, M.; Hakeem, K.R. Development of an efficient micropropagation system for *Tecoma stans* (L.) Juss. ex Kunth using thidiazuron and effects on phytochemical constitution. *Vitr. Cell. Dev. Biol.-Plant* **2019**, *55*, 442–453. [CrossRef]
46. Siddique, I.; Anis, M. Direct plant regeneration from nodal explants of *Balanites aegyptiaca* L. (Del.): A valuable medicinal tree. *New For.* **2009**, *37*, 53–62. [CrossRef]
47. Javed, S.; Anis, M.; Khan, P.; Aref, I. In vitro regeneration and multiplication for mass propagation of *Acacia ehrenbergiana* Hayne: A potential reclaiment of denude arid lands. *Agrofor. Syst.* **2013**, *87*, 621–629. [CrossRef]
48. Ahmad, N.; Shahid, A.; Javed, S.B.; Khan, M.I.; Anis, M. Micropropagation of *Vitex* spp. through in vitro manipulation: Current status and future prospectives. *J. Appl. Res. Med. Aromat. Plants* **2015**, *2*, 114–123. [CrossRef]
49. Elmongy, M.S.; Cao, Y.; Zhou, H.; Xia, Y. Root development enhanced by using indole-3-butyric acid and naphthalene acetic acid and associated biochemical changes of in vitro azalea microshoots. *J. Plant Growth Regul.* **2018**, *37*, 813–825. [CrossRef]
50. Ahmad, A.; Anis, M. *Meta*-topolin improves in vitro morphogenesis, rhizogenesis and biochemical analysis in *Pterocarpus marsupium* Roxb.: A potential drug-yielding tree. *J. Plant Growth Regul.* **2019**, *38*, 1007–1016. [CrossRef]
51. Ahmad, N.; Anis, M. Rapid clonal multiplication of a woody tree, *Vitex negundo* L. through axillary shoots proliferation. *Agrofor. Syst.* **2007**, *71*, 195–200. [CrossRef]

52. Faisal, M.; Singh, S.; Anis, M. In vitro regeneration and plant establishment of *Tylophora indica* (Burm. F.) Merrill: Petiole callus culture. *Vitr. Cell. Dev. Biol.-Plant* **2005**, *41*, 511–515. [CrossRef]
53. Shahzad, A.; Faisal, M.; Anis, M. Micropropagation through excised root culture of *Clitoria ternatea* and comparison between in vitro—Regenerated plants and seedlings. *Ann. Appl. Biol.* **2007**, *150*, 341–349. [CrossRef]
54. Larkin, P.J.; Scowcroft, W.R. Somaclonal variation—A novel source of variability from cell cultures for plant improvement. *Theor. Appl. Genet.* **1981**, *60*, 197–214. [CrossRef] [PubMed]
55. Winter, P.; Kahl, G. Molecular marker technologies for plant improvement. *World J. Microbiol. Biotechnol.* **1995**, *11*, 438–448. [CrossRef]
56. Faisal, M.; Alatar, A.A.; Ahmad, N.; Anis, M.; Hegazy, A.K. An Efficient and Reproducible Method for in vitro Clonal Multiplication of *Rauvolfia tetraphylla* L. and Evaluation of Genetic Stability using DNA-Based Markers. *Appl. Biochem. Biotechnol.* **2012**, *168*, 1739–1752. [CrossRef]
57. Qahtan, A.A.; Faisal, M.; Alatar, A.A.; Abdel-Salam, E.M. High-Frequency Plant Regeneration, Genetic Uniformity, and Flow Cytometric Analysis of Regenerants in *Ruta chalepensis* L. *Plants* **2021**, *10*, 2820. [CrossRef]
58. Yusuf, A.; Nikhilesh, S.; Rao, P.S. Micropropagation of *Vitex negundo* L. and Random Amplified Polymorphic DNA Analysis of Micropropagated Plants. *J. Sustain. For.* **2012**, *31*, 267–282. [CrossRef]
59. Purohit, S.; Jugran, A.K.; Bhatt, I.D.; Palni, L.M.S.; Bhatt, A.; Nandi, S.K. In vitro approaches for conservation and reducing juvenility of *Zanthoxylum armatum* DC: An endangered medicinal plant of Himalayan region. *Trees* **2017**, *31*, 1101–1108. [CrossRef]
60. Thu, H.T.M.; Naing, A.H.; Jeong, H.Y.; Kim, C.K. Regeneration of Genetically Stable Plants from In Vitro Vitrified Leaves of Different Carnation Cultivars. *Plants* **2020**, *9*, 950. [CrossRef]
61. Clapa, D.; Hârța, M. Establishment of an Efficient Micropropagation System for *Humulus lupulus* L. cv. Cascade and Confirmation of Genetic Uniformity of the Regenerated Plants through DNA Markers. *Agronomy* **2021**, *11*, 2268. [CrossRef]

horticulturae

MDPI

Article

Regeneration of African Violet in Response to Light Quality

Zohreh Aslami [1], Masood Ghasemi Ghehsareh [1,*], Sayyed Mohammad Ehsan Mahdavi [2] and Silvana Nicola [3,*]

1 Department of Horticultural Science, College of Agriculture, Shahrekord University, Shahrekord 88186344141, Iran; aslami1@yahoo.com
2 Department of Horticultural Science, College of Agriculture, Shiraz University, Shiraz 7144165186, Iran; smemahdavi@gmail.com
3 Department of Agricultural, Forest and Food Sciences (DISAFA), University of Turin (UNITO), 10095 Grugliasco, Italy
* Correspondence: mghasemi1352@sku.ac.ir (M.G.G.); silvana.nicola@unito.it (S.N.)

Abstract: Light-emitting diode (LED) technology is a form of artificial lighting that offers precise control over spectral composition, creating specific conditions for plant growth and development. However, the influence of various LED wavelengths on the regeneration characteristics in African violet (AV) has not been extensively explored. This study aims to investigate the changes in the regeneration traits of AV when exposed to different LED light colors within controlled conditions. In this study, AV leaf cuttings were prepared and subjected to white, red, blue, and red + blue light colors for a period of three months in a growth chamber. Afterward, they were transferred to the laboratory for further analysis. The results indicated that the AVs treated with red + blue colors exhibited the most significant improvement in several morpho-physiological traits of both the roots and shoots. The highest total biomass (2.96 g), shoot fresh weight (1.76 g), root dry weight (0.14 g), root volume (3.10 cm^3), and shoot length (1.60 cm) were observed in this treatment group. Furthermore, the highest levels of photosynthetic pigments, such as chlorophyll a, chlorophyll b, and carotenoids (0.14, 0.12, and 3.80 mg g^{-1} f.w., respectively), were predominantly observed in the red + blue treatment group. In conclusion, this study introduces a novel methodology for optimizing lighting conditions to enhance the regeneration of African violets, shedding light on the potential for improving AV regeneration practices.

Keywords: light-emitting diodes; propagation; light quality; photosynthesis; root growth

Citation: Aslami, Z.; Ghehsareh, M.G.; Mahdavi, S.M.E.; Nicola, S. Regeneration of African Violet in Response to Light Quality. *Horticulturae* **2024**, *10*, 78. https://doi.org/10.3390/horticulturae10010078

Academic Editors: Konstantinos Bertsouklis, Epaminondas Kartsonas, Angela Carra and Rosario Paolo Mauro

Received: 13 December 2023
Revised: 7 January 2024
Accepted: 10 January 2024
Published: 12 January 2024

1. Introduction

The African violet (AV), *Saintpaulia ionantha* H. Wendl., is a prevalent houseplant native to East Africa, characterized by its appealing flowers and attractive foliage, belonging to the *Gesneriaceae* family [1]. This species was initially named "ionantha", signifying "having violet-like flowers" in Greek, by Hermann Wendland in 1883 [2]. Over the years, AV has established itself as a significant ornamental plant in the floral industry, thanks to its visual appeal, shade tolerance, and the ability to bloom under artificial lighting conditions [1,2]. More importantly, AV has seen a significant rise in economic status, making it one of the best-selling plants in Europe and the United States [3]. However, certain crucial attributes need improvement, such as enhancing its quality concerning the ability to withstand transportation, tolerate cold temperatures, resist diseases and pests, and develop vibrant flower colors [3]. It features relatively short fleshy stems, roughly rounded leaves with scalloped edges, and small clusters of flowers surrounded by foliage [2]. Under suitable conditions, AV can bloom nearly year-round [2]. Among its specific requirements, adequate lighting is of paramount importance in the growth stages of AV [1]. Although AV is typically considered a low-light-friendly houseplant, insufficient light can disrupt its flowering patterns [2]. Therefore, it is crucial to consider the application of various colors of light with specific wavelengths, as they can impact the quality and health of AV from its vegetative phase to maturity [1].

AV can be propagated by seeds, leaf cuttings, and crown division, but most cultivars are propagated by leaf cuttings in commercial production. For this purpose, rooting hormones are not needed [4]. In vegetative propagation with leaf cuttings, a new shoot system must be initiated from an adventitious bud and new adventitious roots [5]. In order for the cells to form adventitious roots and buds, they must undergo development and differentiation. Given that certain cells and plant organs exhibit a more pronounced expression of new meristem growth points, the propagator must establish optimal conditions to foster the regeneration of the root or stem systems [5]. In addition to managing mother plants, the treatment of cuttings and manipulation of environmental conditions are effective for its success [5]. Light is an influential factor in forming adventitious roots and buds on cuttings [6].

Light serves as a crucial energy source that significantly affects various physiological and biochemical processes in plants. In essence, it is evident that light plays a pivotal role in all aspects of plant growth and development, spanning from seed germination to the ripening stages [7]. According to the quality, intensity, and duration of light, some signaling pathways can be activated or deactivated in plants, thereby influencing biological processes related to their morphogenesis [8]. Light quality affects photosynthesis, chlorophyll formation, root and stem length, flower bud formation, seed germination, and the rooting of cuttings [9,10]. In the commercial propagation of plants, it is necessary to use artificial light sources, especially in controlled conditions, due to the inherent limitations of natural light, including seasonal variations, weather fluctuations, and geographical latitude, which can impede plants' access to the optimal wavelengths necessary for their efficient growth [7].

Light-emitting diodes (LEDs) are new artificial lights, which have made it possible to control the quality of light [11]. LED technology has been increasingly used in protected agriculture nowadays, due to its capacity for precise control over light spectrum, intensity, and timing [12]. LED lighting encompasses the visible light ranges between 400 and 700 nm, i.e., photosynthetically active radiation (PAR), including blue (B)-violet light (400–490 nm), yellow-green (G) light (490–550 nm), and red (R) light (660 nm). Among these wavelengths, R and B wavelengths play a central role in the photomorphogenesis of plants, and achieving an optimal ratio between them (R:B ratio) is important. R light affects all stages of the plant life cycle, largely through a complex network of phytochrome photoreceptors [7]. B light promotes plant growth and development owing to two types of photoreceptors, namely cryptochromes and phototropins. Regardless of R and B, chloroplasts display the specific absorption of other colors by exclusive photoreceptors contributing to PAR [13].

Photomorphogenesis is a developmental response of plants to light correlated to photoreceptor proteins and signaling pathways [14]. It takes part in a myriad of morphological, physiological, and molecular activities of plants. Light shapes plant architecture, including the elongation of stem and coleoptile cells, the branching of shoots, and the expansion and enlargement of leaves [15]. Regarding the effect of light quality on root and shoot regeneration, studies have been conducted on some plants and different results have been obtained. The use of R light increased in vitro *Tripterospermum* rooting, but B light prevented it [16]. Kurilčik et al. [17] observed that B light added to R and far red (FR) light affected the rhizogenesis of *Chrysanthemum* micro-cuttings. The B light component was found to inhibit the rooting rate, but it increased the ratio of the fresh and dry weight of the explants [17]. Cavallaro et al. [18] reported the highest pineapple shoot proliferation under R light and the lowest under white (W) LED light. Kwon et al. [19] demonstrated that a combination of R + B LED yielded an optimal rate of in vitro regeneration in *Populus euramericana*, compared to B LED, R LED, or fluorescent light. Bello-Bello et al. [20] investigated the effect of five light treatments, including fluorescent, W, B, R, and R + B LEDs, on the shoot proliferation and growth of the vanilla plant (*Vanilla planifolia* Andrews) under in vitro conditions. Shoot proliferation was the highest in fluorescent light, W LED, and R + B combination. Dewir et al. [21] recorded that the cultivation of *Spathiphyllum cannifolium* under R LED yielded a higher shoot multiplication rate than under B or R + B LEDs. It is also noteworthy that, under R LED, the plant height was maximal in *Gerbera jamesonii* [22]. However, Kostadinova et al. [23] recorded that the shoot length in *Pyrus communis* 'OHF333' was greater under white and B LED. Lotfi [24], investigating the effect of R, B, and equal R + B LED lights on the growth of *Pyrus communis* 'Arbi' explants under in vitro conditions, showed that the shoot height was better under R light. According to the studies mentioned above, obtaining information on using LEDs as a light source for plants is essential, especially in the propagation stage. Furthermore, due to the short size of AVs growing on multi-story shelves, determining the optimal artificial light quality for such conditions becomes inevitable. However, according to our information, no report has been published about the effect of light quality on the regeneration and growth of adventitious shoots/buds in the leaf cuttings of any plant. Therefore, this experiment investigated the effect of different LED light spectra on root and bud regeneration in African violet leaf cuttings. In this investigation, we hypothesized that each light color of the LEDs would significantly enhance the specific regeneration and relevant morpho-physiological traits of AV. The objectives were to (i) investigate the role of LEDs in enhancing the regeneration parameters of AV, particularly in the roots, shoots, and adventitious buds, (ii) compare the effects of white, red, blue, and red + blue light colors on AV regeneration, (iii) assess the accumulation of photosynthetic pigments in AV, and (iv) establish an optimal lighting condition for regenerating AV.

2. Materials and Methods

2.1. Plant Materials and Conditions

2.1.1. Specimen Species

The common propagation method of AV is leaf cutting. Several steps were carried out before transferring the cuttings into specific lighting conditions. In the initial step, a number of micro-propagated AVs were purchased from the Agricultural Biotechnology Research Institute, Isfahan, Iran. In the second step, these micro-propagated AVs were acclimated to the Research Greenhouse of the Department of Horticultural Sciences, Shahrekord University, Shahrekord, Iran (50°49′ E and 32°21′ N—altitude 2125 m a.s.l.). They were planted in plastic pots (i.e., a 1:1 mix of coco coir and perlite) and grown for nearly four months, with daytime and nighttime temperatures averaging 25 $\pm$ 2 and 18 $\pm$ 2 °C, respectively, and a relative humidity range of 60 $\pm$ 5%. In the third step, leaf-cutting materials of AV (approximately 4.5 $\times$ 3 cm, with 2 cm of petiole) were prepared from the mature leaves of the stock plants in the middle of winter, on January 21st. In the fourth step, before placing the leaf cuttings in the growing media (i.e., a 1:1 mix of coco coir and perlite) of the growth chambers, they were treated with a systemic benzimidazole fungicide, known as 'Benomyl' (Zagro Europe GmbH, Rheinfelden, Germany) at a concentration of 1 mg L^{-1} in distilled water for 10 min. This treatment was employed to protect the cuttings from pathogens.

2.1.2. Lighting Treatments

After disinfecting the AV leaf cuttings of the mature leaves with the aforesaid fungicide, they were transferred into four plant growth chambers. The experiment was carried out in two two-story cabinets, with each floor containing a chamber measuring 120 $\times$ 80 cm. Each chamber had a light panel at the top and a planting box equipped with a heating cable at the bottom. All three spectra of light included (1) white (W) [a green (G) (480–660 nm): blue (B) (410–460 nm) ratio of 0.33, i.e., 25 and 75%, respectively], considered the control, (2) red (R) (600–650 nm), (3) B (430–500 nm), and (4) R + B (a R:B ratio of 1) wavelengths (Figure 1). The lights were installed in the chambers and their design was such that it was possible to turn on and off and adjust the intensity of each color of light separately. Therefore, it was possible to use each of the white, blue, and red colors individually and any desired combination or intensity of them.

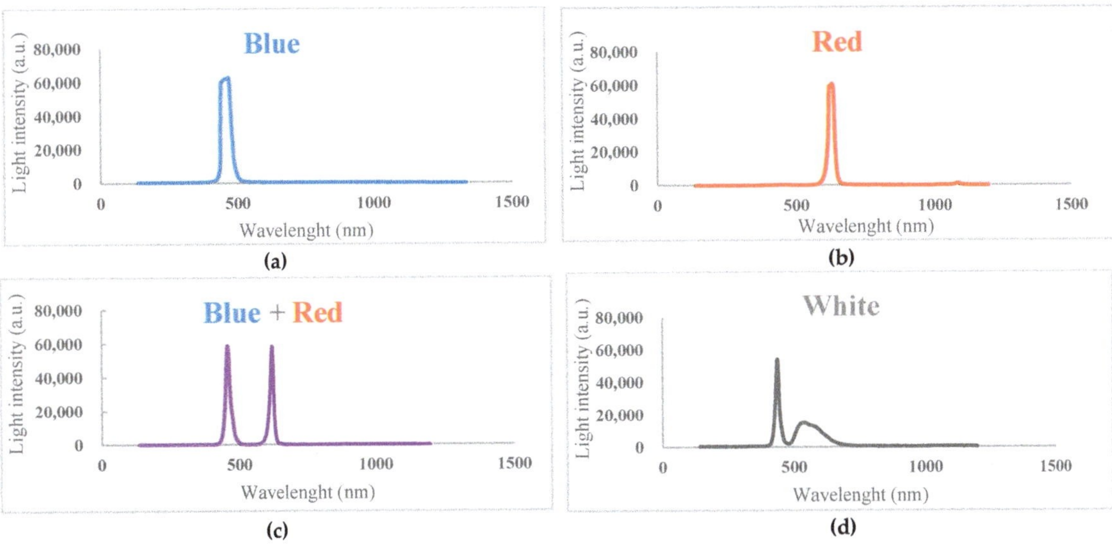

Figure 1. Light spectrum of different LED colors used in the experiment, measured using a spectrometer (Spectrometer V900, Optical Physics Technologists Co., Kashan, Iran).

This study was conducted based on a completely randomized experiment with four light treatments: 100% white, 100% red, 100% blue, and a 50% red + 50% blue combination. Each treatment had its own chamber where the desired light intensity and qualities were adjusted (Figure 2). The light intensity in each chamber/treatment was set at 80 μmol m^{-2} s^{-1}, measured by a PAR quantum sensor (MQ-500: Full-Spectrum Quantum Meter, Apogee Instruments, Inc., Logan, UT, USA). The reason why this light intensity was selected is that the optimal light intensity for micro-propagating of AV is just 70–100 μmol m^{-2}, in accordance with the previous studies [25]. Each treatment consisted of three replicates with 10 leaf cuttings in each one, which were planted in rows with a spacing of 5 cm between leaf cuttings and 10 cm between rows. The propagation substrate consisted of a mixture of coir and perlite in a 1:1 (v/v) ratio, which had been autoclaved for 20 min.

The light panels of the growth chambers were equipped with 3-Watt high-power 252 LEDs (Epistar Group, Xiamen, China), including 84 W, 84 R, and 84 B LEDs, evenly distributed at a distance of 50 cm from the planting beds. The regulation of light intensity, photoperiod, temperature, humidity, and irrigation for each chamber was as follows: (1) the intensity of each lighting treatment was adjusted using a dimmer and a PAR meter; (2) the photoperiod was set to 12 h light from 7 a.m. to 7 p.m. using a timer; (3) the internal chamber temperature was maintained at 21 ± 3 °C; (4) the root-zone temperature was controlled by a thermostat and maintained at 20 ± 2 °C; (5) the relative humidity range was regulated to 80 ± 5% by a humidity meter; (6) the irrigation system applied to the cuttings was an intermittent mist control system with nozzles sized at 0.5 mm and running timer of 1 min per every 2.5 h. The growing media were watered up to field capacity (FC).

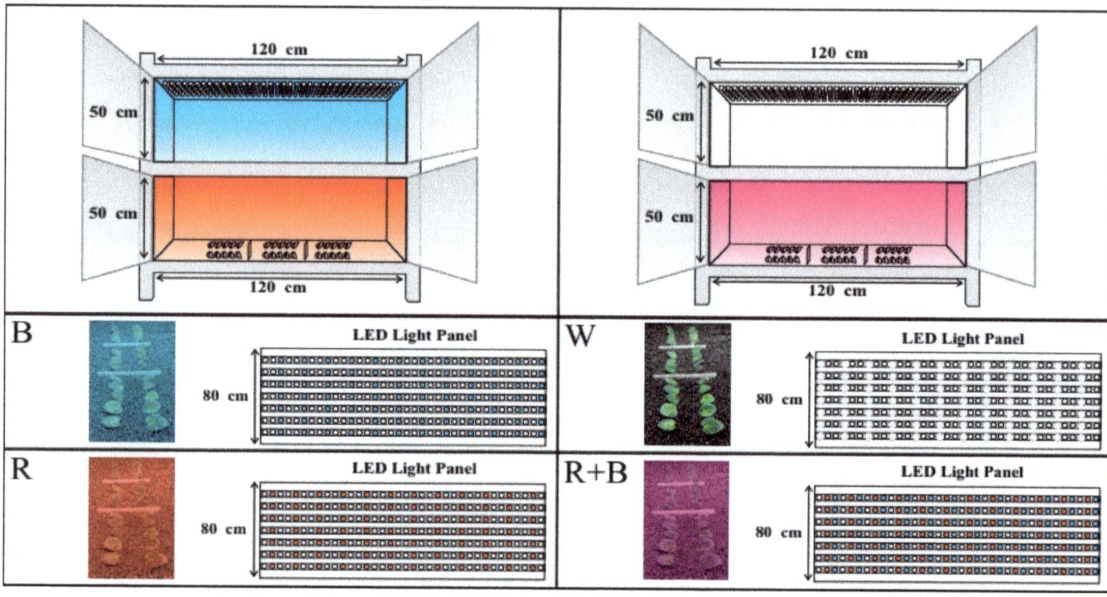

Figure 2. The schematic image of the chambers used to apply different LED light treatments and the layout of African violet leaf cuttings.

2.2. Measurements

2.2.1. Growth Parameters

Three months after planting, which took place in the middle of winter, all the leaf cutting materials subjected to the light treatments were removed from their growing media for the analysis of AV regeneration characteristics. By doing this, the length of roots (RL), the volume of roots (RV), the fresh weight of roots (RFW), the dry weight of roots (RDW), the length of shoots (SL), the number of adventitious buds/shoots (SN), the number of leaves (LN), the fresh weight of shoots (SFW), and the dry weight of shoots (SDW) were measured separately. The adventitious buds/shoots and leaves were counted for each treatment. To determine the length of roots and shoots, an electronic digital vernier caliper (Model: 0–150 mm, Guanglu Instruments Co., Ltd., Guilin, China) was used. The root volume was determined based on Archimedes' principle using the water displacement method, immersing each root in water in a graduated cylinder and measuring the volume [26]. The fresh and dry weights of roots and shoots were measured by an electronic weighing scale (Model: GR-200, A&D Co., Ltd., Tokyo, Japan). The roots and shoots were dried in an oven (FD 56, Binder GmbH, Tuttlingen, Germany) at 70 °C for 48 h. Furthermore, it is worth noting that plants typically produce thin and elongated stems with low dry weight in low-light conditions [27]. Therefore, the following parameters were calculated using measurement ratios: total biomass (TB), shoot length to shoot dry wight ratio (SL:SDW ratio), and root to shoot ratio (Ro:Sh ratio).

2.2.2. Photosynthetic Pigments Content

The chlorophyll content (Chl) (including *a*, *b*, and total) and the carotenoid content (Car) were measured based on the method described by Lichtenthaler [28], with slight modifications. To extract the photosynthetic pigments, 500 mg of fresh leaf blade was ground with a mortar and pestle with 5 mL of 80% acetone. Then, each homogenous treatment was transferred to a 10 mL falcon, and the volume of each falcon was increased to 10 mL with acetone. The extracted solution was centrifuged at 4000 rpm for 5 min. After that, the absorbance of the separated supernatant was read spectrophotometrically (T60

UV-Visible Spectrophotometer, PG Instruments Ltd., Lutterworth, UK) at 663 nm for Chl *a*, 646 nm for Chl *b*, and 470 nm for Car. The equations for determining them quantitatively (μg mL^{-1}) are as follows:

$$\text{Chl } a\ (\mu\text{g mL}^{-1}) = (12.25 \times A_{663}) - (2.79 \times A_{646})$$
$$\text{Chl } b\ (\mu\text{g mL}^{-1}) = (21.50 \times A_{646}) - (5.10 \times A_{663})$$
$$\text{Chl } a{+}b_{\ (\text{total})}\ (\mu\text{g mL}^{-1}) = \text{Chl } a + \text{Chl } b$$
$$\text{Car}\ (\mu\text{g mL}^{-1}) = [(1000 \times A_{470}) - (1.82 \times \text{Chl } a) - (85.02 \times \text{Chl } b)] \div 198$$

where A is optical density, and their formulations were converted from μg mL^{-1} to mg g^{-1} by multiplying each equation yield and V/1000 W, V and W here were the final solution volume and the fresh leaf weight, respectively.

2.3. Experimental Design and Statistical Analysis

The research was laid out in a completely randomized design (CRD) experiment with three replicates and ten leaf cuttings per each replication. Data were analyzed using the SAS® software (version 9.3; SAS Institute Inc., Cary, NC, USA) employing the one-way analysis of variance (ANOVA). The *p*-value was considered in accordance with the least significant difference (LSD) test at the level of 5%. Furthermore, the loading plot and score plot for the parameters under analysis were generated using the Principal Component Analysis (PCA) utilizing Minitab® software (version 19.2020.1; Minitab LLC., State College, PA, USA).

3. Results

The results of the statistical analyses are reported in the following paragraphs, excluding the total chlorophyll (Chl *t*) content and the root to shoot ratio (Ro:Sh ratio) as they had no differences among the LED light treatments at the 5% level.

3.1. Growth Responses of Roots to Light Colors

The analysis indicated that the greatest amount of root length (RL) was in the AVs treated with B light, measuring 14 cm, followed by R + B lights at 12.70 cm, with no significant difference between them. In contrast, the lowest RL was observed in the AVs treated with R light, measuring 7.80 cm, which did not significantly differ from the W treatments at 9.10 cm (Table 1). Regarding the root volume (RV), the highest (3.10 cm^3) and lowest (1.50 cm^3) RVs were observed in the R + B and R treatments, respectively. The latter did not significantly differ from the B and W treatments (Table 1). The greatest root fresh weight (RFW) was obtained in the R + B (1.20 g), followed by the W treatments (1.03 g), showing no significant difference. Conversely, the lowest RFW was in the AV plants treated under R light (0.65 g), and then under B light (0.79 g), with no significant difference observed between these two treatments (Table 1). The greatest RDW, measuring 0.14 g, was observed in the plants treated with R + B light, significantly higher that all the other treatments (Table 1). Overall, the data analysis regarding the root attributes reveals that the combination of R and B lights could enhance the roots growth of AV in most cases; however, R light alone did not significantly impact the roots of the AVs.

Table 1. Effect of different LED light spectra on root growth indices in African violets.

Light Spectra	RL (cm)	RV (cm^3)	RFW (g)	RDW (g)
White	9.10 ± 0.62 ^{b†}	2.20 ± 0.25 ^{b}	1.03 ± 0.12 ^{ab}	0.09 ± 0.01 ^{b}
Red	7.80 ± 0.34 ^{b}	1.50 ± 0.18 ^{b}	0.65 ± 0.05 ^{c}	0.06 ± 0.03 ^{b}
Blue	14.00 ± 0.82 ^{a}	2.00 ± 0.15 ^{b}	0.79 ± 0.13 ^{bc}	0.09 ± 0.03 ^{b}
Red + Blue	12.70 ± 091 ^{ab}	3.10 ± 0.35 ^{a}	1.20 ± 0.13 ^{a}	0.14 ± 0.01 ^{a}
Significance	***	**	*	*

Abbreviations: RL, root length; RV, root volume; RFW, root fresh weight; RDW, root dry weight. † Means having different letters are significantly different at 5% level from LSD test. Data are presented as the mean ± standard error. Not significant (ns), *, **, *** indicate significance at $p \leq 0.05$, $p \leq 0.01$, and $p \leq 0.001$, respectively.

3.2. Leaf and Shoot Growth Dynamics

The results showed that the highest adventitious bud/shoot number (SN) was counted in AV plants grown under R + B or W light spectra, both recording 5.50. The lowest SN was noted in AV plants grown under R light spectrum (3.80), and B light (3.90), with no significant difference observed (Table 2, Figure 3). Additionally, the analysis of the leaf number (LN) indicated that treatments with B light resulted in 3.40 leaves, significantly more than all the other treatments. Specifically, 1.60 leaves were observed for R + B light, 1.30 for W light, and merely 0.50 leaves for R light, all statistically different from one another (Table 2). Regarding the shoot length (SL), considering the AV's typical rosette growth, part of its total length comprises petioles. There was a notable increase in the SL in the AVs treated with R + B light, measuring 1.60 cm, while the shortest SL was seen in the AVs treated with W and B light at 1.30 cm, and R light at 1.40 cm, with no significant difference (Table 2).

Figure 3. The effect of different LED light spectrums on African violet adventitious bud formation.

For the shoot fresh weight (SFW), the R + B treatments exhibited the highest (1.76 g), and R light the lowest (0.97 g), which were significantly different, while the B and W light treatments fell in between, differing from both the highest and lowest value. No significant difference was observed between the W and B treatments (Table 2). A consequent similar trend was observed for the shoot dry weight (SFW), with the highest (0.13 g) and lowest (0.04 g) recorded in the R + B and R treatments, respectively, while B and W fell in the middle, not significantly different from each other (Table 2). Hence, applying R + B spectra might enhance the shoot characteristics and root attributes compared to other LED treatments, while it appears that the R light alone does not positively impact the shoot growth and developmental traits in AVs.

Table 2. Effect of different LED light spectra on shoot growth indices in African violets.

Light Spectra	SN (Count)	LN (Count)	SL (cm)	SFW (g)	SDW (g)
White	$5.50 \pm 0.17^{a\dagger}$	1.30 ± 0.08^{c}	1.30 ± 0.02^{b}	1.24 ± 0.09^{b}	0.08 ± 0.00^{b}
Red	3.80 ± 0.10^{b}	0.50 ± 0.05^{d}	1.40 ± 0.05^{b}	0.97 ± 0.06^{c}	0.04 ± 0.00^{c}
Blue	3.90 ± 0.17^{b}	3.40 ± 0.03^{a}	1.30 ± 0.08^{b}	1.33 ± 0.08^{b}	0.09 ± 0.00^{b}
Red + Blue	5.50 ± 0.35^{a}	1.60 ± 0.06^{b}	1.60 ± 0.03^{a}	1.76 ± 0.07^{a}	0.13 ± 0.00^{a}
Significance	***	***	*	***	***

Abbreviations: SN, adventitious bud/shoot number; LN, leaf number; SL, shoot length; SFW, shoot fresh weight; SDW, shoot dry weight. † Means having different letters are significantly different at 5% level from LSD test. Data are presented as the mean ± standard error. Not significant (ns), *, *** indicate significance at $p \leq 0.05$ and $p \leq 0.001$, respectively.

3.3. Photosynthetic Pigments and Vegetative Characteristics

Under specific LED light color treatments, the chlorophyll (Chl) content exhibited significant variations. Surprisingly, the highest Chl *a* content was observed in the AVs treated with R light (0.15 mg g^{-1} f.w.), which did not significantly differ from the R + B (0.14 mg g^{-1} f.w.) and W (0.12 mg g^{-1} f.w.) treatments. However, the lowest Chl *a* content (0.07 mg g^{-1} f.w.) was measured in the B treatment of the AVs (Table 3). Contrary to Chl *a*, the results indicated that the content of Chl *b* (0.09 mg g^{-1} f.w.) was lowest when the plants were exposed to R light. The highest Chl *b* content (0.13 mg g^{-1} f.w.) was observed in the W and B treatments, with no significant difference between them (Table 3). The Chl *a*/*b* ratio followed a similar pattern to Chl *a*. The AVs treated with the R light showed a ratio at 1.55, while the lowest value was assessed in the AVs treated with the B color, at 0.57 (Table 3).

Table 3. Effect of different LED light spectra on photosynthetic pigments and vegetation indices in African violets.

Light Spectra	Chl *a* (mg g^{-1} f.w.)	Chl *b* (mg g^{-1} f.w.)	Chl *a*/*b* (mg g^{-1} f.w.)	Chl *t* (mg g^{-1} f.w.)	Car (mg g^{-1} f.w.)	TB (g)	Ro:Sh	SL:SDW Ratio
White	0.12 ± 0.02 [a†]	0.13 ± 0.01 [a]	0.98 ± 0.27 [bc]	0.25 ± 0.03	3.20 ± 0.17 [ab]	2.27 ± 0.09 [b]	1.08 ± 0.13	16.50 ± 0.97 [b]
Red	0.15 ± 0.00 [a]	0.09 ± 0.00 [b]	1.55 ± 0.10 [a]	0.24 ± 0.01	2.70 ± 0.19 [b]	1.62 ± 0.09 [c]	1.05 ± 0.13	36.90 ± 4.99 [a]
Blue	0.07 ± 0.00 [b]	0.13 ± 0.01 [a]	0.57 ± 0.06 [c]	0.20 ± 0.02	2.60 ± 0.18 [b]	2.12 ± 0.20 [b]	0.95 ± 0.24	14.92 ± 1.32 [b]
Red + Blue	0.14 ± 0.01 [a]	0.12 ± 0.00 [ab]	1.14 ± 0.52 [ab]	0.26 ± 0.04	3.80 ± 0.44 [a]	2.96 ± 0.08 [a]	1.05 ± 0.05	12.52 ± 0.80 [b]
Significance	*	*	**	ns	*	***	ns	***

Abbreviations: Chl, chlorophyll; Chl *t*, total chlorophyll; Car, carotenoids; TB, total biomass; Ro:Sh, the root to shoot ratio; SL, shoot length; SDW, shoot dry weight. † Means having different letters are significantly different at 5% level from LSD test. Data are presented as the mean ± standard error. Not significant (ns), *, **, *** indicate significance at $p \leq 0.05$, $p \leq 0.01$, and $p \leq 0.001$, respectively.

Parallel to Chl, the content of carotenoids (Car) was assayed. The results indicated that the AVs treated with the R + B lights reached a peak of 3.80 mg g^{-1} f.w., and the B and R lights reduced the Car content to 2.60 mg g^{-1} f.w., with no significant difference between them (Table 3). The total biomass (TB) in the AVs treated with the R + B lights (2.96 g) was significantly higher than in the other treatments. With R light, the plants had the lowest TB at 1.62 g (Table 3). Concerning the shoot length to shoot dry wight ratio (SL:SDW ratio), the plants treated with R light exhibited a notably increased ratio at 36.90 cm g^{-1} compared to the other treatments, which had lower values. However, these values were not significantly different from each other, ranging down to 12.52 cm g^{-1} in the plants grown under R + B light (Table 3).

3.4. Principal Component Analysis (PCA) of Root Growth, Shoot Growth, and Pigments

A PCA was performed on the data from all the AV leaf cuttings, taking into account the different LED wavelength treatments. The biplot graph illustrating this analysis is shown in Figure 4. The first two components (PCs) together accounted for over 90% of the total variance, with PC1 and PC2 explaining 59% and 32.4% of the variance, respectively. Therefore, these PCs can be effectively used to evaluate the relationships between the traits and treatments. The percentage of variance explained and the eigenvalues associated with each component (four components in total) are detailed in Table 4.

Table 4. Eigenvalues associated with each component in the principal components analysis.

	PC1	PC2	PC3	PC4
Eigenvalue	10.038	5.510	1.452	0.000
Proportion	0.590	0.324	0.085	0.000
Cumulative	0.590	0.915	1.000	1.000

In this study, several correlations and patterns were observed. PC1 showed positive correlations with the RFW, RDW, RV, SFW, SDW, TB, SN, and Chl *b*. Conversely, PC1 had negative correlations with the SL:SDW and Ro:Sh. On the other hand, PC2 was found to have positive correlations with the LN, RL, and Chl *b*, and negative correlations with the Chl *a*, Chl *t*, Chl *a/b*, Car, SL, and SN. In terms of the LED light treatments, the R, B, and R + B treatments were clearly differentiated and clustered based on PC1 and PC2. The R + B and W treatments were quite similar and positioned on the positive side of PC1, specifically in the lower right quadrant, and were clustered with the SN, RFW, and Car. On the other hand, the B light treatments were situated on the positive side of PC2, in the upper right quadrant, near the y-axis, and clustered with LN and RL. In contrast, the R treatments appeared on the negative side of PC1, in the lower left quadrant, and were correlated with the SL:SDW, Ro:Sh, and Chl *a/b* (Figure 4).

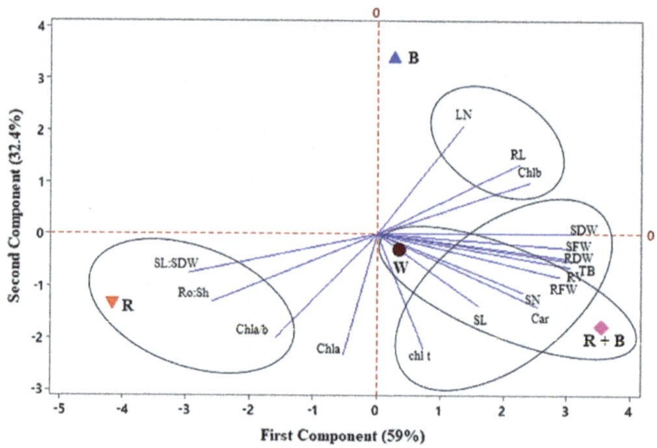

Figure 4. Biplot of principal component analysis for measured traits and treatments of African violets under different LED light conditions. Abbreviations: Chl, chlorophyll; Chl *t*, total chlorophyll; Car, carotenoids; LN, leaf number; RFW, root fresh weight; RDW, root dry weight; SFW, fresh weight of shoots; RL, root length; RV, root volume; SFW, shoot fresh weight; SDW, shoot dry weight; SN, adventitious bud/shoot number; SL, shoot length; TB, total biomass; Ro:Sh, the root to shoot ratio.

4. Discussion

In this study, the highest RL measurements were observed in the cuttings exposed to the B and R + B light colors, while the lowest measurements were recorded for the R and then W light. These findings align with the existing literature suggesting that B light stimulates the RL, while R light inhibits it. Previous studies in other plant species, such as *Symphyotrichum novi-belgii* var. *novi-belgii*, have shown similar results [29]. The primary reason for this phenomenon in roots may be linked to the biosynthesis and transport of auxin and gibberellin and their interaction with photoreceptors in plants [30]. For instance, B light has been found to increase root growth by reducing gibberellin levels and increasing auxin in Norway spruce (*Picea abies* (L.) Karst.) [31]. Conversely, R light facilitates the transfer of auxin from leaves to roots [32]. Studies have revealed that Cryptochrome1 (CRY1) positively regulates primary root elongation but negatively impacts lateral root development, particularly under a higher B light intensity [33]. Additionally, the present investigation found that the R + B LED lights increased levels of the RV, RFW, and RDW in the AVs, while the lowest values were observed in the R treatments. The AVs treated with W light showed a higher RV, RFW, and RDW compared to those treated with R light. Figure 1 indicates that the W LED lamps used in this experiment consisted of blue-violet (400–470 nm with a pick at 440 nm) and blue-green (655–485 nm with a pick at 557 nm) spectrum, resulting in similar impacts on the RV, RFW, and RDW as the B treatments in AV. Similar findings have been reported in *Malus domestica* Borkh., where the B and W lights did not significantly differ in the total RL, RV, and lateral root numbers [34].

The results revealed that the SFW and SDW were higher in the AVs treated with the R + B lights than in those treated with the R light, consistent with numerous previous studies. For instance, Miao et al. [35] in *Cucumis sativus* L. and Lim et al. [36] in *Gerbera jamesonii* cv. 'Shy Pink', reported the highest fresh and dry weight in the combination of R + B. Generally, R light initiates photosynthesis, while B light promotes the process [35,37]. The short-term effectiveness of B light is less than that of R light under limited light conditions [38]. Conversely, long-term exposure to R light can lead to the abnormal functioning of photosystem II (PSII), reducing the rate of photosynthesis [38]. Thus, the accumulation of dry and fresh matters can vary under different light conditions due to these reasons.

This investigation revealed that the SN in the AVs treated with W and R + B lights was significantly higher than those treated with B and R alone. Firstly, various pieces of literature support this phenomenon, highlighting the increased SN in many propagated plants when exposed to R + B wavelengths in comparison to other pure colors [19,20,36,39]. Bello-Bello et al. [20] in vanilla (*Vanilla planifolia* Andrews), Hung et al. [39] in *Vaccinium corymbosum*, Kwon et al. [19] in *Populus euramericana.*, and Lim et al. [36] in *Gerbera jamesonii* cv. 'Shy Pink' found that the combination of B and R lights could enhance the regeneration of shoots compared to using B or R LED lights under in vitro conditions. Secondly, as depicted in Figure 1, the W LED light here comprised two spectra: 75% blue-violet (400–470 nm) and 25% blue-green (655–485 nm). This underlines the significance of green (G) wavelengths in the regeneration of AV shoots. In this context, Kaewjampa and Shimasaki [40] demonstrated that the use of interval lighting with G LEDs could stimulate shoot proliferation and formation in *Cymbidium waltz 'Idol'* under in vitro conditions. The experiment clearly demonstrated the R + B treatment consistently produced the highest SL measurements in the AVs, while no significant differences were observed between the W, R, and B wavelengths. This aligns with previous research in blueberry [39], banana [41], and vanilla [20], where the increasing effect of R + B light on the shoot length was also observed. Conflicting reports exist on the effects of light quality on the SL, with R light generally increasing elongation and B light inhibiting growth [36,42]. However, there are exceptions, such as lettuce [43] and petunia [44], where the positive and negative effects of B and R lights on SL were observed, respectively. R light can individually enhance the SL by causing the excessive elongation of internodes in stems, known as red light syndrome [45], which is not considered a desirable trait [46]. On the other hand, the promotion of shoot elongation due to the shade-avoidance responses in plants often occurs when B light is combined with lower PHY activity (e.g., pure B and impure B created by adding low-level far-red light) [47,48]. However, in the present study, where the AVs treated with B showed the highest measurements of the LN, SFW, and SDW compared to the R-treated AVs, it appears that the increase in the SL of the AVs is more closely related to photosynthesis and less to shade-avoidance responses. Nonetheless, the highest ratio of SL:SDW was related to the R treatment, which indicates the effect of an unfavorable increase in the shoot length due to the individual R light (red light syndrome). Therefore, maintaining a balance between B and R wavelengths is crucial for the optimal growth and development of plants, as emphasized in previous studies [42,46].

Regarding photosynthetic pigments, the research revealed that R light (including both R and R + B colors) increased the content of Chl *a*, while B light (comprising W, B, and R + B colors) enhanced the content of Chl *b* in AV. This suggests that AVs, like many other plant species, have the ability to adjust their Chl content in response to different light conditions, specifically in the presence of R or B wavelengths, in order to maintain their health [49]. The highest and lowest measurements of Chl a/b in the AVs were associated with the R and B lights, respectively. In natural lighting, the Chl a/b ratio is typically falls within the range of 2 to 4. Various factors, such as ambient light, biotic or abiotic stresses, and plant growth stages, greatly influence the levels of Chl *a*, Chl *b*, total Chl, and Chl a/b in plants [50,51]. Furthermore, under different light conditions, Chl *a* and Chl *b* can convert into each other [51,52]. For instance, Chl *a* can be converted to Chl *b* when the rate of Chl synthesis is low in darkness [51], or Chl *b* can be converted back to Chl *a* in the chloroplast [50,52]. Apart from Chl, the increase in the Car levels in the AVs exposed to R + B colors compared to B and R colors individually aligns with the findings reported by Shin et al. [53], who observed enhanced Car and Chl synthesis in *Doritaenopsis* hort. (*Orchidaceae*) plants exposed to R + B LEDs. The AVs treated with W light showed the highest accumulation of Car, consistent with the general understanding that B wavelengths tend to increase the Car content in many plants. Previous research has also demonstrated that light supplementation (R:B = 7:2) in the morning improved Chl and Car contents in the leaves of *Solanum lycopersicum* L., further supporting the findings of the present study [54].

Ultimately, the PCA revealed the impact of the different lighting treatments on the root and shoot regeneration and growth. Consistent with the findings of the mean comparisons, the R + B treatment demonstrated a positive influence on various root and shoot growth indices, such as the RL, RV, RFW, RDW, SN, SL, SFW, SDW, TB, and Car content. Additionally, R light positively affected the SL:SDW, Chl *a*, Chl *a/b*, but negatively affected several growth indices, particularly the LN and RL. Furthermore, both the R + B and W treatments exhibited similar effects on root and shoot regeneration, growth, and the biosynthesis of Car and Chl *a*.

5. Conclusions

In conclusion, the light quality significantly influences photosynthesis and various aspects of plant growth and development. Therefore, determining the optimal light spectrum for each plant species is crucial. LED lighting, especially that emitting blue, red, and white light wavelengths, offers the potential to tailor wavelengths for individual species. The present study on African violet leaf cuttings unveiled that a blend of blue and red light had a more positive effect on root and shoot regeneration and growth compared to other lights. Furthermore, plants treated separately with white and blue light showed better performance than those exposed to red light. This suggests that blue light might have a more significant role in stimulating African violets' regeneration than red light, contrasting with reports suggesting a higher ratio of red to blue light for overall plant growth. The outcomes of this research are anticipated to shed more light on how different light wavelengths impact plant-regenerative processes.

Author Contributions: Conceptualization, M.G.G. and S.N.; methodology, M.G.G.; formal analysis, M.G.G. and S.M.E.M.; investigation, M.G.G., Z.A. and S.M.E.M.; data curation, Z.A.; writing—original draft preparation, M.G.G. and S.M.E.M.; writing—review and editing, S.N.; supervision, M.G.G.; funding acquisition, S.N. All authors have read and agreed to the published version of the manuscript.

Funding: This research was funded by Shahrekord University, Iran, Ref. 2021,141/5323, and partly by the Italian Ministry of Education and Research, Project MUR-PRIN 2020, Ref. 2020ELWM82.

Data Availability Statement: All data are presented in the article.

Conflicts of Interest: The authors declare no conflicts of interest.

References

1. da Silva, J.A.T.; Zeng, S.; Wicaksono, A.; Kher, M.M.; Kim, H.; Hosokawa, M.; Dewir, Y.H. In Vitro Propagation of African Violet: A Review. *S. Afr. J. Bot.* **2017**, *112*, 501–507. [CrossRef]
2. Mason, J.; Cole, G.; Beermann, M.; Fraser, A. *Gesneriads: African Violets, Gloxinias, Streptocarpus and Others*; ACS Distance Education: Robina MDC, QLD, Australia, 2017.
3. Winkelmann, T.; Grunewaldt, J. Regeneration of Plants from Protoplasts of *Saintpaulia ionantha* H. Wendl.(African Violet). In *Plant Protoplasts and Genetic Engineering VII*; Springer: Berlin/Heidelberg, Germany, 1996; pp. 141–149.
4. Ghehsareh, M.G.; Kafi, M. *Scientific and Practical Floriculturee*; Esfahan Golbon Press: Isfahan, Iran, 2015; Volume 1. (In Persian)
5. Hartmann, H.; Kester, D.; Davies, F.; Geneve, R.; Wilson, S. *Plant Propagation: Principles and Practices*, 9th ed.; Pearson: London, UK, 2017.
6. Marcenaro, S.; Voyiatzi, C.; Lercari, B. Photocontrol of in Vitro Bud Regeneration: A Comparative Study of the Interaction between Light and IAA in a Wild Type and an Aurea Mutant of Lycopersicon Esculentum. *Physiol. Plant.* **1994**, *91*, 329–333. [CrossRef]
7. Fang, S.; Lang, T.; Cai, M.; Han, T. Light Keys Open Locks of Plant Photoresponses: A Review of Phosphors for Plant Cultivation LEDs. *J. Alloys Compd.* **2022**, *902*, 163825. [CrossRef]

8. Zheng, L.; He, H.; Song, W. Application of Light-Emitting Diodes and the Effect of Light Quality on Horticultural Crops: A Review. *HortScience* **2019**, *54*, 1656–1661. [CrossRef]
9. Barta, D.J.; Tibbitts, T.W.; Bula, R.J.; Morrow, R.C. Evaluation of Light Emitting Diode Characteristics for a Space-Based Plant Irradiation Source. *Adv. Space Res.* **1992**, *12*, 141–149. [CrossRef] [PubMed]
10. Bula, R.J.; Morrow, R.C.; Tibbitts, T.W.; Barta, D.J.; Ignatius, R.W.; Martin, T.S. Light-Emitting Diodes as a Radiation Source for Plants. *HortScience Publ. Am. Soc. Hortic. Sci.* **1991**, *26*, 203–205. [CrossRef]
11. Dutta Gupta, S.; Agarwal, A. Artificial Lighting System for Plant Growth and Development: Chronological Advancement, Working Principles, and Comparative Assessment. *Light Emit. Diodes Agric. Smart Light.* **2017**, 1–25. [CrossRef]
12. Stamford, J.D.; Stevens, J.; Mullineaux, P.M.; Lawson, T. LED Lighting: A Grower's Guide to Light Spectra. *HortScience* **2023**, *58*, 180–196. [CrossRef]
13. Folta, K.M.; Carvalho, S.D. Photoreceptors and Control of Horticultural Plant Traits. *HortScience* **2015**, *50*, 1274–1280. [CrossRef]
14. Paradiso, R.; Proietti, S. Light-Quality Manipulation to Control Plant Growth and Photomorphogenesis in Greenhouse Horticulture: The State of the Art and the Opportunities of Modern LED Systems. *J. Plant Growth Regul.* **2022**, *41*, 742–780. [CrossRef]
15. Schäfer, E.; Nagy, F. *Photomorphogenesis in Plants and Bacteria: Function and Signal Transduction Mechanisms*, 3rd ed.; Springer Science & Business Media: Dordrecht, The Netherlands, 2006; ISBN 1402038119.
16. Moon, H.K.; Park, S.-Y.; Kim, Y.W.; Kim, C.S. Growth of Tsuru-Rindo (*Tripterospermum japonicum*) Cultured in Vitro under Various Sources of Light-Emitting Diode (LED) Irradiation. *J. Plant Biol.* **2006**, *49*, 174–179. [CrossRef]
17. Kurilčik, A.; Miklušytė-Čanova, R.; Dapkūnienė, S.; Žilinskaitė, S.; Kurilčik, G.; Tamulaitis, G.; Duchovskis, P.; Žukauskas, A. In Vitro Culture of Chrysanthemum Plantlets Using Light-Emitting Diodes. *Cent. Eur. J. Biol.* **2008**, *3*, 161–167. [CrossRef]
18. Cavallaro, V.; Avola, G.; Fascella, G.; Pellegrino, A.; Ierna, A. Effects of Spectral Quality and Light Quantity of LEDs on In Vitro Shoot Development and Proliferation of *Ananas comosus* L. Merr. *Agronomy* **2023**, *13*, 1072. [CrossRef]
19. Kwon, A.-R.; Cui, H.-Y.; Lee, H.; Shin, H.; Kang, K.-S.; Park, S.-Y. Light Quality Affects Shoot Regeneration, Cell Division, and Wood Formation in Elite Clones of *Populus euramericana*. *Acta Physiol. Plant.* **2015**, *37*, 1–9. [CrossRef]
20. Bello-Bello, J.J.; Martínez-Estrada, E.; Caamal-Velázquez, J.H.; Morales-Ramos, V. Effect of LED Light Quality on in Vitro Shoot Proliferation and Growth of Vanilla (*Vanilla planifolia* Andrews). *Afr. J. Biotechnol.* **2016**, *15*, 272–277. [CrossRef]
21. Dewir, Y.H.; Chakrabarty, D.; Hahn, E.J.; Paek, K.Y. A Simple Method for Mass Propagation of Spathiphyllum Cannifolium Using an Airlift Bioreactor. *In Vitro Cell. Dev. Biol.* **2006**, *42*, 291–297. [CrossRef]
22. Pawłowska, B.; Żupnik, M.; Szewczyk-Taranek, B.; Cioć, M. Impact of LED Light Sources on Morphogenesis and Levels of Photosynthetic Pigments in *Gerbera jamesonii* Grown in Vitro. *Hortic. Environ. Biotechnol.* **2018**, *59*, 115–123. [CrossRef]
23. Kostadinova, S.; Mollov, I.; Dzhambazov, B.; Naimov, S.; Vassilev, K.; Georgiev, B. Preliminary Study on the Effect of LED Light and Cytokinin on the Growth of Pear Plants In Vitro. In Proceedings of the 5th Balkan Scientific Conference on Biology, Plovdiv, Bulgaria, 15–16 April 2021; p. 1.
24. Lotfi, M. Effects of Monochromatic Red and Blue Light-Emitting Diodes and Phenyl Acetic Acid on in Vitro Mass Production of Pyrus Communis 'Arbi'. *J. Hortic. Postharvest Res.* **2022**, *5*, 119–128. [CrossRef]
25. Dewir, Y.H.; El-Mahrouk, M.E.-S.; Al-Shmgani, H.S.; Rihan, H.Z.; Teixeira da Silva, J.A.; Fuller, M.P. Photosynthetic and Biochemical Characterization of in Vitro-Derived African Violet (*Saintpaulia ionantha* H. Wendl) Plants to Ex Vitro Conditions. *J. Plant Interact.* **2015**, *10*, 101–108. [CrossRef]
26. Siswantoro, J.; Prabuwono, A.S.; Abdulah, A. Volume Measurement of Food Product with Irregular Shape Using Computer Vision and Monte Carlo Method: A Framework. *Procedia Technol.* **2013**, *11*, 764–770. [CrossRef]
27. Salokhe, V.M.; Sharma, A.K. *Greenhouse Technology and Applications*; Agrotech Publishing Academy: Udaipur, India, 2006; ISBN 8183210570.
28. Lichtenthaler, H.K. Chlorophylls and Carotenoids: Pigments of Photosynthetic Biomembranes. In *Methods in Enzymology*; Elsevier: Amsterdam, The Netherlands, 1987; Volume 148, pp. 350–382, ISBN 0076-6879.
29. Schroeter-Zakrzewska, A.; Kleiber, T. The Effect of Light Colour and Type of Lamps on Rooting and Nutrient Status in Cuttings of Michaelmas Daisy. *Bulg. J. Agric. Sci.* **2014**, *20*, 1426–1434.
30. Li, C.-X.; Xu, Z.-G.; Dong, R.-Q.; Chang, S.-X.; Wang, L.-Z.; Khalil-Ur-Rehman, M.; Tao, J.-M. An RNA-Seq Analysis of Grape Plantlets Grown In Vitro Reveals Different Responses to Blue, Green, Red LED Light, and White Fluorescent Light. *Front. Plant Sci.* **2017**, *8*, 78. [CrossRef] [PubMed]
31. OuYang, F.; Mao, J.-F.; Wang, J.; Zhang, S.; Li, Y. Transcriptome Analysis Reveals That Red and Blue Light Regulate Growth and Phytohormone Metabolism in Norway Spruce [*Picea abies* (L.) Karst.]. *PLoS ONE* **2015**, *10*, e0127896. [CrossRef]
32. Meng, L.; Song, W.; Liu, S.; Dong, J.; Zhang, Y.; Wang, C.; Xu, Y.; Wang, S. Light Quality Regulates Lateral Root Development in Tobacco Seedlings by Shifting Auxin Distributions. *J. Plant Growth Regul.* **2015**, *34*, 574–583. [CrossRef]
33. Zeng, J.; Wang, Q.; Lin, J.; Deng, K.; Zhao, X.; Tang, D.; Liu, X. Arabidopsis Cryptochrome-1 Restrains Lateral Roots Growth by Inhibiting Auxin Transport. *J. Plant Physiol.* **2010**, *167*, 670–673. [CrossRef] [PubMed]
34. Li, Z.; Chen, Q.; Xin, Y.; Mei, Z.; Gao, A.; Liu, W.; Yu, L.; Chen, X.; Chen, Z.; Wang, N. Analyses of the Photosynthetic Characteristics, Chloroplast Ultrastructure, and Transcriptome of Apple (*Malus domestica*) Grown under Red and Blue Lights. *BMC Plant Biol.* **2021**, *21*, 483. [CrossRef] [PubMed]
35. Miao, Y.; Chen, Q.; Qu, M.; Gao, L.; Hou, L. Blue Light Alleviates 'Red Light Syndrome'by Regulating Chloroplast Ultrastructure, Photosynthetic Traits and Nutrient Accumulation in Cucumber Plants. *Sci. Hortic.* **2019**, *257*, 108680. [CrossRef]

36. Lim, M.-J.; Murthy, H.N.; Song, H.-Y.; Lee, S.-Y.; Park, S.-Y. Influence of White, Red, Blue, and Combination of LED Lights on In Vitro Multiplication of Shoots, Rooting, and Acclimatization of *Gerbera jamesonii* Cv.'Shy Pink'Plants. *Agronomy* **2023**, *13*, 2216. [CrossRef]
37. Meng, X.; Wang, Z.; He, S.; Shi, L.; Song, Y.; Lou, X.; He, D. LED-Supplied Red and Blue Light Alters the Growth, Antioxidant Status, and Photochemical Potential of in Vitro-Grown *Gerbera jamesonii* Plantlets. *Hortic. Sci. Technol.* **2019**, *37*, 473–489. [CrossRef]
38. Hogewoning, S.W.; Wientjes, E.; Douwstra, P.; Trouwborst, G.; Van Ieperen, W.; Croce, R.; Harbinson, J. Photosynthetic Quantum Yield Dynamics: From Photosystems to Leaves. *Plant Cell* **2012**, *24*, 1921–1935. [CrossRef]
39. Hung, C.D.; Hong, C.-H.; Kim, S.-K.; Lee, K.-H.; Park, J.-Y.; Nam, M.-W.; Choi, D.-H.; Lee, H.-I. LED Light for in Vitro and Ex Vitro Efficient Growth of Economically Important Highbush Blueberry (*Vaccinium corymbosum* L.). *Acta Physiol. Plant.* **2016**, *38*, 1–9. [CrossRef]
40. Kaewjampa, N.; Shimasaki, K. Effects of Green LED Lighting on Organogenesis and Superoxide Dismutase (SOD) Activities in Protocorm-like Bodies (PLBs) of Cymbidium Cultured in Vitro. *Environ. Control Biol.* **2012**, *50*, 247–254. [CrossRef]
41. Trivedi, A.; Sengar, R.S. Effect of Various Light-Emittimg Diodes on Growth and Photosynthetic Pigments of Banana (*Musa acuminata*) CV. Grande Naine in Vitro Plantlets. *Int. J. Chem. Stud.* **2017**, *5*, 1819–1821. [CrossRef]
42. Runkle, E. Light Wavebands & Their Effects on Plants. Prepr. 2015. Available online: https//gpnmag.com/article/light-wavebands-and-their-effects-plants (accessed on 10 December 2023).
43. Okamoto, K.; Yanagi, T.; Kondo, S. Growth and Morphogenesis of Lettuce Seedlings Raised under Different Combinations of Red and Blue Light. In Proceedings of the II Workshop on Environmental Regulation of Plant Morphogenesis 435, Wellesbourne, UK, 8–10 May 1996; pp. 149–158.
44. Fukuda, N.; Ajima, C.; Yukawa, T.; Olsen, J.E. Antagonistic Action of Blue and Red Light on Shoot Elongation in Petunia Depends on Gibberellin, but the Effects on Flowering Are Not Generally Linked to Gibberellin. *Environ. Exp. Bot.* **2016**, *121*, 102–111. [CrossRef]
45. Trouwborst, G.; Hogewoning, S.W.; van Kooten, O.; Harbinson, J.; van Ieperen, W. Plasticity of Photosynthesis after the 'Red Light Syndrome'in Cucumber. *Environ. Exp. Bot.* **2016**, *121*, 75–82. [CrossRef]
46. Kim, S.-J.; Hahn, E.-J.; Heo, J.-W.; Paek, K.-Y. Effects of LEDs on Net Photosynthetic Rate, Growth and Leaf Stomata of Chrysanthemum Plantlets in Vitro. *Sci. Hortic.* **2004**, *101*, 143–151. [CrossRef]
47. Kong, Y.; Zheng, Y. Phototropin Is Partly Involved in Blue-Light-Mediated Stem Elongation, Flower Initiation, and Leaf Expansion: A Comparison of Phenotypic Responses between Wild Arabidopsis and Its Phototropin Mutants. *Environ. Exp. Bot.* **2020**, *171*, 103967. [CrossRef]
48. Kong, Y.; Stasiak, M.; Dixon, M.A.; Zheng, Y. Blue Light Associated with Low Phytochrome Activity Can Promote Elongation Growth as Shade-Avoidance Response: A Comparison with Red Light in Four Bedding Plant Species. *Environ. Exp. Bot.* **2018**, *155*, 345–359. [CrossRef]
49. Zhou, B.; Li, Y. Phytochrome and Light Signal Transduction in Plants. *Plant Physiol. Commun.* **2006**, *42*, 134.
50. Pyke, K. *Plastid Biology*; Cambridge University Press: Cambridge, UK, 2009; ISBN 0521885019.
51. Tanaka, A.; Tanaka, Y.; Takabe, T.; Tsuji, H. Calcium-Induced Accumulation of Apoproteins of the Light-Harvesting Chlorophyll Ab-Protein Complex in Cucumber Cotyledons in the Dark. *Plant Sci.* **1995**, *105*, 189–194. [CrossRef]
52. Kusaba, M.; Ito, H.; Morita, R.; Iida, S.; Sato, Y.; Fujimoto, M.; Kawasaki, S.; Tanaka, R.; Hirochika, H.; Nishimura, M. Rice NON-YELLOW COLORING1 Is Involved in Light-Harvesting Complex II and Grana Degradation during Leaf Senescence. *Plant Cell* **2007**, *19*, 1362–1375. [CrossRef] [PubMed]
53. Shin, K.S.; Murthy, H.N.; Heo, J.W.; Hahn, E.J.; Paek, K.Y. The Effect of Light Quality on the Growth and Development of in Vitro Cultured Doritaenopsis Plants. *Acta Physiol. Plant.* **2008**, *30*, 339–343. [CrossRef]
54. Wang, S.; Meng, X.; Tang, Z.; Wu, Y.; Xiao, X.; Zhang, G.; Hu, L.; Liu, Z.; Lyu, J.; Yu, J. Red and Blue LED Light Supplementation in the Morning Pre-Activates the Photosynthetic System of Tomato (*Solanum lycopersicum* L.) Leaves and Promotes Plant Growth. *Agronomy* **2022**, *12*, 897. [CrossRef]

horticulturae

MDPI

Article

In Vitro Germination and Propagation of *Dyckia brevifolia*, An Ornamental and Endangered Bromeliad

Konstantinos Bertsouklis * and Konstantina-Panagiota Panagaki

Laboratory of Floriculture and Landscape Architecture, Department of Crop Science, School of Plant Sciences, Agricultural University of Athens, 75 Iera Odos Street, GR-11855 Athens, Greece; panagakitina@gmail.com
* Correspondence: kber@aua.gr; Tel.: +30-210-5294558

Abstract: *Dyckia brevifolia* is an endangered plant used for ornamental purposes. As no references to the in vitro propagation of the species exist, the present study aims at investigating the possibility of an efficient micropropagation protocol. Seeds collected from mother plants were germinated at high percentages (84–86%) at a range of 15–25 °C, without any pre-treatment, and demonstrated their highest germination speed index (191.51) at 25 °C. In vitro-grown seedlings were used as the starting material for micropropagation on solid, or liquid, MS medium, supplemented with a variety of concentrations of cytokinins (BA, KIN or 2IP). Shoots and leaves were used as starting explants. Liquid media supplemented with BA or 2IP at 1.0 mg L^{-1} led to high multiplication rate and 2.7, or 2.3, lateral shoots were regenerated while on 2IP a high percentage (77.5%) of rooting occurred at the same time. Rooted microshoots were acclimatised ex vitro at 100% and acclimatised plants were transplanted in pots where they grew with a survival rate of 100% after two months. The in vitro propagation protocol presented in this study could enhance the large-scale propagation use of *D. brevifolia* as an ornamental plant and, simultaneously, contribute to the ex-situ conservation of the species.

Keywords: acclimatisation; Bromeliaceae; GSI; liquid culture; micropropagation; tissue culture

Citation: Bertsouklis, K.; Panagaki, K.-P. In Vitro Germination and Propagation of *Dyckia brevifolia*, An Ornamental and Endangered Bromeliad. *Horticulturae* **2022**, *8*, 390. https://doi.org/10.3390/horticulturae8050390

Academic Editor: Sergio Ruffo Roberto

Received: 30 March 2022
Accepted: 25 April 2022
Published: 28 April 2022

1. Introduction

Bromeliads are widely used in floriculture. Their attractive rosettes, range of colours of flowers, and graceful foliage, make them ideal ornamental plants. Still, information on their reproductive strategies remains scant [1,2]. Certain members of Bromeliaceae are often viewed as a prime example of adaptive radiation [3] as they can be found in a wide range of environments, from mesic to xeric, from terrestrial to epiphytic, and from sea level areas to high elevation ones. Their key adaptations include leaf succulence, spiny leaves, which gives them herbivore protection, and a crassulacean acid metabolism (CAM), which allows them to inhabit rocky, semi-arid environments [4,5]. However, bromeliads are relatively difficult to propagate, with one plant producing no more than a few individual plantlets characterised by slow growth [6]. Thus, in the recent past, the development of plant tissue culture-based techniques has been gaining ground since such techniques have proved valuable tools in developing new strategies for the bromeliads' mass propagation [7].

Dyckia is a genus that includes 170 species (either terrestrial or saxicolous, tankless perennial plants) and is one of the most species-rich genera in the Pitcairnioideae subfamily [8]. Many of the Dyckia members are of high ornamental interest [9]. According to recent studies, the Dyckia genus is one of the three Bromeliaceae genera which have formed a well-supported xerophytic clade [4,10,11], and includes species which are native to and thrive in environments with poor soil, limited water supply, and high sunlight exposure [12]. *Dyckia brevifolia* Baker (aka 'sawblade') is a species endemic to Southern Brazil [13]. It has yellow flowers on a 30-cm-long spike. It forms about 30 leaves, with fierce spiny margins (20 cm long), grouped in a dense rosette. Leaves are erect in the centre and recurve outwards, growing to a height of 40–50 cm [14]. *D. brevifolia* is a heliophyte that tolerates full sunlight, but can also adapt to a river's flow, whether such adaptation entails its submergence during floods or its dehydration during periods of low tide [15]. What is more, the morphoanatomical aspects of *D. brevifolia* carry both xeromorphic and hydromorphic characteristics which, in a rheophytic environment, give rise to adaptations during periods of both low and high water [16]. At present, the species, a member of the Brazilian consolidated ornamentals [17], can be found in an area no broader than a mere two hectares [1] and is faced with extinction as a result of the human activities [13] taking place in the Itajaí-Açu River Basin for construction of a hydroelectric plant [18]. It is similar to *Dyckia distachya*, another rare bromeliad included in the official list of Brazilian species threatened with extinction [19,20]. The special characteristics described above add such a high ornamental value to the species that a study of seed germination, in tandem with the development of in vitro propagation methods, could enhance its exploitation as an ornamental plant and lead to the establishment of strategies for its conservation in situ and ex situ alike.

In Greece, *D. brevifolia* is used as a landscape or potted plant, suitable for landscape architecture. It can also be used as a xerophyte. In vitro propagation of the bromeliad can take place using explants derived from the leaves' basal region since application of plant growth regulators can activate the vascular elements present in the explants [6]. This type of explant has been successfully used in other bromeliad in vitro systems [21–24]. For instance, in the case of *Dyckia macedoi* [9], in vitro morphogenesis was carried out starting from leaf explants and, in the case of *Dyckia distachya*, in vitro morphogenesis began with flower stalk segments [7]. In the cases of *Dyckia sulphurea* Koch [25] and *Dyckia maritima* Baker [26], the role of starting material towards the establishment of a successful micropropagation technique was played by shoot explants. Pompelli [27] as well as Pompelli and Guerra [28] used seeds as starting material for in vitro propagation of *Dyckia distachya* Hassler and so did Silva et al. [29] for *Dyckia agudensis*. The use of explants of seedling origin for in vitro cultures initiation does encourage a high proliferation rate [30], but the fact that Bromeliad seeds lose their viability quite rapidly [31] must be taken into account. Propagation by seed enhances genetic diversity which, in turn, contributes to the selection of genotypes of exceptional commercial value [32–34], has little environmental impact [35], and is followed by efficient clonal propagation strategies necessary in meeting the increased demands of the ornamental plant market. Proper molecular methods should be employed, in tandem with the study of the genetic variability in natural populations, for the assessment of genetic diversity and the relationship between individuals aiming at the preservation of natural variation. Cases of ex vitro seed germination of *D. brevifolia* have indeed been reported [36,37]. However, to the best of our knowledge, there are no studies researching in vitro propagation of *D. brevifolia*.

All *Dyckia* species show promising potential for the ornamental plant market [37]. Moreover, an efficient protocol for seed and clonal propagation could facilitate the introduction of suitable clones of *D. brevifolia* into nursery production and supporting floriculture use. The use of seedlings as starting plant material has proved quite effective for other species [30,38], enhancing those species' high genetic diversity and making it possible to select clones suitable for the floricultural industry as well as for breeding programs. The present study examines the in vitro germinability of *D. brevifolia* seeds and the use of in vitro-grown seedlings as starting material in order to investigate the effect of explant type, cytokinins, and medium phase of in vitro morphogenesis and micro-shoot rooting.

2. Materials and Methods

2.1. Plant Material

Seeds were collected fully mature, at Gryllis Water Lilies Nurseries (Marathon, Attica, Greece, 38°08′02.7″ N 23°56′33.7″ E) from potted mother plants. They were transferred to the Laboratory of Floriculture and Landscape Architecture, Agricultural University of Athens, Athens. They were stored dry, in the dark, in 9-cm, unsealed, plastic Petri dishes between filter paper, at T = 25 °C, and a relative humidity of 30%. After three months of collection, in-vitro germination and propagation took place (Figure 1).

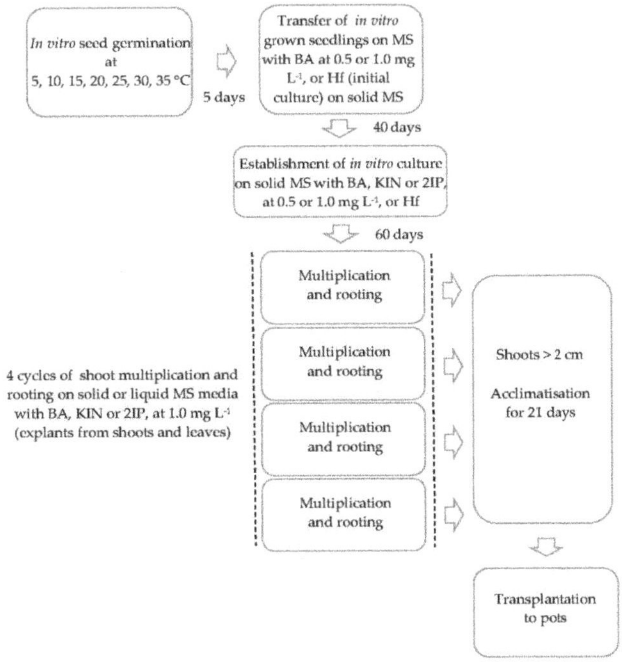

Figure 1. Schematic representation of *Dyckia brevifolia* in vitro propagation procedures starting from in vitro germinated seeds.

2.2. *In Vitro Germination*

Before initiating germination treatments, seeds were surface-sterilized with 20% (v/v) commercial bleach (4.6% *w/v* sodium hypochlorite) for 10 min and rinsed three times for three min per time with sterile distilled water. Seeds were sown in 9-cm plastic Petri dishes containing Hf, half-strength Murashige and Skoog (MS) medium [39]. Petri dishes were incubated at 5, 10, 15, 20, 25, 30, and 35 °C. In compliance with the rules of the International Seed Testing Association [40], germination was defined as the appearance of a radicle that would be at least 2 mm long and would be recorded every 1 d. T_{50} was defined as time taken by cumulative germination to reach 50% of its maximum [41]. One hundred seeds were used per treatment (five Petri dishes per treatment/20 seeds per Petri dish). The germination speed index (GSI) was calculated using the formula proposed by Maguire [42]:

$$GSI = G1/N1 + G2/N2 + \ldots + Gn/Nn$$

in which: G1, G2 and Gn = number of normal seedlings, computed during the first, second and last counts; and N1, N2, Nn = number of sowing days during the first, second and last counts.

2.3. *Micropropagation*

2.3.1. Establishment of Initial Cultures

Ten days after the completion of seed germination, young seedlings (plantlets) were transferred to Hf medium (8 g L^{-1} agar) and to MS medium supplemented with 6-benzyladenine (BA) (0.5 or 1.0 mg L^{-1}) for in vitro cultures. Forty (40) days later, the plantlets were sub-cultured on solid medium supplemented with BA, kinetin (KIN), and

6-γ-γ-(dimethylallylamino)-purine (2IP) at 0.5 and 1.0 mg L^{-1}, or Hf medium. The root system of young plantlets was excised before culturing on a new MS medium in both previous stages.

2.3.2. Effect of Explant Type and Medium Type on Shoot Multiplication

During the multiplication stage, lateral shoots and leaves excised from the base of plantlet stems were cultured on solid or liquid, Hf MS medium; or supplemented with BA, KIN and 2IP at 1.0 mg L^{-1}. Explants were sub-cultured on the same medium as that on which each had originated. A total of four subcultures followed.

2.3.3. In-Vitro Culture Conditions and Data Collection

In vitro cultures of the initial and establishment phases were carried out in 60 mL glass vessels (three explants per vessel); and covered with plastic wrap (Sanitas; Sarantis S.A., Athens, Greece). In vitro liquid, or solid, cultures of the multiplication phase were conducted in Magenta GA-7 vessels (77 mm × 77 mm × 77 mm, Sigma-Aldrich) (four explants per vessel). All cultures were maintained at 25 °C, with a 16 h photoperiod at 37.5 μmol m^{-2} s^{-1} provided by cool-white, fluorescent lamps. All solid media were solidified with 8 g L^{-1} agar and the pH of all media was adjusted to 5.7–5.8 before addition of the agar and autoclaving (121 °C for 20 min). Data on the initial cultures were collected after 40 days of culture. In compliance with previous studies [30], data on the establishment and multiplication phase were collected after 60 days of culture. Data were also collected on shoot proliferation percentages, shoot numbers per explant, shoot lengths, and number of leaves per shoot. To obtain the proliferation potential of the cultures, the "multiplication index" (MI) of each culture was calculated by multiplying the percentage of explants that produced shoots by the mean number of shoots per responding explant. Rooting percentages and root numbers and lengths were recorded during the rooting experiments.

2.3.4. Ex Vitro Acclimatisation

Rooting and shooting took place at the same time. Rooted microshoots 2.0–2.5 cm long, on various media, separated from shoot-clusters and thoroughly rinsed under running tap water in order to remove the medium before being transferred to 500 mL containers (eight plantlets per container), on peat (pH 5.5–6.5, Klasmann-Delimann Gmbh, Geeste, Germany) and perlite (particles diameter 1–5 mm, Perloflor, Isocon S.A., Athens, Greece) substrate 1:1 (v/v). All containers were covered with transparent plastic wrap (Sanitas) to maintain humidity. Pots were then placed for one week in a growth chamber (25 °C and 16-h cool white fluorescent light 37.5 μmol.m^{-2} s^{-1}/8-h dark photoperiod). Next, pots were uncovered for one week more. Afterwards, pots were transferred to a heated glasshouse (37°58′58.0″ N, 23°42′19.2″ E) and placed on a greenhouse bench for another 7 days. At the end of that period, data on acclimatisation were recorded. The plants were transplanted singly in 500 mL plastic pots with peat: perlite (1/1, v/v) and fertilized monthly with 2.0 g L^{-1} complete water-soluble fertilizer (Nutrileaf 60, 20-20-20; Miller Chemical and Fertilizer Corp., Hanover, PA, USA). The final survival rate was checked two months later.

2.4. Statistical Analysis

Our statistical analysis used the completely randomized design (CRD) method. The significance of the results was tested by one- or two-way analysis of variance (ANOVA). Data on percentages were arcsine-transformed prior to statistical analysis. The treatment means were compared via use of the Student's t-test at $p \leq 0.05$ (JMP 14.0 software, SAS Institute Inc., Cary, NC, USA, 2013). As shown on the data tables, the number of replicates per treatment differed between experiments. Experiments on the multiplication phase involved four subcultures. Data for the purposes of our statistical analysis were pooled.

3. Results

3.1. Seed Germination

The disinfection method of seeds resulted in 100% of success. The germination percentages were high at 15, 20, and 25 °C, i.e., 86.00, 89.00, and 84.00%, respectively (Table 1; Figures 2 and 3), with no statistical differences. Germination at 25 °C was completed in 3 days. At 30 °C the germination percentage was 35.00%. Cardinal germination temperatures were at 15 and 30 °C (86 and 35% germination, respectively). T_{50} was completed in 2 days at 25 °C. Seeds for that treatment had a higher germination speed index (191.51) (Table 1).

Table 1. In vitro germination of *Dyckia brevifolia* seeds, T_{50} and germination speed index (GSI) at temperatures shown, after six months of storage at room temperature.

Temperature (°C)	Germination (%) ± SD *	T_{50} (Days)	GSI
5	0	-	-
10	0	-	-
15	86.00 ± 8.21 a	15	25.24 c
20	89.00 ± 5.47 a	5	135.33 ab
25	84.00 ± 11.81 a	2	191.51 a
30	35.00 ± 16.62 b	4	84.17 b
35	0	-	-

* Different letters in the same column indicate significant differences. Mean (±SE) separation in columns by Student's *t*-test at $p \leq 0.05$, $n = 5$, 20 seeds/Petri dish (total 100 seeds per treatment).

(A) **(B)**

Figure 2. Germinated seeds on MS, Hf, of *Dyckia brevifolia* at 25 °C after 3 (**A**) and 20 (**B**) days of culture in Petri dishes. Bars represent 1.0 cm of length.

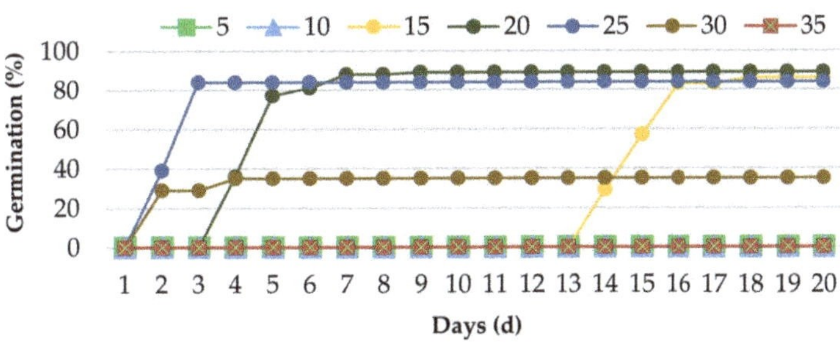

Figure 3. Germination time course curves of *Dyckia brevifolia* seeds as affected by temperature. Five replicates of 20 seeds per treatment were used.

3.2. Micropropagation

3.2.1. Initial Culture and Establishment

D. brevifolia plantlets deriving from in vitro grown seedlings were transferred on MS medium supplemented with BA at 0.5 or 1.0 mg L^{-1} or Hf medium to continue their growth (initial culture). There was no difference between the treatments, and the 40-day plantlets were 0.7 cm in height, and showed vigorous growth, of five to six leaves. In the subsequent establishment stage, the survival of plantlets was 95–100%. The plantlets were more vigorous, with more leaves than previously, i.e., during the initial phase (Figure 4A). They were similar in height (0.6–0.8 cm) in most of the media tested, with the exception of the medium containing 0.5 mg L^{-1} BA on which the growth of shoots was the shortest one (0.5 cm). The number of leaves was higher (8.4 leaves) on a Hf medium and presented no difference from the MS medium supplemented with KIN or 2IP at 1.0 mg L^{-1} (Table 2).

Figure 4. In vitro regeneration of *Dyckia brevifolia*: (**A**) plantlets growing and root formation during initial transfer of young seedlings after forty days on solid MS-Hf medium; (**B**) first shoots forming on shoot explants after two weeks of culture on solid MS medium containing BA 1 mg L^{-1}; (**C**) multiple shoot formation on a shoot explant after six weeks on a solid MS medium with 1 mg L^{-1} BA; (**D**) multiple shoot and spontaneous root formation on a shoot explant after six weeks of culture on a liquid MS medium with 1 mg L^{-1} 2IP; (**E**) first shoots forming on leaf- explants after two weeks of culture on a solid MS medium containing BA 1 mg L^{-1}; (**F**) multiple shoot and spontaneous root formation on a shoot explant after six weeks of culture on a solid MS medium with 1 mg L^{-1} 2IP; (**G**) plantlets during ex vitro acclimatisation; (**H**) two-month-old plants showing vigorous growth. Bars represent a length of 1 cm.

Table 2. Effect of cytokinin type and concentration on shoot proliferation from shoot explants excised from shoots produced on MS medium (Hf or containing BA at 0.5 or 1.0 mg).

Cytokinin	Concentration (mg L^{-1})	Survival (%)	Growth (cm)	Number of Leaves
Hf †	-	95.0 a	0.7 ab	8.4 a
BA	0.5	100.0 a	0.5 c	6.5 c
	1.0	97.5 a	0.6 abc	6.4 c
KIN	0.5	97.5 a	0.8 a	6.8 bc
	1.0	95.0 a	0.6 abc	7.3 ab
2IP	0.5	100.0 a	0.7 abc	6.1 c
	1.0	100.0 a	0.7 abc	7.6 ab
$F_{\text{one-way ANOVA}}$		NS	***	***
F_{cyt}		NS		
F_{conc}		NS		
$F_{cyt \times conc}$		NS	*	*

Different letters in the same column indicate significant differences. NS, *, *** Nonsignificant or significant at $p \leq 0.05$, $p \leq 0.001$, respectively, n = 40–50. † Hf (hormone free) treatment was excluded for 2-way ANOVA.

3.2.2. Multiplication Stage

Shoot Explants

The multiplication stage comprised a total of four subcultures so as to ascertain whether growth would be optimized in: (a) solid MS containing BA, KIN, or 2IP, at 1.0 mg L^{-1}, or Hf; (b) liquid MS containing BA, KIN, or 2IP, at 1.0 mg L^{-1}, or Hf. The two-way analysis showed that the liquid media were superior to the solid ones in terms of growth (height of plantlets), leaf number/main shoot, number of lateral shoots and MI (Table 3). The plantlets showed sturdy growth and were vigorous with lateral shoots (Figure 4B,C,D). The type of cytokinin used also played a significant, positive role. The maximum MI (1.93 and 1.97) was recorded on liquid media supplemented with 2IP or BA at 1.0 mg L^{-1} (Table 3, Figure 4B). The percentage of lateral shoot formation was higher on solid and liquid media containing BA (83 and 73%, respectively) and on liquid media with 2IP (84%) (Figure 4D). The higher lateral shoot number (2.7) was observed in the Hf liquid medium with BA (Table 3, Figure 4C). More leaves were formed on Hf MS, and the higher growth was estimated when explants were cultured on MS containing KIN or 2IP (2.3 and 2.2 cm, respectively).

Table 3. Effect of cytokinin type and concentration on shoot proliferation from shoot explants, excised from plantlets on solid and liquid media containing BA, KIN, or 2IP, at 1.0 mg L^{-1}, or Hf.

MS	Cytokinin	Stem Growth (cm)	Number of Leaves/Main Shoot	Formation of Lateral Shoots (%)	Lateral Shoot Number	MI	Number of Leaves/Lateral Shoot
Solid	Hf †	0.9 b	8.5 cd	0.0 c	0.0 c	-	0.0 c
	BA	0.8 b	6.2 e	83.0 a	1.8 ab	1.50 ab	4.6 b
	KIN	0.9 b	7.7 de	0.0 c	0.0 c	-	0.0 c
	2IP	0.7 b	6.1 e	54.0 b	2.0 ab	1.08 b	4.9 b
Liquid	Hf †	1.7 ab	12.1 a	8.0 b	1.5 bc	0.11 b	4.9 b
	BA	1.9 ab	10.2 bc	73.0 a	2.7 a	1.97 a	4.9 b
	KIN	2.3 a	11.6 ab	63.0 ab	1.9 ab	1.19 b	6.1 a
	2IP	2.2 a	10.5 bc	84.0 a	2.3 ab	1.93 a	6.0 a
$F_{\textit{one-way}\ \text{ANOVA}}$		***	***	**	***	*	***
F_{med}		***	***		***	**	***
F_{cyt}		NS	**		***	**	***
$F_{cyt \times med}$		NS	NS	*	NS	NS	NS

Different letters in the same column indicate significant differences. NS, *, **, *** Non-significant or significant at $p \leq 0.05$, $p \leq 0.01$, $p \leq 0.001$, respectively, n= 50–60, † Hf = hormone-free.

Leaf Explants

Bud proliferation started at the base of both shoot and leaf explants 2 weeks after culture initiation (Figure 4B,E). Regarding leaf explants, shoots originated directly from protuberances which had formed at the leaf blade's cut without any intermediate callus phase. Two-way analysis revealed that solid media were more effective than liquid media, as lateral shoot formation was higher (Table 4). On the other hand, explants cultured in liquid media, gave rise to plantlets of greater growth. Multiple shoots formed in all media, but the lateral shoot formation was higher on solid MS supplemented with BA, KIN or 2IP (41.0, 35.0 and 44.5%, respectively). Lateral shoot number stood at 7.4 and the MI was higher on liquid media containing BA, registering 1.9 (Table 4).

Table 4. Shoot proliferation from leaf-origin explants excised from shoots produced either on Hf MS medium or on MS containing BA at 0.5 or 1.0 mg L^{-1}, on solid and liquid media containing BA, KIN or 2IP at 1.0 mg L^{-1} or Hf.

MS	Cytokinin	Formation of Lateral Shoots (%)	Growth (cm)	Lateral Shoot Number	MI	Number of Leaves/Shoot
Solid	Hf †	20.0 b	1.0 ab	1.4 c	0.28 c	8.8 a
	BA	41.0 a	0.5 c	1.4 c	0.57 c	7.0 a
	KIN	35.0 a	0.8 bc	1.5 c	0.53 c	8.0 a
	2IP	44.5 a	0.9 b	1.5 c	0.67 bc	8.9 a
Liquid	Hf †	17.0 b	1.2 ab	1.0 c	0.17 c	7.4 a
	BA	26.0 ab	1.0 ab	7.4 a	1.90 a	6.6 a
	KIN	22.5 b	1.3 a	2.6 b	0.59 bc	7.6 a
	2IP	27.5 b	1.0 ab	3.3 b	0.90 b	6.3 a
$F_{one\text{-}way}$ ANOVA		***	**	***	**	NS
F_{med}		***	*			
F_{cyt}		NS	NS			
$F_{cyt \times med}$		NS	NS	*	*	*

Different letters in the same column indicate significant differences. NS, *, **, *** Non-significant or significant at $p \leq 0.05$, $p \leq 0.01$, $p \leq 0.001$, respectively, n = 50–60. † Hf = hormone-free.

3.2.3. In Vitro Rooting and Ex Vitro Acclimatisation

Shoot-origin plantlets produced on establishment media, rooted spontaneously at a high percentage (95%) on Hf medium, or supplemented with 0.5–1.0 mg L^{-1} KIN or 2IP (Table 5). Rooted plantlets produced 2.1 and 2.3 roots per plantlet of 1.2 and 1.1 cm length on Hf and MS containing 0.5 mg L^{-1} KIN, respectively. During the multiplication phase, rooting was continuous. Shoot- and leaf-origin plantlets rooted at a higher percentage (94.0 and 74.5%, respectively) on a solid MS medium (Table 6, Figure 4F). The percentage of rooting was high on liquid HF or MS medium supplemented with 1 mg L^{-1} 2IP (74.5 and 77.5%, respectively) for shoot-origin shoots. Leaf-origin shoots rooted at a lower percentage (Table 6). Rooted microshoots were acclimatised ex vitro at 100% (Figure 4G). Acclimatised plants were transplanted in pots. Their survival rate while growing was 100% after two months (Figure 4H). Lateral sprouts were formed at the base of some plantlets during the acclimatisation phase (Figure 5).

Table 5. In vitro rooting of shoots during first subculture on MS supplemented BA, KIN or 2IP at 0.5 or 1.0 mg L^{-1}, or Hf.

Cytokinin	Concentration (mg L^{-1})	Rooting (%)	Root Number	Root Length (cm)
Hf †	-	95.0 a	2.1 a	1.2 a
BA	0.5	20.0 b	1.0 b	0.5 c
	1.0	13.0 b	1.0 b	0.5 c
KIN	0.5	93.0 a	2.3 a	1.1 ab
	1.0	81.0 a	1.6 ab	0.9 bc
2IP	0.5	100.0 a	1.3 b	0.9 bc
	1.0	79.0 ab	1.6 ab	0.8 c
$F_{\text{one-way ANOVA}}$		**	**	**
F_{cyt}				NS
F_{conc}				**
$F_{cyt \times conc}$		*	*	NS

Different letters in the same column indicate significant differences, † Hf = hormone-free. NS, *, ** Non-significant or significant at $p \leq 0.05$, $p \leq 0.01$, respectively, n = 50–60.

Table 6. In vitro rooting of shoots derived from shoots or leaves as affected by the type of MS (solid or liquid) and of cytokinin during multiplication stage.

		Shoot-Origin			Leaf-Origin		
MS	Cytokinin	Rooting (%)	Root Number	Root Length (cm)	Rooting (%)	Root Number	Root Length (cm)
Solid	Hf †	94.0 a	2.2 b	1.1 c	73.5 a	2.0 a	1.2 a
	BA	0.0 c	0.0 c	0.0 c	50.0 ab	1.0 a	0.5 a
	KIN	47.0 b	1.4 b	1.0 c	18.0 b	1.7 a	1.0 a
	2IP	43.0 b	1.8 b	0.9 c	52.5 ab	2.0 a	0.9 a
Liquid	Hf †	74.5 ab	4.2 a	1.7 b	62.5 ab	2.4 a	1.4 a
	BA	0.0 c	0.0 c	0.0 c	0.0 b	0.0 b	0.0 b
	KIN	18.3 b	3.9 a	2.3 a	0.0 b	0.0 b	0.0 b
	2IP	77.5 ab	3.8 a	2.4 a	0.0 b	0.0 b	0.0 b
$F_{\textit{one-way}\ \text{ANOVA}}$		***	***	***	*	NS	NS
F_{med}			***				
F_{cyt}			NS				
$F_{cyt \times med}$		***	NS	*	**	*	*

Different letters in the same column indicate significant differences. NS, *, **, *** Non-significant or significant at $p \leq 0.05$, $p \leq 0.01$, $p \leq 0.001$, respectively, n = 50–60. † Hf = hormone-free.

Figure 5. Lateral sprouts formed at the base of one of the plantlets during acclimatisation. Bar represents length of 1 cm.

4. Discussion

The small size of Bromeliaceae seeds and their diminished reproduction capacity [43,44] has resulted in scientists' turning to alternative propagation by means of tissue culture-based techniques, starting with young tissues of in vitro grown seedlings. With regard to the temperature range for seed germination of the *D. brevifolia*, the present study is the first one to investigate and define cardinal temperatures as ranging from 15 °C to 30 °C. At temperatures of 20–25 °C, *D. brevifolia* exhibited maximum germination in a brief period of time. The present study's finding has been confirmed by Paula et al. [36] who recently evaluated the effect of liquid nitrogen during cryopreservation of *D. brevifolia* seeds and determined a germination percentage of 95% at 25 °C. At the same time, Moresco et al. [37] have also cited a high germination rate. Such references lead to the thought that fast germination of *D. brevifolia* may prove an adaptive strategy of taking advantage of the short period of rainfall and water availability that *D. brevifolia* has at its disposal in its natural habitat. That strategy has been reported as successful for *Haloxylon recurvum* Bunge ex. Boiss, for *H. salicornicum* Bunge ex. Boiss [45], and for *Limonium axillare* (Forssk.) Boiss [46]. The study also indicates that *D. brevifolia* seeds have no dormancy period as reported for *D. distachya*, another Dyckia species [28]. With regard to germination percentages of two other Dyckia species, it has been observed that *D. distachya* seeds germinate at higher percentages when the temperature range is 20–30 °C [47], in the presence of light. Similarly, seeds of *Dyckia tuberosa* have shown a higher germination percentage at a range of 30–35 °C [48], with cardinal temperatures for germination ranging from 10 °C to 40 °C. The lower optimum temperature for germination of *D. brevifolia* may be explained by the major impact of the environment on seed germinability during seed production [49]. Regarding the optimum temperatures for germination of Mediterranean species overall, the usual range is between 15–20 °C [50–52]. The range of germination of *D. brevifolia*, which is similar to that of other Mediterranean species, could enhance its use as an ornamental plant, as is the case with *D. encholirioides*, another endangered species, which has no limitations when germinated in vitro, in contrast to its germination in a natural environment where there are certain obstacles that could reduce its germination rate [53,54]. It could also lead to the introduction of new varieties/types into the Mediterranean Basin countries: plants grown from seeds are characterised by high quality that could lead to the production of disease- and/or virus-free plants enjoying a prolonged lifespan [55]. In vitro propagated plants should be indexed as being free of viruses and virus-like diseases through enzyme-linked immunosorbent assay (ELISA) and molecular methods [56]. Moreover, seed propagation is the basis for production of commercially attractive agronomic and horticultural plants [57]. Thus, using seeds as the starting material could meet the increasing demand for ornamental succulent plants [58].

Many areas around the planet face strong anthropogenic pressure [59]. Inevitably as well as regrettably, several species will be included in the official lists of threatened species. It is quite the paradox that, although certain species are faced with extinction in their natural habitat, those very species are widely used in ornamental horticulture and by landscape designers. Cases in point, *Origanum dictamnus*, a plant endemic to Crete, Greece, [60] or *Neoregelia cruenta*, an endemic bromeliad found exclusively on the sandy parts of the coastal plain of Rio de Janeiro [22,61]. In vitro techniques are of paramount importance for conservation purposes. In view of that, an efficient micropropagation protocol for *D. brevifolia* could enhance as well as accelerate efforts to that direction. In the present study, initial culture and establishment phases were successful for explants derived from young seedlings, providing an effective method of production of in vitro-grown plantlets. An effective plant tissue culture protocol is based on the selection of the appropriate explant type [62]. The present study has shown that shoot-origin explants can lead to a successful, direct-shoot regeneration on solid or liquid media supplemented with BA at 1.0 mg L^{-1}. Of particular interest is the ability of the explants to produce lateral shoots with a high rate of spontaneous rooting on liquid media with 2IP (1.0 mg L^{-1}) at the same time. Regarding leaf origin-explants, shoots originating directly from protuberances located at the cut end of the leaf blade, without any intermediate callus phase, produced multiple shoots both in liquid and solid media. Bud and protuberance formation at the base of leaf explants prior to bud development has been described in the case of *Dyckia macedoi* [9].

The relevant literature indicates that successful propagation systems of other bromeliads do exist. Those systems are based on tissue culture through liquid media on which the number of leaf-derived buds could reach double the number of buds as produced on solid media with the same composition [63]. Similar base leaf regeneration in bromeliads was also observed on the base of leaf sheaths removed from greenhouse-grown seedlings of *Puya tuberosa* [64]. In the present study, the shoot production that took place by leaf-derived explants on a liquid MS medium supplemented with 1.0 mg L^{-1} BA was exceptionally high (7.4 shoots/explant). On the other hand, the percentage of lateral shoot formation was a mere 26.0%, since the task of excising a specific leaf explant that could be used as a suitable explant successfully proved particularly daunting.

The ultimate survival of acclimatised plantlets is crucial for a successful in vitro propagation protocol. In the present study, the percentage of survival after two months reached a full 100%, an acclimatisation rate which is quite high and observed also in the case of *D. distachya* [27], and of *D. agudensis* [29]. The present experimentation resulted in an efficient plant propagation method which could be used either for commercial propagation of selected clones or for in vitro and ex situ conservation programs.

5. Conclusions

In conclusion, the present study investigated a fully reliable procedure for propagation on *D. brevifolia* starting from a small quantity of plant material. In quite a short period, three-month-old seeds of *D. brevifolia* germinated profusely, at 15 to 25 °C. Cardinal temperatures for germination were defined at 15 °C and 30 °C. Regarding micropropagation, seedling-origin shoot explants responded more eagerly than single leaves to in vitro culture. Nevertheless, both exhibited high shoot multiplication on liquid MS medium supplemented with 1.0 mg L^{-1} 2IP, with simultaneous rooting. Microshoots rooted abundantly on Hf, half-strength MS medium and were successfully established at ex vitro conditions.

In many countries, ornamental bromeliads fetch a high market price as they are sought after by the fields of floriculture and landscape architecture. To our knowledge, this is the first report on in vitro propagation of *D. brevifolia*. This experimental procedure leads to the production of a high number of individuals, independently of the natural vegetative cycle of *D. brevifolia* in the wild. This type of propagation may facilitate both the needs for increased demand that floriculture or ornamental horticulture face and in producing new specimens for in vitro and ex situ conservation purposes. Starting with young tissue taken from in vitro germinated seeds is essential for the preservation of the genetic diversity that threatened species are in need of. Future studies on genetic stability could evaluate the use of regenerated plants for reintroduction in those plants' natural habitats.

Author Contributions: Conceptualization, K.B.; methodology, K.B. validation, K.B.; formal analysis, K.B.; investigation, K.B.; resources, K.B. and K.-P.P.; data curation, K.B. and K.-P.P.; writing—original draft preparation, K.B.; writing—review and editing, K.B.; visualization, K.B.; supervision, K.B.; project administration, K.B.; funding acquisition, K.B. All authors have read and agreed to the published version of the manuscript.

Funding: This research received no external funding.

Institutional Review Board Statement: Not applicable.

Informed Consent Statement: Not applicable.

Acknowledgments: The authors would like to thank the plant nursery Gryllis Water Lilies (POBOX Z2-04, Vranas—Marathon Attica, Greece) for donating the seeds of the species investigated.

Conflicts of Interest: The authors declare no conflict of interest.

References

1. Martinelli, G. The Bromeliads of the Atlantic Forest. *Sci. Am.* **2000**, *282*, 86–93. [CrossRef] [PubMed]
2. Zanella, C.M.; Janke, A.; Palma-Silva, C.; Kaltchuk-Santos, E.; Pinheiro, F.G.; Paggi, G.M.; Soares, L.E.S.; Goetze, M.; Büttow, M.V.; Bered, F. Genetics, evolution and conservation of Bromeliaceae. *Genet. Mol. Biol.* **2012**, *35*, 1020–1026. [CrossRef] [PubMed]
3. Benzing, D.H. *Bromeliaceae: Profile of an Adaptive Radiation*; Cambridge University Press: Cambridge, UK, 2000. [CrossRef]
4. Givnish, T.J.; Barfuss, M.H.J.; Ee, B.V.; Riina, R.; Schulte, K.; Horres, R.; Gonsiska, P.A.; Jabaily, R.S.; Crayn, D.M.; Smith, J.A.C.; et al. Phylogeny, adaptive radiation, and historical biogeography in Bromeliaceae: Insights from an eightlocus plastid phylogeny. *Am. J. Bot.* **2011**, *98*, 872–895. [CrossRef] [PubMed]
5. Silvestro, D.; Zizka, G.; Schulte, K. Disentangling the effects of key innovations on the diversification of Bromelioideae (Bromeliaceae). *Evolution* **2014**, *68*, 163–175. [CrossRef]
6. Hosoki, T.; Asahira, T. In Vitro propagation of bromeliads in liquid culture. *HortScience* **1980**, *15*, 603–604.
7. Daquinta, M.; Almeida, A.P.; Guerra, M.P. In Vitro morphogenesis of immature flower and buds of flower stalk in *Dyckia distachya*. *J. Bromel. Soc.* **1998**, *49*, 72–76. Available online: https://isb.emnuvens.com.br/iheringia/article/view/162 (accessed on 4 March 2022).
8. Krapp, F.; de Barros Pinangé, D.S.; Benko-Iseppon, A.M.; Leme, E.M.; Weising, K. Phylogeny and evolution of Dyckia (Bromeliaceae) inferred from chloroplast and nuclear sequences. *Plant Syst. Evol.* **2014**, *300*, 1591–1614. [CrossRef]
9. Mercier, H.; Kerbauy, G.B. Micropropagation of *Dyckia macedoi*–an endangered endemic Brazilian bromeliad. *Bot. Gard. Microprop. News* **1993**, *1*, 70–72.
10. Givnish, T.J.; Millam, K.C.; Berry, P.E.; Sytsma, K.J. Phylogeny, adaptive radiation, and historical biogeography of Bromeliaceae inferred from ndhF sequence data. *Aliso* **2007**, *23*, 3–26. [CrossRef]
11. Rex, M.; Schulte, K.; Zizka, G.; Peters, J.; Vásquez, R.; Ibisch, P.L.; Weising, K. Phylogenetic analysis of *Fosterella* L.B. Sm. (Pitcairnioideae, Bromeliaceae) based on four chloroplast DNA regions. *Mol. Phylogenet. Evol.* **2009**, *51*, 472–485. [CrossRef]
12. Givnish, T.J.; Barfuss, M.H.J.; Van Ee, B.; Riina, R.; Schulte, K.; Horres, R.; Gonsiska, P.A.; Jabaily, R.S.; Crayn, D.M.; Smith, J.A.C.; et al. Adaptive radiation, correlated and contingent evolution, and net species diversification in Bromeliaceae. *Mol. Phylogenetics Evol.* **2014**, *71*, 55–78. [CrossRef]
13. Santos-Silva, F.; Saraiva, D.P.; Monteiro, R.F.; Pita, P.; Mantovani, A.; Forzza, R.C. Invasion of the South American dry diagonal: What can the leaf anatomy of Pitcairnioideae (Bromeliaceae) tell us about it? *Flora* **2013**, *208*, 508–521. [CrossRef]
14. Synge, P.M. *Dictionary of Gardening*; Clarendon Press: Oxford, UK; Cumberlege: London, UK, 1981; p. 725.
15. Reitz, R. Bromeliaceas e a malaria—bromelia endémica. In *Flora Ilustrada Catarinense. Parte I. Fasciculo Bromelia*; Reitz, R., Ed.; Herbário Barbosa Rodrigues: Itajaí, Brazil, 1983; p. 518.
16. Lobo, G.M.; de Souza, T.V.; Voltolini, C.H.; Reis, A.; Santos, M. Leaf epidermis of the rheophyte *Dyckia brevifolia* Baker (Bromeliaceae). *Sci. World J.* **2013**, *2013*, 307593. [CrossRef]
17. da Silva Sousa, R.P.; Costa, W.S.; Matos, P.e.S.; Carvalho, A.S.; Martins, F.D.; Torres, K.R. Ornamental potential of species from the ferruginous Campo rupestre of the Carajás National Forest, Brazilian Amazon. *Comun. Sci.* **2012**, *12*, e3260. [CrossRef]
18. Rogalski, J.M.; Reis, A.; Rogalski, M.; Montagna, T.; Dos Reis, M.S. Mating System and Genetic Structure Across All Known Populations of *Dyckia brevifolia*: A Clonal, Endemic, and Endangered Rheophyte Bromeliad. *J. Hered.* **2017**, *108*, 299–307. [CrossRef]
19. Ibama. Reconhecer Como Espécies da Flora Brasileira Ameaçadas de Extinção Aquelas Constantes do Anexo i a Esta Instrução Normativa 2014, n. 443, de 17 de Dezembro de 2014. Available online: http://www.ibama.gov.br/sophia/cnia/legislacao/MMA/PT0443-171214.pdf (accessed on 4 March 2022).
20. Strehl, T. Periodically submersed bromeliads. *Bromélia* **1994**, *3*, 19–21.
21. Mercier, H.; Kerbauy, G.B. Micropropagation of ornamental bromeliads (Bromeliaceae). In *Biotechnology in Agriculture and Forestry*; Bajaj, Y.P.S., Ed.; Springer: Berlin, Germany, 1997; pp. 43–57.
22. Carneiro, L.; Araújo, R.; Brito, G.; Fonseca, M.H.P.B.; Costa, A.; Crocomo, O.J.; Mansur, E. In Vitro regeneration from leaf explants of *Neoregelia cruenta* (R. Graham) L.B. Smith, an endemic bromeliad from Eastern Brazil. *Plant Cell Tissue Organ Cult.* **1998**, *55*, 79–83. [CrossRef]
23. Alves, G.M.; Guerra, M.P. Micropropagation for mass propagation and conservation of *Vriesea friburgensis* var. *paludosa* from microbuds. *J. Bromel. Soc.* **2001**, *51*, 202–212.
24. Pompelli, M.F.; Fernandes, D.; Guerra, M.P. Somatic embryogenesis in *Dyckia distachia* Hassler (Bromeliaceae)—An endangered bromeliad from South Brazil. *Propag. Ornam. Plants* **2005**, *5*, 192–198.
25. Murashige, T. Plant propagation through tissue culture. *Annu. Rev. Plant Phys.* **1974**, *25*, 135–165. Available online: http://dx.doi.org/10.1146/annurev.pp.25.060174.001031 (accessed on 4 March 2022). [CrossRef]
26. Silva, A.L.L.; Franco, E.T.H.; Dornelles, E.B.; Gesing, J.P.A. Micropropagation of *Dyckia maritima* Baker-Bromeliaceae. *Iheringia Sér. Botânica* **2008**, *63*, 135–138.
27. Pompelli, M.F. Morfogênese In Vitro, Métodos Demicropropagação e Conservação de Germoplasma de *Dyckia distachya* Hassler. Master's Thesis, Universidade Federal de Santa Catarina, Florianópolis, Brazil, 2002; 93p.
28. Pompelli, M.F.; Guerra, M.P. Micropropagation enables the mass propagation and conservation of *Dyckia distachya* Hassler. *Crop Breed. Appl. Biotechnol.* **2005**, *5*, 117–126. [CrossRef]

29. Silva, A.L.L.; Dornelles, E.B.; Bisognin, D.A.; Franco, E.T.H.; Horbach, M.A. Micropropagation of *Dyckia agudensis* Irgang & Sobral—An extinction threatened bromeliad. *Iheringia Sér. Botânica* **2007**, *62*, 39–43.
30. Papafotiou, M.; Bertsouklis, K.F.; Trigka, M. Micropropagation of *Arbutus unedo*, *A. andrachne*, and their natural hybrid, *A.* × *andrachnoides* from seedling explants. *J. Hortic. Sci. Biotechnol.* **2013**, *6*, 768–775. [CrossRef]
31. Zotz, G. A longer story than expected: Seeds of several species (Tillandsioideae) remain viable for up to two years. *J. Bromel. Soc.* **2013**, *63*, 83–86. Available online: https://go.gale.com/ps/i.do?p=AONE&u=anon~{}898338be&id=GALE\T1\textbar{}A610341224&v=2.1&it=r&sid=bookmark-AONE&asid=b882c762 (accessed on 4 March 2022).
32. Sarasan, V.; Kite, G.C.; Sileshi, G.W.; Stevenson, P.C. Applications of phytochemical and *in vitro* techniques for reducing over-harvesting of medicinal and pesticidal plants and generating income for the rural poor. *Plant Cell Rep.* **2011**, *30*, 1163–1172. [CrossRef]
33. Generoso, A.L.; Carvalho, V.S.; Walter, R.; Campbell, G.; Araújo, L.S.; Santana, J.G.S.; Cunha, M. Mature-embryo culture in the cryopreservation of passion fruit (*Passifora edulis* Sims) seeds. *Sci. Hortic.* **2019**, *256*, 108638. [CrossRef]
34. Silva, S.S.S.; Souza, E.H.; Souza, F.V.D.; Max, D.A.S.; Rossi, M.L.; Costa, M.A.P.C. Post-seminal development and cryopreservation of endemic or endangered bromeliads. *An. Acad. Bras. Cienc.* **2021**, *93*, e20191133. [CrossRef]
35. Silva, L.F.; de Souza, D.C.; Resente, L.V.; Gonçalves, W.M. Manejo de recursos genéticos vegetais. *An. da Acad. Pernambucana de Ciência Agronômica* **2018**, *15*, 109–126. Available online: http://www.journals.ufrpe.br/index.php/apca/article/view/1824 (accessed on 4 March 2022).
36. de Paula, J.C.B.; Men, G.B.; Biz, G.; Júnior, W.A.R.; de Faria, R.T. Cryopreservation of seeds from endangered Brazilian bromeliads-*Dyckia brevifolia* Baker and *D. delicata* Larocca & Sobral. *Rev. Bras. de Ciências Agrárias* **2020**, *15*, 1–8. [CrossRef]
37. Moresco, V.P.; Omura, M.S.; de Paula, J.C.B.; Furlan, F.F.; Takahashi, L.S.A. Physiological potential of Dyckia spp. bromeliad seeds under different temperatures. *Ciências Agrárias* **2021**, *42*, 2639–2650. [CrossRef]
38. Vlachou, G.; Papafotiou, M.; Bertsouklis, K.F. Studies on Seed Germination and Micropropagation of *Clinopodium nepeta*: A medicinal and aromatic plant. *HortScience* **2019**, *54*, 1558–1564. [CrossRef]
39. Murashige, T.; Skoog, F. A revised medium for rapid growth and bioassays with tobacco tissue cultures. *Physiol. Plant.* **1962**, *15*, 473–497. [CrossRef]
40. International Seed Testing Association. International rules for seed testing. *Seed Sci. Technol.* **1999**, *27*, 333.
41. Soltani, A.; Galeshi, S.; Zeinali, E.; Latifi, N. Genetic variation for and interrelationships among seed vigor traits in wheat from the Caspian Sea coasts of Iran. *Seed Sci. Technol.* **2001**, *29*, 653–662.
42. Maguire, J.D. Speed of germination-aid in selection and evaluation for seedling emergence and vigor. *Crop. Sci.* **1962**, *2*, 176–177. [CrossRef]
43. Kowalski, V.K.; Tardivo, R.S.; Oliveira, F.M.C.; Mourão, K.S.M. Morphology and anatomy of seedlings of Bromeliaceae from the perspective of ecophysiological types. *Flora* **2021**, *285*, 151959. [CrossRef]
44. Malda, G.; Suzán, H.; Backhaus, R.A. In vitro culture as a potential method for the conservation of endangered plants possessing crassulacean acid metabolism. *Sci. Hort.* **1999**, *81*, 71–87. [CrossRef]
45. Sharma, T.P.; Sen, D.N. A new report on abnormally fast germinating seeds of *Haloxylon* spp.: An ecological adaptation to saline habitat. *Curr. Sci.* **1989**, *58*, 382–385. [CrossRef]
46. Mahmoud, A.; El Sheikh, A.M.; Baset, S.A. Germination of two halophytes: *Halopeplis perfoliata* and *Limonium axillare* from Saudi Arabia. *J. Arid. Environ.* **1983**, *6*, 87–98. [CrossRef]
47. Wiesbauer, M.B.; Hmeljevski, K.V.; Zimmermann, T.G.; dos Reis, M.S.; Reis, A.; de Souza, S.L. Reintrodução de *Dyckia distachya* Hassler nas áreas de Influência das Hidrelétricas de Itá e Machadinho. *V Congresso de Inovação Tecnológica em Energia Elétrica*. 2009, Volume 20011, p. 202006. Available online: https://www.cgti.org.br/publicacoes/wp-content/uploads/2016/03/Reintroduc%CC%A7a%CC%83o-de-Dyckia-distachya-Hassler-nas-a%CC%81reas-de-Influe%CC%82ncia-das-Hidrele%CC%81tricas-de-Ita%CC%81-e-Machadinho.pdf (accessed on 4 March 2022).
48. Vieira, D.C.; Socolowski, F.; Takaki, M. Germinação de sementes de *Dyckia tuberosa* (Vell.) Beer (Bromeliaceae) sob diferentes temperaturas em luz e escuro. *Rev. Bras. Bot.* **2007**, *30*, 183–188. [CrossRef]
49. Penfield, S.; MacGregor, D.R. Effects of environmental variation during seed production on seed dormancy and germination. *J. Exp. Bot.* **2017**, *68*, 819–825. [CrossRef] [PubMed]
50. Thanos, C.A.; Doussi, M.A. Ecophysiology of seed germination in endemic labiates of Crete. *Isr. J. Plant Sci.* **1995**, *43*, 227–237. [CrossRef]
51. Kadis, C.; Georghiou, K. Seed dispersal and germination behavior of three threatened endemic labiates of Cyprus. *Plant Spec. Biol.* **2010**, *25*, 77–84. [CrossRef]
52. Bertsouklis, K.; Papafotiou, M. Seed germination of *Arbutus unedo*, *A. andrachne* and their natural hybrid *A. andrachnoides* in relation to temperature and period of storage. *HortScience* **2013**, *48*, 347–351. [CrossRef]
53. Pompelli, M. Germinação de *Dyckia encholirioides* var *encholirioides* (Bromeliaceae, Pitcairnioideae). *Rev. Floresta E Ambiente* **2004**, *13*, 1–9.
54. Pompelli, M.F.; Fernandes, D.; Guerra, M.P. Germination of *Dyckia encholirioides* (Gaudichaud) Mez var. *encholirioides* under saline conditions. *Seed Sci. Technol.* **2006**, *34*, 759–763. [CrossRef]
55. Adams, C.R.; Early, M.P. *Principles of Horticulture*; Elsevier Butterworth-Heinemann Publication: Burlington, MA, USA, 2004.

56. Cruz-Cruz, C.A.; González-Arnao, M.T.; Engelmann, F. Biotechnology and Conservation of Plant Biodiversity. *Resources* **2013**, *2*, 73–95. [CrossRef]
57. Hartmann, H.T.; Kester, D.E.; Geneve, R.L. *Hartmann & Kester's Plant Propagation Principles and Practices*; (No. 631.53 H2555p Ej.1 025385); Prentice Hall: Hoboken, NJ, USA, 2011; 2555p.
58. Cabahug, R.A.M.; Nam, S.Y.; Lim, K.B.; Jeon, J.K.; Hwang, Y.J. Propagation techniques for ornamental succulents. *Korean Flower Assoc.* **2018**, *26*, 90–101. [CrossRef]
59. Le Roux, J.J.; Hui, C.; Castillo, M.L.; Iriondo, J.M.; Keet, J.H.; Khapugin, A.A.; Médail, F.; Rejmánek, M.; Theron, G.; Yannelli, F.A.; et al. Recent Anthropogenic Plant Extinctions Differ in Biodiversity Hotspots and Coldspots. *Curr. Biol.* **2019**, *29*, 2912–2918. [CrossRef]
60. Kougioumoutzis, K.; Kokkoris, I.P.; Panitsa, M.; Strid, A.; Dimopoulos, P. Extinction Risk Assessment of the Greek Endemic Flora. *Biology* **2021**, *10*, 195. [CrossRef]
61. Leme, E.M.C. Considerations on a new Neoregelia from Brazil. *J. Bromel. Soc.* **1983**, *33*, 118–120.
62. Tsay, H.S.; Lee, C.Y.; Agrawal, D.C.; Basker, S. Influence of ventilation closure, gelling agent and explant type on shoot bud proliferation and hyperhydricity in *Scrophularia yoshimurae*—A medicinal plant. *Vitr. Cell. Dev. Biol. Plant* **2006**, *42*, 445–449. Available online: https://www.jstor.org/stable/20461600 (accessed on 4 March 2022). [CrossRef]
63. Guerra, M.P.; Vesco, L.L.D. Strategies for the Micropropagation of Bromeliads. In *Protocols for In Vitro Propagation of Ornamental Plants. Methods in Molecular Biology (Methods and Protocols)*; Jain, S., Ochatt, S., Eds.; Humana Press: Totowa, NJ, USA, 2010; Volume 589. [CrossRef]
64. Varadarajan, G.S.; Varadarajan, U.; Locy, R.D. Application of tissue culture techniques to maintain a rare species, *Puya tuberosa*. *J. Bromel. Soc.* **1993**, *43*, 112–118.

horticulturae

Article

In Vitro Propagation of the Mount Parnitha Endangered Species *Sideritis raeseri* subsp. *Attica*

Konstantinos Bertsouklis *, Panagiota Theodorou and Paraskevi-Evangelia Aretaki

Laboratory of Floriculture and Landscape Architecture, Department of Crop Science, School of Plant Sciences, Agricultural University of Athens, 75 Iera Odos Street, GR-11855 Athens, Greece
* Correspondence: kber@aua.gr; Tel.: +30-210-5294558

Abstract: Over the past few decades, both wildfires and human-sparked fires have ravaged Mount Parnitha, destroying the mountain's unique natural ecosystem, applying pressure to its flora, and subjecting the vulnerable populations of *Sideritis raeseri* subsp. *attica* to excessive stress. The present study aims to establish an efficient micropropagation method starting from in vitro-grown seedlings. The in vitro germination study carried out during the production of seedlings revealed a higher germination rate (34.0% and 37.0%, respectively) at 20.0 °C and 25.0 °C. The in vitro-derived seedlings studied were used as the starting material for the establishment of various media. Murashige and Skoog (MS) media, hormone-free and containing 0.5 mg L^{-1} 6-benzyladenine (BA), led to the satisfactory (84.0–89.0%) establishment of plantlets. During the multiplication phase, the study used BA in conjunction with α-naphthaleneacetic acetic acid and four different cytocinins (BA; kinetin (KIN); 6-(γ-γ-dimethylallylamino) purine; zeatin) at low concentrations (0.5 mg L^{-1}). During the second subculture, a high multiplication index (7.3 and 6.4, respectively) was found for the hormone-free MS medium and the MS medium containing KIN at 0.5 mg L^{-1}. Hyperhydricity took place on the media supplemented with hormones. Rooting occurred on the half-strength MS medium (51.0%). After two months, the plants' survival rate stood at 100.0%, as did their ex vitro acclimatisation rate, which also registered at 100.0%. The present results could encourage not only the introduction of *S. raeseri* subsp. *attica* into the industry of floriculture as a new ornamental plant but also its ex vitro conservation.

Keywords: in vitro germination; ex vitro acclimatisation; GSI; micropropagation; tissue culture

Citation: Bertsouklis, K.; Theodorou, P.; Aretaki, P.-E. In Vitro Propagation of the Mount Parnitha Endangered Species *Sideritis raeseri* subsp. *Attica*. *Horticulturae* **2022**, *8*, 1114. https://doi.org/10.3390/horticulturae8121114

Academic Editor: Luigi De Bellis

Received: 19 October 2022
Accepted: 24 November 2022
Published: 27 November 2022

1. Introduction

Mount Parnitha is one of the ten national parks of Greece, located 35.0 km north of the metropolitan city of Athens [1]. It is part of the Natura 2000 Network (established by the European Union, Directive 92/43/EEC), a network of sites that are home to rare habitats and species. Mount Parnitha comprises three vegetation zones with 1093 plant species; ninety-three of these are Greek endemics [2]. Since ancient times, the vegetation of Parnitha has been under enormous human pressure brought about by anthropogenic factors, urbanisation, and fires. It is clear that, over time, the types of pressures and their temporal patterns, in conjunction with invasions and urbanisation [3], may eventually take their toll on Parnitha, leading to the extinction of some of the mountain's unique flora. Since 1913, there have been 438 fires on Mount Parnitha, which have ravaged the entire forest approximately six times (Figure 1A) [4–6].

Sideritis L. is one of the three largest genera of the Lamiaceae family. It includes more than 150 species of shrubs and annual or perennial herbs that are found mostly in the Mediterranean Basin [7] and it has long been known as an integral part of Greek flora. Its natural populations are threatened by cultivation in neighbouring fields with cultivars of unrelated resources [8]. The genus is used in homoeopathic medicine as a herbal tea and in folk medicine for the preparation of a variety of remedies. In Greece, there are only two taxa, i.e., *S. scardica* and *S. raeseri* Boiss. & Heldr. subsp. *raeseri*, which are cultivated on a small scale for the preparation of mountain tea. Recently, the cultivation of *S. syriaca* subsp. *syriaca* and *S. clandestine* (Bory & Chaub.) Hayek subsp. *clandestina* has also begun [9].

Figure 1. (**A**) The slopes of Mount Parnitha after the widespread fires of 2007, 15 years ago; (**B**) mother plants of *Sideritis raeseri* subsp. *attica* on Parnitha; (**C**) flowering phase in July; (**D**) mother plants during seed collection.

S. raeseri Boiss. & Heldr. subsp. *attica* (Heldr.) Papan. & Kokkini [10] was placed on the list of vulnerable species [11] but its extinction risk (Figure 1B) has recently been updated, placing it on the list of the endangered species found on Mount Parnitha [12]. Small populations of the species can also be found on Mounts Kithairon and Pateras [13]. A hardy perennial that grows up to 80.0 cm, *S. raeseri* subsp. *attica* sprawls on rocky slopes at elevations over 1000.0 m. Its yellow flowers bloom from June to August on short, woody stems 30.0 cm long (Figure 1C,D). The species shows the potential for being a promising introduction as an ornamental plant into the landscape architecture and floriculture industries, as it is quite suitable for thematic gardens, rocky places, and degraded areas.

The members of the *Sideritis* genus have been studied for their antioxidants and their valuable components [14–16]. Moreover, they could be used as a natural source of antioxidants in the food industry and medicine [17,18]. The Cretan *S. syriaca* subsp. *syriaca* shows promising medicinal and cosmetic potential [19] and is grown locally in Crete as a pilot crop [20,21]. As to *S. raeseri*, commercial nursery growers provide farmers with plantlets that have been established mainly asexually and collected from populations found in the wild. Still, the origin and properties of this type of material remain uncharacterised [21]. The increased interest of nurseries and farmers in the propagation of aromatic plants in tandem with the intensive harvesting practices during the species' flowering stage constitutes a threat to the distribution of *S. raeseri*.

To the best of our knowledge, apart from a study of *S. raeseri* Boiss & Heldr subsp. *raeseri* starting from adult mother plants [22], there are no reports in the literature on the propagation of the present subspecies. *S. raeseri* subsp. *attica* has attractive, ornamental characteristics that make it an ideal candidate for commercial exploitation. It is precisely those characteristics that could be exploited in synergy with the development of in vitro propagation methods, leading to the introduction of suitable methods for the plant's conservation, in situ and ex situ alike. In this task, biotechnological methods are essential to developing successful in situ conservation strategies, as well as ex situ collections aimed at plant preservation [23–26]. Not only does seed propagation for in vitro use enhance genetic diversity leading to the selection of genotypes of high commercial value [27] but it also barely impacts the environment as it requires a small number of seeds to produce many plants and does not involve harvesting plant parts, especially from endangered species [28]. Furthermore, in vitro seed propagation allows for quick shoot multiplication by small quantities of initial explant material derived from plants in the wild through germinated seeds and makes propagation independent of seasons. The practice of seed propagation accompanied by efficient clonal propagation strategies can meet the increased demands of the ornamental plant market. Moreover, it presents a challenge for researchers, including the authors of the present study, who wish to both exploit and conserve the *Sideritis* species via studies on in vitro propagation.

Therefore, the aim of the present study was to suggest an efficient protocol for in vitro propagation of *S. raeseri* subsp. *attica* starting from in vitro-grown seedlings. The effects of the nutrient medium and auxin concentration on in vitro rooting were tested. In addition, an assessment of seed germinability was carried out. Finally, the acclimatisation rate was tested targeting the exploitation of the species and its introduction as an ornamental plant into the floriculture industry.

2. Materials and Methods

2.1. Plant Material

In July 2020, we visited Mount Parnitha in Attica, Greece (38°09′59.3892″ N, 23°42′58.9932″ E), where we proceeded to collect fully matured seeds on the same day. The fully matured seeds came from native plants in the wild (Figure 1C,D). Our harvest was then transferred to our laboratory at the Agricultural University of Athens, where the seeds were stored in paper bags in the dark at room temperature, where the temperature stood at 25.0 °C and the relative humidity at 30.0%. In vitro germination and propagation did not take place until three months had elapsed after seed storage.

2.2. In Vitro Germination

We accomplished the seeds' surface sterilisation with 20.0% (*v*/*v*) commercial bleach (4.6% *w/v* sodium hypochlorite) for 10.0 min and then rinsed them three times (3.0 min/time) with sterile, distilled water. After disinfection, the seeds were sown in plastic Petri dishes (9.0 cm) with half-strength Murashige and Skoog (MS) media [29]. The seed cultures were incubated at 5.0, 10.0, 15.0, 20.0, 25.0, 30.0, and 35.0 °C. Germination was defined as the appearance of a radicle at least 2.0 mm long and could be recorded daily in compliance with the rules of the International Seed Testing Association [30]. T_{50} (days) was defined as the days needed to reach 50% of the final germination percentage [31]. We used sixty seeds per treatment (four Petri dishes per treatment/15 seeds per Petri dish) and employed the Maguire formula [32], which measures the germination speed index (GSI) as

$$GSI = G1/N1 + G2/N2 + \ldots + Gn/Nn$$

in which G1, G2, and Gn = the number of normal seedlings measured during the first, second, and last counts; and N1, N2, and Nn = the number of sowing days during the first, second, and last counts.

2.3. Micropropagation

2.3.1. Establishment of Initial Cultures

Five days after we had finalised the seed germination stage, we transferred the seedlings (young plantlets) grown in vitro to a hormone-free MS medium (8.0 g L^{-1} agar), where they remained for forty days until the time we moved on to the next stage, the establishment of initial cultures. We initiated a subculture on a hormone-free medium, as well as on a medium containing 6-benzyladenine (BA) at 0.5 mg L^{-1}, with or without 4.0 g L^{-1} polyvinylpyrrolidone (PVP).

2.3.2. Multiplication

During the multiplication stage, node explants were cultured on a hormone-free MS medium or supplemented with BA at 0.1, 0.5, and 1.0 mg L^{-1} and α-naphthaleneacetic acetic acid (NAA) at 0.01 and 0.1 mg L^{-1} in various combinations (first subculture). The effect of four different cytokinins, i.e., BA, kinetin (KIN), 6-(γ-γ-dimethylallylamino) purine (2iP), and zeatin (ZEA), at 0.5 mg L^{-1} was tested on the second subculture. The MS medium and McCown woody plant medium (WPM) [33] were tested during the multiplication stage to control hyperhydricity using node explants from shoots deriving from the first and second subcultures.

2.3.3. In Vitro Rooting

Microshoots that were 1.5–2.0 cm long and produced by sub-culturing on various media during the multiplication stage were transferred for rooting onto half-strength MS media or half-strength WPM media (MS/2 and WPM/2, respectively) containing 0.0, 0.5, 1.0, or 2.0 mg L^{-1} indole-3-butyric acid (IBA). A completely randomised two-factor factorial design was used for each plant species (i.e., two types of nutrient medium × four IBA concentrations).

2.3.4. In Vitro Culture Conditions and Data Collection

In vitro cultures were carried out in Magenta GA-7 vessels (7.2 cm × 7.2 cm × 10.0 cm, Sigma-Aldrich), with four explants per vessel. The cultures were maintained at 25.0 °C. Cool-white, fluorescent lamps provided a 16 h photoperiod at 37.5 μmol m^{-2} s^{-1}. All media contained 30.0 g L^{-1} sucrose; were solidified with 8.0 g L^{-1} agar; had their pH adjusted to 5.7–5.8; and were autoclaved at 121.0 °C for 20.0 min. The collection of data took place after forty days of culture for the shoot regeneration and after twenty days for the rooting experiments. Data were collected on the shoot proliferation percentage, shoot number per explant, length of the shoots, number of nodes per shoot, lateral shoots, lateral shoot length, and number of nodes. The percentage (%) of hyperhydricity and callus were also recorded. The proliferation potential of the cultures, in other words, the multiplication index (MI) of each culture, was calculated by multiplying the percentage of explants that produced shoots by the mean number of shoots and mean shoot length per responding explant, then dividing by 0.6 (the length of each explant used for the sub-culture). Rooting percentages and root numbers and lengths were recorded during the rooting experiments.

2.3.5. Ex Vitro Acclimatisation

All rooted microshoots measuring 1.5–2.0 cm in length were thoroughly rinsed under running tap water. They were then placed into containers (500.0 mL, eight plantlets/container) that included both peat (pH 5.5–6.5, Klasmann-Delimann GmbH, Geeste, Germany) and perlite (particles diameter 1.0–5.0 mm, Perloflor, Isocon S.A., Athens, Greece) substrate 1:1 (v/v). In order to control the humidity, we covered all containers with plastic wrap (Sanitas; Sarantis S.A., Athens, Greece). Next, the containers were moved into a growth chamber for seven days (chamber temperature: 25.0 °C; 16 h cool-white, fluorescent light 37.5 μmol m^{-2} s^{-1}/8.0 h dark photoperiod). Next, the containers stayed uncovered for a period of seven days and then were transferred onto a heated glasshouse bench for another seven days (37°58′58.0″ N, 23°42′19.2″ E). It was at the end of this period that data on acclimatisation were recorded. After acclimatisation, the plants were transplanted into 500.0 mL plastic pots containing peat:perlite (1/1, v/v) and were fertilised monthly with 2.0 g L^{-1} complete water-soluble fertiliser (Nutrileaf 60, 20-20-20; Miller Chemical and Fertilizer Corp., Hanover, PA, USA). The last step, which took place two months later, involved the calculation of the plants' survival rate.

2.4. Statistical Analysis

For the purposes of our statistical analysis, we used a completely randomised design method. The experiments of seed germination had 4 replications of 15 seeds each, i.e., 60 seeds per treatment. The seed germination results for each treatment are shown as the mean ± SE (standard error mean). As shown in the in vitro propagation data tables, the number of replicates per treatment differed among the experiments.

The significance of the results was tested by one-way ANOVA. In the case of in vitro rooting, two-way ANOVA was performed to find the statistically significant differences and possible interactions between the medium (MS/2, WPM/2) and IBA concentration (0.0, 0.5, 1.0, or 2.0 mg L^{-1}). The treatment means were compared using Tukey's test at $p \leq 0.05$ (JMP 14.0 software, SAS Institute Inc., Cary, NC, USA, 2013). Data on percentages were arcsine-transformed prior to the statistical analysis.

3. Results

3.1. Seed Germination

The disinfection method of the seeds resulted in a 90.0% success rate. The germination percentages of 34.0% and 37.0% proved to be similar at 20.0 °C and 25.0 °C, respectively (Table 1; Figures 2 and 3). The germination percentages at the cardinal germination temperatures of 15.0 °C and 30.0 °C were 17.0% and 7.0%, respectively, with no difference. T_{50} was completed in 16 days at 20 °C and 20 days at 25 °C; the seeds in those treatments had a higher GSI, which stood at 24.3 and 20.3, respectively (Table 1). Germination was completed in 32.0–35.0 days.

Table 1. In vitro germination of *Sideritis raeseri* subsp. *attica* seeds, T_{50}, times for full germination, and germination speed indexes (GSI) at the temperatures shown.

Temperature (°C)	Germination (%) ± SE †	T_{50} †† (Days)	Time for Full Germination (Days)	GSI
10.0	0.0 ± 0.0 c	-	-	-
15.0	17.0 ± 4.6 b	21.0	35.0	9.0 b
20.0	34.0 ± 5.3 a	16.0	33.0	24.3 a
25.0	37.0 ± 7.5 a	20.0	32.0	20.3 a
30.0	7.0 ± 6.5 b	30.0	32.0	1.7 b
35.0	0.0 ± 0.0 c	-	-	-
$F_{\text{one-way ANOVA}}$	*			*

† (SE): standard error; †† (T_{50}): time for 50% germination; *: significant at $p \leq 0.05$; mean (±SE) separation in columns by Tukey's test at $p \leq 0.05$; mean values followed by the same letter are not significantly different at $p \leq 0.05$; n = 4, 15 seeds/Petri dish (total 60 seeds per treatment).

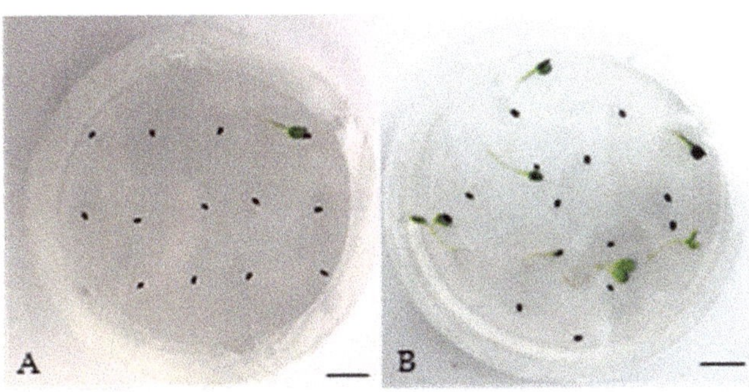

Figure 2. *Sideritis raeseri* subsp. *attica* seeds' germination in Petri dishes containing half-strength Murashige and Skoog media (MS) at 20.0 °C under 16.0 h light/8.0 h darkness: (**A**) after 9.0 days and (**B**) after 30.0 days of incubation. Bars represent a length of 1.0 cm.

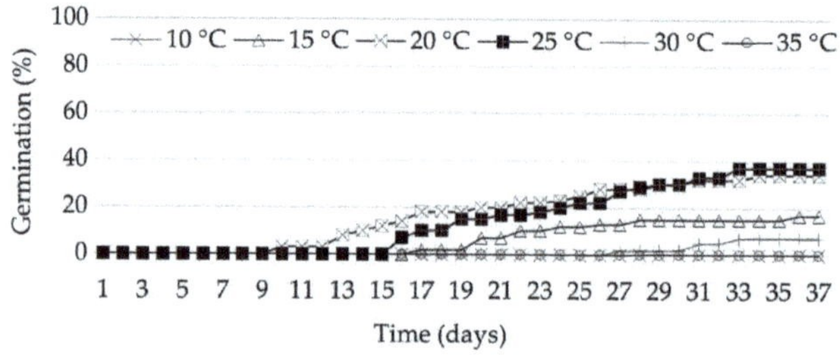

Figure 3. Germination time course curves of *Sideritis raeseri* subsp. *attica* seeds affected by temperature. Four Petri dishes (15 seeds/treatment) were used.

3.2. Micropropagation

3.2.1. Establishment of Initial Cultures

S. raeseri subsp. *attica* plantlets deriving from in vitro-grown seedlings were transferred to a hormone-free medium for further growth (initial culture). Forty days after the transfer, the plantlets had risen to 2.0–2.5 cm in height and showed vigorous growth of 5.0–7.0 leaves (Figure 4A,B). The establishment of initial cultures on MS media that was either hormone-free or supplemented with 0.5 mg L^{-1} was successful during a subsequent stage. The shooting percentage was exceptionally high and stood at over 84.0% (Table 2). Two shoots of similar heights were formed (1.1–1.4 cm) on the hormone-free media and the media supplemented with BA and presented no differences. On the other hand, on the media containing PVP, the shoots were the shortest, growing no higher than 0.6 cm (Figure 5A). A small number of short lateral shoots were formed. As to the plantlets themselves, 12.5–13.0% of them grew roots on the MS media that was hormone-free and the MS media that was supplemented with BA (Table 2, Figure 5A).

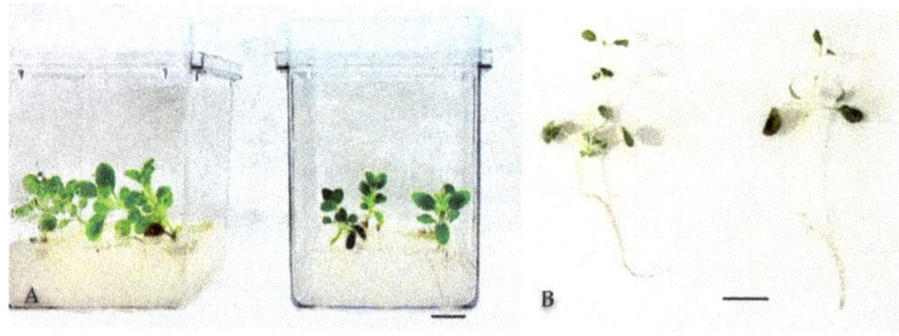

Figure 4. Forty-day-old young seedlings of *Sideritis raeseri* subsp. *attica* grown in vitro on half-strength MS media without any hormones (MS/2) (**A**) before and (**B**) during the subculture. Bars represent a length of 1.0 cm.

Table 2. Establishment stage: initial cultures of node explants excised from in vitro-grown, young seedlings of *Sideritis raeseri* subsp. *attica* on MS media without hormones (Hf) or supplemented with 6-benzyladenine (BA) at 0.5 mg L^{-1}, with or without polyvinylpyrrolidone (PVP) at 4.0 mg L^{-1}.

BA (mg L^{-1})	PVP (g L^{-1})	Shooting (%)	Shoot Number	Shoot Length (cm)	Node Number	Hyperhydricity	MI †	Lateral Shoot Number	Length of Lateral Shoots (cm)	Node Number of Lateral Shoots	Rooting (%)	Root Number	Root Length (cm)
-	-	89.0 a	2.2 a	1.1 a	4.2 a	-	3.6 ab	0.6 a	0.1 a	0.4 a	12.5 a	15.0 a	2.6 a
0.5	-	84.0 a	2.1 a	1.4 a	4.5 a	19.0 a	4.1 a	0.2 a	0.1 a	0.1 b	13.0 a	18.0 a	1.5 a
0.5	4.0	95.0 a	2.0 a	0.6 b	3.3 b	12.0 a	1.9 b	0.5 a	0.1 a	0.1 b	-	-	-
$F_{one\text{-}way\ ANOVA}$		ns	ns	**	**	ns	**	ns	ns	*	ns	ns	ns

† (MI): multiplication index = shooting (%) × mean shoot number per explant × mean shoot length/0.6; *, **, significant at $p \leq 0.05$, $p \leq 0.01$, respectively; F values represented by ns indicate no significant differences at $p \leq 0.05$; mean separation in columns by Tukey's test at $p \leq 0.05$; mean values followed by the same letter are not significantly different at $p \leq 0.05$; n = 45–60.

Figure 5. In vitro regeneration of *Sideritis raeseri* subsp. *attica*: (**A**) plantlets growing and root formation during the establishment of an initial culture of young seedlings after 40.0 days on various MS media; (**B**) multiplication stage and shoots forming on shoot explants after 40.0 days of culture on various MS media; (**C**) in vitro root formation after 20.0 days of culture on half-strength MS or WPM media (MS/2 or WPM/2, respectively) containing indole-3-butyric acid (IBA) at 0.0, 0.5, 1.0, or 2.0 mg L^{-1}; (**D**) plantlets during ex vitro acclimatisation on 1:1 T:P and two-month-old plants showing vigorous growth. Bars represent a length of 1 cm and the numbers represent the concentrations of phytohormones. (MS): Murashige and Skood medium; (Hf): hormone-free medium; (MS/2): half-strength MS.

3.2.2. Multiplication

Two subcultures were carried out on MS media containing (a) low concentrations of BA as well as NAA and (b) various cytokinins (BA, KIN, 2iP, ZEA) at 0.5 mg L^{-1}. Another subculture on a hormone-free MS medium and WPM media followed in order to eliminate the percentage of hyperhydricity recorded during the previous two subcultures. The first subculture presented a shooting percentage of over 78.0% in all treatments (Table 3; Figure 5B); no differences were observed in the lengths of the formed shoots and number of nodes and the shoot number registered higher (3.6 shoots/explant) on the 0.5/0.01 BA/NAA (mg L^{-1}) media. A small number of short, lateral shoots were formed and a higher MI (6.3) was calculated on the MS medium with 0.5/0.01 BA/NAA (mg L^{-1}) due to the formation of a higher number of shoots. Shoots were hyperhydric at a percentage of over 23.0% in all hormone-containing media with no differences.

Table 3. Effect of various concentrations of BA and α-naphthaleneacetic acetic acid (NAA) on MS media on shoot proliferation of *Sideritis raeseri* subsp. *attica* during the first subculture.

Hormone (mg L^{-1})		Shooting (%)	Shoot Number	Shoot Length	Node Number	Hyperhy-Dricity	Callus Formation	MI †	Lateral Shoot Number	Length of Lateral Shoots (cm)	Node Number of Lateral Shoots
BA	NAA										
-	-	81.0 a	2.1 b	1.1 a	3.9 a	6.0 b	23.0 c	3.1 b	0.3 b	0.2 a	0.5 a
0.1	-	78.0 a	2.4 b	1.2 a	3.2 a	32.0 ab	50.0 b	3.7 b	0.2 b	0.1 a	0.3 a
0.5	-	87.0 a	2.8 ab	0.9 a	3.2 a	23.0 ab	55.0 b	3.7 b	0.2 b	0.1 a	0.2 a
0.5	0.01	87.0 a	3.6 a	1.2 a	3.4 a	31.0 ab	54.0 b	6.3 a	0.6 a	0.1 a	0.3 a
0.5	0.1	92.0 a	2.6 ab	1.0 a	3.0 a	50.0 a	85.0 a	4.0 b	0.7 a	0.2 a	0.2 a
1.0	0.1	78.0 a	3.3 ab	0.8 a	4.5 a	20.0 ab	94.0 a	3.4 b	0.0 b	0.0 a	0.0 a
$F_{\text{one-way ANOVA}}$		ns	*	ns	ns	*	*	*	*	ns	ns

† (MI): multiplication index = shooting (%) × mean shoot number per explant × mean shoot length/0.6; *, significant at $p \leq 0.05$; *F* values represented by ns indicate no significant differences at $p \leq 0.05$; mean separation in columns by Tukey's test at $p \leq 0.05$; mean values followed by the same letter are not significantly different at $p \leq 0.05$; n = 45–50.

We used four different cytokines (BA, KIN, 2iP, ZEA) supplemented at low concentrations (0.5 mg L^{-1}) in the case of the second subculture. The shooting percentage was quite high, standing at over 78.5% regardless of the medium (Table 4, Figure 5B). The shoot production was higher on the MS medium supplemented with BA, with a 4.3 formation of shoots/explants, whereas the shoot length was higher (1.8 cm) on the hormone-free or KIN-supplemented media (Table 4). The range of hyperhydricity percentages was similar to the first subculture, i.e., 8.0–52.0%. It was higher for plantlets derived on 2iP-containing medium (52.0%). With regard to the MI, the hormone-free medium (7.3) and the media supplemented with BA or KIN (6.2 and 6.4, respectively) proved to be the most suitable.

Table 4. Effect of hormone-free MS medium and media supplemented with four different cytokinins, i.e., BA, kinetin (KIN), 6-(γ-γ-dimethylallylamino) purine (2iP), and (zeatin) ZEA, at 0.5 mg L^{-1} on shoot proliferation of node explants excised from in vitro-produced shoots during the first subculture of *Sideritis raeseri* subsp. *attica* seeds (second subculture).

Cytokinin (mg L^{-1})	Shooting (%)	Shoot Number	Shoot Length	Node Number	Hyperhy-Dricity	Callus Formation	MI †	Lateral Shoot Number	Length of Lateral Shoots (cm)	Node Number of Lateral Shoots
-	86.5 a	2.8 b	1.8 a	4.7 a	8.0 b	-	7.3 a	0.5 a	0.2 a	0.7 a
BA	78.5 a	4.3 a	1.1 b	4.2 ab	39.0 ab	62.0 ab	6.2 a	0.1 b	0.1 a	0.1 b
KIN	89.5 a	2.4 b	1.8 a	4.9 a	16.0 b	48.0 b	6.4 a	0.1 b	0.1 a	0.1 b
2iP	73.5 a	3.6 ab	1.2 b	4.0 b	52.0 a	63.0 a	5.3 ab	0.1 b	0.1 a	0.3 b
ZEA	81.5 a	3.4 ab	1.0 b	3.9 b	34.0 ab	88.0 a	4.6 b	0.1 b	0.1 a	0.3 b
$F_{\text{one-way ANOVA}}$	ns	***	***	***	*	*	*	**	ns	*

† (MI): multiplication index = shooting (%) × mean shoot number per explant × mean shoot length/0.6; *, **, *** significant at $p \leq 0.05$, $p \leq 0.01$ $p \leq 0.001$, respectively; *F* values represented by ns indicate no significant differences at $p \leq 0.05$; mean separation in columns by Tukey's test at $p \leq 0.05$; mean values followed by the same letter are not significantly different at $p \leq 0.05$; n = 45–50.

The subcultures on the hormone-free MS and WPM-supplemented media were successful across the board in terms of shooting formation, which stood at 100.0% (Table 5). The shoot numbers decreased and 1.9–2.0 shoots/explant of 1.5–1.7 cm in length formed. The MI registered at 5.0–5.4 (Table 5). Further, it was observed that the culture of explants on the hormone-free media thoroughly eliminated hyperhydricity (Table 5, Figure 5B).

Table 5. Effect of two different media, MS and McCown woody plant medium (WPM) without hormones, on the elimination of hyperhydration during the multiplication phase of *Sideritis raeseri* subsp. *attica*.

Medium	Shooting (%)	Shoot Number	Shoot Length (cm)	Node Number	Hyperhy-Dricity	Callus Formation	MI †	Lateral Shoot Number	Length of Lateral Shoots (cm)	Node Number of Lateral Shoots
MS	100 a	1.9 a	1.7 a	4.3 a	-	84.0	5.4 a	0.1 a	0.0 a	0.2 a
WPM	100 a	2.0 a	1.5 a	4.0 a	-	-	5.0 a	0.1 a	0.0 a	0.1 a
$F_{\text{one-way ANOVA}}$	ns	ns	ns	ns	-	-	ns	ns	ns	ns

† (MI): multiplication index = shooting (%) × mean shoot number per explant × mean shoot length/0.6; *F* values represented by ns indicate no significant differences at $p \leq 0.05$; mean separation in columns by Tukey's test at $p \leq 0.05$; mean values followed by the same letter are not significantly different at $p \leq 0.05$; n = 45–50.

Finally, a small number of short lateral shoots also formed during all multiplication phases (Tables 3–5; Figure 5A,B)

3.2.3. In Vitro Rooting and Ex Vitro Acclimatisation

Spontaneous rooting of the microshoots derived during the establishment of the initial culture took place at low percentages (12.5%) on the MS media that were either hormone-free or supplemented with 0.5 mg L^{-1} (Table 2, Figure 5A). Microshoots also rooted spontaneously during the multiplication stage at higher percentages (Figure 4B, data not shown). Two-way ANOVA showed that the interaction of the main factors, medium type, and IBA concentration (0.0, 0.5, 1.0, or 2.0 mg L^{-1}) was significant during the in vitro rooting phase (Table 6). A higher rooting percentage (51.0%) of plantlets was observed on the MS/2 medium containing 0.0 mg L^{-1} IBA (51.0%). Plantlets were observed forming longer roots (0.9 cm). The MS/2 media supplemented with 2.0 (mg L^{-1}) IBA showed a reduction in root formation (14.0%). However, the root number was higher on the WPM/2 containing 2.0 mg L^{-1} IBA (Table 6, Figure 5C). Rooted microshoots were successfully (100.0%) acclimatised ex vitro (Figure 5D). Once the plants were acclimatised, they were transplanted into pots. Their survival rate after the transfer was again 100.0% (Figure 5D).

Table 6. In vitro rooting of microshoots derived in the multiplication stage affected by the concentration of IBA (0.0, 0.5, 1.0, or 2.0 mg L^{-1}) and the type of medium (half-strength MS or half-strength WPM).

Treatments	Rooting (%)	Root Number	Root Length (cm)
IBA (mg L^{-1})			
0.0 (control)	- †	-	-
0.5	-	-	-
1.0	-	-	-
2.0	-	-	-
Medium			
MS/2	-	-	-
WPM/2	-	-	-

Table 6. *Cont.*

Treatments		Rooting (%)	Root Number	Root Length (cm)
Interaction (IBA × Medium)				
0.0 mg L^{-1} (control) ×	MS/2	51.0 a	2.3 ab	0.9 a
	WPM/2	32.0 ab	0.4 b	0.4 ab
0.5 mg L^{-1} ×	MS/2	23.0 ab	2.6 ab	0.3 b
	WPM/2	20.0 ab	2.7 ab	0.4 ab
1.0 mg L^{-1} ×	MS/2	20.0 ab	2.4 ab	0.3 ab
	WPM/2	20.0 ab	3.7 ab	0.3 ab
2.0 mg L^{-1} ×	MS/2	14.0 b	2.1 ab	0.2 b
	WPM/2	28.0 ab	6.6 a	0.4 ab
F_{IBA}		ns	ns	ns
F_{medium}		ns	ns	ns
$F_{IBA \times med}$		*	*	*

† When interactions are significant, factors are not considered and mean values are not shown; ns: non-significant; * denotes significant differences between means at $p < 0.05$, indicated by different letters in columns; mean separation in columns by Tukey's test at $p \leq 0.05$; $n = 45$–50.

4. Discussion

The present study's main aim focused on the regeneration of the endangered *S. raeseri* subsp. *attica* starting with young, in vitro-grown seedlings. Nevertheless, while focusing on regeneration, we carried out a preliminary assessment of the species' germinability as previous studies of other *Sideritis* species found a strong correlation between germination and the environmental conditions [34,35]. In the present study, the germination rate was low and did not exceed 40.0%. Very low germination rates were also recorded for *S. perfoliate* and *S. erythabtha*, whereas *S. stricta* seeds did not germinate at all [36]. The germination rates for *S. athoa* were 70.0–72.0% on the solid MS/2 at 20.0 °C and 25.0 °C [37]. As for *S. chamaedryfolia* and *S. pungens*, the germination rates registered at 74.0% and 99.0%, respectively [34]. Shtereva et al. [38] achieved very impressive germination rates (up to 98.0%) for *S. scardica* after pre-treatment (mechanical scarification) and treatment with gibberellic acid. The present study defined the cardinal temperatures as ranging from 15.0 to 30.0 °C, whereas a range of 20.0–25.0 °C was found to be the optimal temperature range for seed germination, as temperatures higher or lower than these in the range were found to decrease the germination rate to all-too-low percentages. To confirm our findings, Papafotiou and Kalantzis [37] indicated the same optimal temperature range for *S. athoa* and Estrelles et al. [34] reported the same temperature range as being optimal for *S. chamaedryfolia* and *S. pungens*, which are found in Iberian habitats. The usual range for the germination of Mediterranean species is between 15.0 and 20.0 °C [39–41]. However, Picciau et al. [42] found that the germination of species growing at higher altitudes necessitates no pre-treatment at relatively high temperatures (20.0 °C), a finding also confirmed by the present study. The impact of the environment during seed production proved powerful [39]. Not only does the environment influence the seed size, germination rate, and viability but it also causes variations even within the same genotype [43–45]. Endemic species are also closely correlated with the characteristics of their habitats [34] and native plants have their own strategies for optimising their evolution in their respective ecosystems [46]. Moreover, it has been found that the inducement of physiological dormancy in seeds of *S. serrata* seedbanks in the soil under high and prolonged thermoperiods is possible [47]. Thermo-dormancy is a secondary dormancy that plays a role in preventing seed germination during occasional summer rainfalls [48,49]. The seeds used in the present study were collected from plants in one day and it may be construed that a proportion of thermodormancy could not be induced during storage. The low germination rate could be attributed to climate change, which is characterised by high temperatures and frequent conditions of water storage. This is a climate issue that negatively impacts seed germination as it reduces seed viability and germinability [50] and leads to the realisation that further studies should focus on the effects of environmental parameters, such as temperature and photoperiod, on seed quality and germination. The ecophysiology of endangered species is far more vulnerable than that of other, non-endangered endemic species.

With regard to the in vitro propagation of *S. raeseri* subsp. *attica*, our study employed PVP, which has been used before in a wide range (from 0.5 to 20 g L^{-1}) for the micropropagation of *S. raeseri* subsp. *raeseri* [51]. It has been found that phenolics are produced often during the establishment stage of in vitro propagation. The production of phenolics has been observed in the case of explants derived from ex vitro-grown plants, as well as in the case of young seedlings, i.e., *Acacia catechu* [52], *Stereospermum personatum* [53], and the *Sideritis* species [36,54]. PVP is an antioxidant commonly used in binding phenolics that prevents the browning of both explants and medium during that stage of in vitro propagation of many species [55,56] such as *Thymus bleicherianus* Pomel [57], *Aloe vera* [58,59], and rose cultivars [60]. It is used at a range of concentrations, usually 0.5 to 5.0 g L^{-1}. Sahini and Gupta [53] used PVP at 1.5 g L^{-1} to control browning in subcultures of stem nodes from in vitro-grown plants of *A. catechu* on BA-supplemented MS media. On the other hand, PVP at times tends to have some undesirable effects due to its absorption of plant growth regulators and nutrients from the plant tissue [59]. Confirming the findings of the last study mentioned above, the present research also found that the use of PVP at low concentrations inhibited shoot length, possibly due to the absorption of auxins. Sarropoulou and Maloupa [51] also used PVP at higher concentrations (10 g L^{-1} PVP), which negatively impacted shoot length. Our own findings lead us to believe that there is no need for further research on the effect of PVP at a lower concentration since no visible production of phenolics emerged.

The micropropagation method presented in this study could help in vitro techniques to address the conservation and exploitation of *Sideritis*. The hormone-free MS medium proved to be efficient for the initial establishment of in vitro cultures, providing shooting formation and elongation, which both increased through subcultures. The supplementing of low concentrations of cytocinins based on references on the Lamiaceae and other xerophytic plants species, which responded better on the MS media containing low concentrations of cytokinins (up to 0.5 mg L^{-1}), also proved beneficial [61]. BA or KIN at 0.5 mg L^{-1} and BA/NAA at 0.5/0.01 (mg L^{-1}) were more effective than the hormone-free MS medium and shoot production was higher. On the other hand, they resulted in the formation of hyperhydric shoots. The shoot length was higher on the hormone-free MS medium and a multiplication cycle on the hormone-free MS medium may have eliminated the hyperhydricity phenomenon. The same medium was also found to be suitable for *S. angustifolia* with a higher elongation [62].

IBA was the most efficient auxin for the in vitro induction of the rooting of *Sideritis stricta* [63], *S. scardica* [38], *S. athoa* [37], and certain other Lamiaceae species [64,65]. As was the case with *S. augustifolia*, in the present study, rooting took place on the hormone-free MS/2 media [62]. Further investigation is required to promote in vitro-induced rooting. Regarding the survival of acclimatised plants, which is the final step of a successful micropropagation protocol, the present study verified that acclimatisation was fully successful (100.0%). An equally high acclimatisation rate was also observed in the cases of *S. raeseri* [51] in Greece and *S. stricta* in Turkey [63].

Seed production is frequently employed as the starting material for commerce-oriented horticultural and aromatic plants and could meet the increasing commercial demand for the *Sideritis* species, which, in this way, could be introduced into the floriculture industry. In the meantime, seed production could result in the establishment of ex situ conservation methods in botanical gardens, arboretums, or other ex situ facilities [66]. It could also help with the reintroduction of the species into its natural habitat, a practice that is effective with very small populations [67]. Bioresource collections could also help to preserve endangered species such as *Sideritis* [68]. It is worth mentioning that *Origanum dictamnus* is another endangered species from Greece, which has found uses in the practices of Greek ornamental horticulture [12]. The conservation of threatened plants and programs aimed at the restoration of natural habitats must be supported by the development of optimal germination protocols.

5. Conclusions

To the best of our knowledge, our study is the first to present an efficient in vitro protocol of *S. raeseri* subsp. *attica*. It is also a preliminary study on the seed germinability of this endangered species. Three-month-old seeds of *S. raeseri* subsp. *attica* germinated, registering average percentages. The cardinal temperatures for germination were defined as 15.0 °C to 30.0 °C. With regard to micropropagation, in vitro culture was established from seedling-origin shoot explants on a hormone-free MS medium. During multiplication, the MS medium supplemented with KIN at 0.5 mg L^{-1} proved to have a high shoot production and low formation of hyperhydric shoots. A multiplication cycle on the hormone-free MS medium eliminated the hyperhydricity phenomenon. Be that as it may, it is evident that more research is required to promote in vitro-induced rooting provided that acclimatisation proves to be 100.0% successful, as it did in our study. The present method, in tandem with suitable genetic markers, could contribute to commercial exploitation and conservation programs alike.

Author Contributions: Conceptualisation, K.B.; methodology, K.B.; validation, K.B.; formal analysis, K.B.; investigation, K.B.; resources, K.B., P.T. and P.-E.A.; data curation, K.B., P.T. and P.-E.A.; writing—original draft preparation, K.B.; writing—review and editing, K.B.; visualisation, K.B. and P.T.; supervision, K.B.; project administration, K.B.; funding acquisition, K.B. All authors have read and agreed to the published version of the manuscript.

Funding: This research received no external funding.

Acknowledgments: The authors would like to thank Apostolos-Emmanouil Mpazanis, MSc, agronomist researcher, for donating the seeds of the species investigated.

Conflicts of Interest: The authors declare no conflict of interest.

References

1. Ganatsas, P.; Daskalakou, E.; Paitaridou, D. First results on early post-fire succession in an *Abies cephalonica* forest (Parnitha National Park, Greece). *IForest Biogeosci. For.* **2012**, *5*, 6–12. [CrossRef]
2. Aplada, E.; Georgiadis, T.; Tiniakou, A.; Theocharopoulos, M. Phytogeography and ecological evaluation of the flora and vegetation of Mt Parnitha (Attica, Greece). *Edinb. J. Bot.* **2007**, *64*, 185–207. [CrossRef]
3. Le Roux, J.J.; Hui, C.; Castillo, M.L.; Iriondo, J.M.; Keet, J.H.; Khapugin, A.A.; Médail, F.; Rejmánek, M.; Theron, G.; Yannelli, F.A.; et al. Recent anthropogenic plant extinctions differ in biodiversity hotspots and coldspots. *Curr. Biol.* **2019**, *29*, 2912–2918. [CrossRef] [PubMed]
4. Dragozi, E.; Gitas, I.Z.; Stavrakoudis, D.G.; Theocharis, J.B. Burned area mapping using support vector machines and the FuzCoC feature selection method on VHR IKONOS imagery. *Remote Sens.* **2014**, *6*, 12005–12036. [CrossRef]
5. Karani, P. National Park Parnes: Present and Future. Ph.D. Thesis, Harokopio University, Athens, Greece, 2008. (In Greek).
6. Chatzimpiros, P.; Roumelioti, N.; Zamba, A.; Hadjibiros, K. Road transport emissions and capacity of forests in the region of Athens for sequestring these emissions: Carbon flow before and after forest fires. *Eur. J. Environ. Sci.* **2016**, *6*, 18–24. [CrossRef]
7. Fraga, B.M. Phytochemistry and chemotaxonomy of *Sideritis* species from the Mediterranean region. *Phytochemistry* **2012**, *76*, 7–24. [CrossRef] [PubMed]
8. Patelou, E.; Chatzopoulou, P.; Polidoros, A.N.; Mylona, P.V. Genetic diversity and structure of *Sideritis raeseri* Boiss. & Heldr. (Lamiaceae) wild populations from Balkan Peninsula. *J. Appl. Res. Med. Arom. Plants* **2020**, *16*, 100241. [CrossRef]
9. Grigoriadou, K.; Krigas, N.; Sarropoulou, V.; Papanastasi, K.; Tsoktouridis, G.; Maloupa, E. In vitro propagation of medicinal and aromatic plants: The case of selected Greek species with conservation priority. *Vitr. Cell Dev. Biol. Plant* **2019**, *55*, 635–646. [CrossRef]
10. Papanikolaou, K.; Kokkini, S. A taxonomic revision of *Sideritis* L. section Empedoclia (RAFIN.) BENTHAM (Labiatae) in Greece. In *Aromatic Plants*; Springer: Dordrecht, The Netherland, 1982; pp. 101–128.
11. Phoitos, D.; Constantinidis, T.; Kamari, G. *The Red Data Book of Rare and Threatened Plants of Greece*; Hellenic Botanical Society: Patras, Greece, 2009; Volume II, pp. 295–297. ISBN 978-960-9407-09-0.
12. Kougioumoutzis, K.; Kokkoris, I.P.; Panitsa, M.; Strid, A.; Dimopoulos, P. Extinction risk assessment of the Greek endemic flora. *Biology* **2021**, *10*, 195. [CrossRef] [PubMed]

13. Constantinidis, T. The flora and vegetation of the mountains Gerania, Pateras and Kitheron (SE Sterea Ellas, Greece). Ph.D. Thesis, National and Kapodistrian University of Athens, Athens, Greece, 1997. (In Greek with an English summary). [CrossRef]
14. Gonzalez-Burgos, E.; Carretero, M.E.; Gomez-Serranillos, M.P. *Sideritis* spp.: Uses, chemical composition and pharmacological activities-a review. *J. Ethnopharmacol.* **2011**, *135*, 209–225. [CrossRef] [PubMed]
15. Lall, N.; Chrysargyris, A.; Lambrechts, I.; Fibrich, B.; Blom Van Staden, B.; Twilley, D.; Nuno de Canha, M.; Oosthuizen, C.B.; Bodiba, D.; Tzortzakis, N. *Sideritis perfoliate* (subsp. *perfoliata*) nutritive value and its potential medicinal properties. *Antioxidants* **2019**, *8*, 521. [CrossRef] [PubMed]
16. Żyżelewicz, D.; Kulbat-Warycha, K.; Oracz, J.; Żyżelewicz, K. Polyphenols and other bioactive compounds of *Sideritis* plants and their potential biological activity. *Molecules* **2020**, *25*, 3763. [CrossRef] [PubMed]
17. Kassi, E.; Papoutsi, Z.; Fokialakis, N.; Messari, I.; Mitakou, S.; Moutsatsou, P. Greek plant extracts exhibit selective estrogen receptor modulator (SERM)-like properties. *J. Agric. Food Chem.* **2004**, *52*, 6956–6961. [CrossRef] [PubMed]
18. Tsaknis, J.; Lalas, S. Extraction and identification of natural antioxidant from *Sideritis euboea* (mountain tea). *J. Agric. Food Chem.* **2005**, *53*, 6375–6381. [CrossRef] [PubMed]
19. Bourgou, S.; Ben Haj Jilani, I.; Karous, O.; Megdiche-Ksouri, W.; Ghrabi-Gammar, Z.; Libiad, M.; Khabbach, A.; El Haissoufi, M.; Lamchouri, F.; Greveniotis, V.; et al. Medicinal-cosmetic potential of the local endemic plants of Crete (Greece), northern Morocco and Tunisia: Priorities for conservation and sustainable exploitation of neglected and underutilized Phytogenetic resources. *Biology* **2021**, *10*, 1344. [CrossRef] [PubMed]
20. Krigas, N.; Tsoktouridis, G.; Anestis, I.; Khabbach, A.; Libiad, M.; Megdiche-Ksouri, W.; Ghrabi-Gammar, Z.; Lamchouri, F.; Tsiripidis, I.; Tsiafouli, M. Exploring the potential of neglected local endemic plants of three Mediterranean regions in the ornamental sector: Value chain feasibility and readiness timescale for their sustainable exploitation. *Sustainability* **2021**, *13*, 2539. [CrossRef]
21. Kloukina, C.; Tomou, E.M.; Krigas, N.; Sarropoulou, V.; Madesis, P.; Maloupa, E.; Skaltsa, H. Non-polar secondary metabolites and essential oil of ex situ propagated and cultivated *Sideritis syriaca* L. subsp. *syriaca* (Lamiaceae) with consolidated identity (DNA Barcoding): Towards a potential new industrial crop. *Ind. Crops Prod.* **2020**, *158*, 112957. [CrossRef]
22. Sarropoulou, V.; Maloupa, E. Effect of the NO donor "sodium nitroprusside" (SNP), the ethylene inhibitor "cobalt chloride" ($CoCl_2$) and the antioxidant vitamin E "α-tocopherol" on In Vitro shoot proliferation of *Sideritis raeseri* Boiss. & Heldr. subsp. *raeseri*. *PCTOC* **2017**, *128*, 619–629. [CrossRef]
23. Krishnan, P.N.; Decruse, S.W.; Radha, R.K. Conservation of medicinal plants of Western Ghats, India and its sustainable utilization through in vitro technology. *Vitr. Cell Dev. Biol.-Plant* **2011**, *47*, 110–122. [CrossRef]
24. Reed, B.M.; Sarasan, V.; Kane, M.; Bunn, E.; Pence, V.C. Biodiversity conservation and conservation biotechnology tools. *Vitr. Cell Dev. Biol.-Plant* **2011**, *47*, 1–4. [CrossRef]
25. Galus, A.; Chenari Bouket, A.; Belbahri, L. In vitro propagation and acclimatization of dragon tree (*Dracaena draco*). *Horticulturae* **2019**, *5*, 64. [CrossRef]
26. Rafiq, S.; Wagay, N.A.; Bhat, I.A.; Kaloo, Z.A.; Rashid, S.; Lin, F.; El-Abedin, T.K.Z.; Wani, S.H.; Mahmoud, E.A.; Almutairi, K.F.; et al. In vitro propagation of *Aconitum chasmanthum* Stapf Ex Holmes: An endemic and critically endangered plant species of the western Himalaya. *Horticulturae* **2021**, *7*, 586. [CrossRef]
27. Sarasan, V.; Kite, G.C.; Sileshi, G.W.; Stevenson, P.C. Applications of phytochemical and in vitro techniques for reducing overharvesting of medicinal and pesticidal plants and generating income for the rural poor. *Plant Cell Rep.* **2011**, *30*, 1163–1172. [CrossRef] [PubMed]
28. Silva, L.F.; de Souza, D.C.; Resente, L.V.; Gonçalves, W.M. Manejo de recursos genéticos vegetais. *An da Acad. Pernambucana de Ciência Agronômica* **2018**, *15*, 109–126. Available online: http://www.journals.ufrpe.br/index.php/apca/article/view/1824 (accessed on 4 March 2022).
29. Murashige, T.; Skoog, F. A revised medium for rapid growth and bioassays with tobacco tissue cultures. *Physiol. Plant.* **1962**, *15*, 473–497. [CrossRef]
30. International Seed Testing Association. International rules for seed testing. *Seed Sci. Technol.* **1999**, *27*, 333.
31. Soltani, A.; Galeshi, S.; Zeinali, E.; Latifi, N. Genetic variation for and interrelationships among seed vigor traits in wheat from the Caspian Sea coasts of Iran. *Seed Sci. Technol.* **2001**, *29*, 653–662.
32. Maguire, J.D. Speed of germination-aid in selection and evaluation for seedling emergence and vigor. *Crop. Sci.* **1962**, *2*, 176–177. [CrossRef]
33. Lloyd, G.; McCown, B. Commercially feasible micropropagation of mountain laurel, *Kalmia latifolia*, by use of shoot-tip culture. *Int. Plant Prop. Soc. Proc.* **1980**, *30*, 421–427.
34. Estrelles, E.; Güemes, J.; Riera, J.; Boscai, U.; Ibars, A.; Costa, M. Seed germination behavior in *Sideritis* from different Iberian habitats. *Not. Bot. Horti Agrobot.* **2010**, *38*, 9–13. [CrossRef]
35. Yankova-Tsvetkova, E.; Yurukova-Grancharova, P.; Vitkova, A. Reproductive biology of the Balkan endemic *Sideritis scardica* (Lamiaceae). *Bot. Serb.* **2013**, *37*, 83–87.
36. Uçar, E.; Turgut, K. In Vitro Propagation of Some Mountain Tea (*Sideritis*) Species. *Ziraat Fakültesi Derg. Akdeniz Üniversitesi* **2009**, *22*, 51–57.
37. Papafotiou, M.; Kalantzis, A. Seed germination and in vitro propagation of *Sideritis athoa*. *Acta Hortic.* **2009**, *813*, 471–476. [CrossRef]

38. Shtereva, L.A.; Vassilevska-Ivanova, R.D.; Kraptchev, B.V. In vitro cultures for micropropagation, mass multiplication and preservation of an endangered medicinal plant *Sideritis scardica Griseb. Bot. Serbica* **2015**, *39*, 11–120. Available online: https://botanicaserbica.bio.bg.ac.rs/arhiva/pdf/2015_39_2_633_full.pdf (accessed on 10 October 2022).
39. Penfield, S.; MacGregor, D.R. Effects of environmental variation during seed production on seed dormancy and germination. *J. Exp. Bot.* **2017**, *68*, 819–825. [CrossRef] [PubMed]
40. Thanos, C.A.; Doussi, M.A. Ecophysiology of seed germination in endemic labiates of Crete. *Isr. J. Plant Sci.* **1995**, *43*, 227–237. [CrossRef]
41. Kadis, C.; Georghiou, K. Seed dispersal and germination behaviour of three threatened endemic labiates of Cyprus. *Plant Spec. Biol.* **2010**, *25*, 77–84. [CrossRef]
42. Picciau, R.; Pritchard., H.W.; Mattana, E.; Bacchetta, G. Thermal thresholds for seed germination in Mediterranean species are higher in mountain compared with lowland areas. *Seed Sci. Res.* **2019**, *29*, 44–54. [CrossRef]
43. Donohue, K. Completing the cycle: Maternal effects as the missing link in plant life histories. *Philos. Trans. R. Soc. B. Biol. Sci.* **2009**, *364*, 1059–1074. [CrossRef] [PubMed]
44. Donohue, K.; De Casas, R.R.; Burghardt, L.T.; Kovach, K.; Willis, C.G. Germination, postgermination adaptation, and species ecological ranges. *Annu. Rev. Ecol. Evol. Syst.* **2010**, *41*, 293–319. Available online: https://www.annualreviews.org/doi/10.1146/annurev-ecolsys-102209-144715 (accessed on 10 October 2022). [CrossRef]
45. Finch-savage, W.E.; Bassel, G.W. Seed vigour and crop establishment: Extending performance beyond adaptation. *J. Exp. Bot.* **2016**, *67*, 567–591. [CrossRef] [PubMed]
46. Baskin, J.M.; Baskin, C.C. Plant population differences in dormancy and germination characteristics of seeds: Heredity or environment? *Am. Midl. Nat.* **1973**, *90*, 493–498. [CrossRef]
47. Copete, M.A.; Herranz, J.M.; Ferrandis, P.; Copete, E. Annual dormancy cycles in buried seeds of shrub species: Germination ecology of *Sideritis serrata* (Labiatae). *Plant Biol.* **2015**, *17*, 798–807. [CrossRef] [PubMed]
48. Crocker, W.M. Mechanics of dormancy in seeds. *Am. J. Bot.* **1916**, *3*, 99–120. [CrossRef]
49. Baskin, J.M.; Baskin, C.C. The seed bank in a population of an endemic plant species and its ecological significance. *Biol. Consev.* **1978**, *14*, 125–130. [CrossRef]
50. Cristaudo, A.; Catara, S.; Mingo, A.; Restuccia, A.; Onofri, A. Temperature and storage time strongly affect the germination success of perennial *Euphorbia species* in Mediterranean regions. *Ecol. Evol.* **2019**, *9*, 10984–10999. [CrossRef] [PubMed]
51. Sarropoulou, V.; Maloupa, E. Effect of polyamine type and concentration on in vitro propagation and ex situ conservation of *Sideritis raeseri* Boiss & Heldr. Subsp. *raeseri. Res. J. Life Sci. Bioinform. Pharmaceut. Chem. Sci.* **2019**, *5*, 583–598. [CrossRef]
52. Sahini, R.; Gupta, S.C. In vitro plantlet regeneration from seedling nodal explants of *Acacia catechu*. *Indian J. Exp. Biol.* **2002**, *40*, 1050–1055. Available online: https://pubmed.ncbi.nlm.nih.gov/12587736/ (accessed on 10 October 2022).
53. Shukla, S.; Shukla, S.K.; Mishra, S.K. In vitro plant regeneration from seedling explants of *Stereospermum personatum* DC: A medicinal tree. *Trees* **2009**, *23*, 409–413. [CrossRef]
54. Aki, C.; Cördük, N. Inhibition of browning problem during micropropagation of *Sideritis trojana* Bornm., an endemic medicinal herb of Turkey, Rom. *Biotechnol. Lett.* **2011**, *16*, 6760–6765.
55. Rout, G.R.; Samantaray, S.; Das, P. Manipulation and propagation of medicinal plants. *Biotechnol. Adv.* **2000**, *18*, 91–120. [CrossRef] [PubMed]
56. Thomas, T.D. The role of activated charcoal in plant tissue culture. *Biotechnol. Adv.* **2008**, *26*, 618–631. [CrossRef] [PubMed]
57. Aicha, N.; Abdelmalek, E.M. Rapid in vitro regeneration and clonal multiplication of *Thymus bleicherianus* pomel, a rare and threatened medicinal and aromatic plant in Morocco. *Med. Aromat. Plants* **2014**, *3*, 145.
58. Liao, Z.; Chen, M.; Tan, F.; Sun, X.; Tang, K. Microprogagation of endangered Chinese aloe. *PCTOC* **2004**, *76*, 83–86. [CrossRef]
59. Hamdeni, I.; Slim, S.; Sanaa, A.; Louhaichi, M.; Boulila, A.; Bettaieb, T. Rosemary essential oil enhances culture establishment and inhibits contamination and enzymatic browning: Applications for in vitro propagation of *Aloe vera* L. *S. Afr. J. Bot.* **2022**, *147*, 1199–1205. [CrossRef]
60. Carelli, B.P.; Echeverrigaray, S. An improved system for the In Vitro propagation of rose cultivars. *Sci. Hortic.* **2002**, *92*, 69–74. [CrossRef]
61. Papafotiou, M.; Martini, A.N.; Vlachou, G. In Vitro propagation as a tool to enhance the use of native ornamentals in archaeological sites of Greece. *Acta Hortic.* **2017**, *1155*, 301–308. [CrossRef]
62. Faria, J.L.C.; Kostenyuk, I.; Segura, J. Isolation, culture and plant regeneration from proto-plasts of *Sideritis angustifolia*. *J. Plant Physiol.* **1998**, *153*, 251–254. [CrossRef]
63. Yavuz, D.Ö. Optimization of regeneration conditions and in vitro propagation of *Sideritis stricta* Boiss & Heldr. *Int. J. Biol. Macromol.* **2016**, *90*, 59–62. [CrossRef] [PubMed]
64. Siddique, I.; Anis, M. An improved plant regeneration system and ex vitro acclimatization of *Ocimum basilicum* L. *Acta Physiol. Plant.* **2008**, *30*, 493–499. [CrossRef]
65. Iyer, P.V.; Pai, J.S. Micropropagation of sweet marjoram (*Majorana hortensis* Moench). *J. Spices Aromat. Crops* **1998**, *7*, 47–49.
66. Chen, G.; Sun, W. The role of botanical gardens in scientific research, conservation, and citizen science. *Plant Divers* **2018**, *40*, 181–188. [CrossRef] [PubMed]

67. Maunder, M. Plant reintroduction: An overview. *Biodivers. Conserv.* **1992**, *1*, 51–61. [CrossRef]
68. Chokheli, V.A.; Dmitriev, P.A.; Rajput, V.D.; Bakulin, S.D.; Azarov, A.S.; Varduni, T.V.; Stepanenko, V.V.; Tarigholizadeh, S.; Singh, R.K.; Verma, K.K.; et al. Recent development in micropropagation techniques for rare plant species. *Plants* **2020**, *9*, 1733. [CrossRef] [PubMed]

horticulturae

MDPI

Article

Somatic Embryogenesis and Flow Cytometric Assessment of Nuclear Genetic Stability for *Sansevieria* spp.: An Approach for In Vitro Regeneration of Ornamental Plants

Caterina Catalano [1,†], Angela Carra [1,†], Francesco Carimi [1,*], Antonio Motisi [1], Maurizio Sajeva [2], Alan Butler [3], Sergio Lucretti [4], Debora Giorgi [4], Anna Farina [4] and Loredana Abbate [1]

1 CNR—Istituto di Bioscienze e BioRisorse, Corso Calatafimi 414, 90129 Palermo, Italy
2 Department of Biological, Chemical and Pharmaceutical Sciences and Technologies (STEBICEF), University of Palermo, Via Archirafi 18, 90123 Palermo, Italy
3 The International Sansevieria Society, P.O. Box 64759, Tucson, AZ 85728-4759, USA
4 ENEA, Italian National Agency for New Technologies, Energy and Sustainable Economic Development, Department for Sustainability, Biotechnologies and Agroindustry Division, Casaccia Research Center, 00123 Rome, Italy
* Correspondence: francesco.carimi@ibbr.cnr.it (F.C.)
† These authors contributed equally to this work.

Abstract: *Sansevieria* Thunb. species are traditionally known as succulent ornamental plants worldwide. They are also cultivated for medicinal, fodder, soil conservation and fiber uses, and for their capacity to reduce environmental pollution. *Sansevieria* sexual propagation is limited by the lack of viable seeds, and reproduction is largely made via vegetative propagation by suckers or cuttings. For these reasons, genetic improvement by conventional breeding is limited. To overcome this problem and to address the increasing demand from customers for novel *Sansevieria* varieties, many commercial companies regularly use in vitro propagation, as is the case in the breeding process of several ornamental plants. In this paper, for the first time, we report a procedure for in vitro somatic embryogenesis and plant regeneration starting from three flower explants for seven different *Sansevieria* genotypes. Regeneration was attempted using stigmas/styles, anther/filament, and ovary which were cultured on a Murashige and Skoog solidified medium under three different plant growth regulator combinations. A good regeneration rate was obtained with all genotypes used under all culture conditions tested from every explant type, with percentages ranging from 0 to 73.3%. "Genetic stability" assessment of regenerated plants in respect to their mother plants was verified through flow cytometry analysis showing a high degree of uniformity, with only *S. parva* exhibiting a different level of DNA fluorescence among in vitro regenerated plants. This is an interesting achievement in the aim to produce true-to-type plants and new variants with desirable characteristics, both of which are desired features in ornamentals improvement.

Keywords: floral explants; in vitro culture; regenerated plants; somaclonal variability; flow DNA ploidy

Citation: Catalano, C.; Carra, A.; Carimi, F.; Motisi, A.; Sajeva, M.; Butler, A.; Lucretti, S.; Giorgi, D.; Farina, A.; Abbate, L. Somatic Embryogenesis and Flow Cytometric Assessment of Nuclear Genetic Stability for *Sansevieria* spp.: An Approach for In Vitro Regeneration of Ornamental Plants. *Horticulturae* **2023**, *9*, 138. https://doi.org/10.3390/horticulturae9020138

Academic Editor: Genhua Niu

Received: 28 November 2022
Revised: 10 January 2023
Accepted: 14 January 2023
Published: 19 January 2023

1. Introduction

Sansevieria Thunb. is a genus of succulent plants with approximately 90 distinct species identified [1], originally found in tropical central Africa and drier parts of eastern and southern Africa [2]. *Sansevieria* plants are cultivated for medicinal, fodder, soil conservation and fiber uses [3]. Recently it has been reported that *Sansevieria* plants have also a high capacity to reduce environmental pollution produced by gases and heavy metals thanks to the ability to absorb hazardous pollutants such as VOCs and CO_2 emissions [4,5]. Unquestionably, *Sansevieria* species are much better known as ornamental plants [6,7]. *Sansevieria*s are liked for their shape, patterns and colors of leaves which varies from dark green, pale green, grayish green or a combination of green and white or yellow with several patterns of the lamina [8]. After the "green wave" in the 70's, *Sansevierias* sales decreased but in recent years, the market has expanded and consumer demand has increased leading to an increase in the culture of *Sansevieria*. Nevertheless, intensive culture could not fulfil the demand due to the slow growing manner of plant propagation. Usually, *Sansevierias* are propagated by suckers and leaf cuttings but both methods need a great amount of starting plant material and a long period of time to produce a competitive number of new specimens [8]. Moreover, sexual propagation is limited by inviable seeds [9]. Due to these reasons, genetic improvement by conventional breeding is limited. Mass propagation through in vitro procedures has been investigated as a solution to address the increasing commercial demand. Recently, direct and indirect organogenesis from leaf explants have been reported for *S. trifasciata* and *S. masoniana* [3,10–12]. Here we report for the first time, to the best of our knowledge, a technique for plant regeneration starting from flower explants. These kinds of explants have been used to regenerate several crops, such as citrus [13], grapevine [14], caper [15] and fennel [16], with clear advantages since flower explants are easy to sterilize, due to the almost complete absence of contamination of their inner tissues. Further benefits of using of young floral tissues are inherent in the wide availability of young explants, which positively influence the regeneration potential [17,18] and in the preservation of the mother plants' integrity (non-destructive procedure).

Ornamental plants are economically important for the horticulture industry and the growth of the sector depends on the creation of new cultivars with commercially attractive features [19]. Plant breeding studies are the main method to improve ornamental plants, and plant tissue culture is the tool used in both fundamental research and commercial applications. As tissue culture technologies sometimes can produce genetic variability, regenerants could bear new interesting features for the ornamental plants industry. For this reason, the exploitation of somaclonal variability has become part of the usual breeding activity of many commercial enterprises. Several ornamental cultivars derived from tissue culture via somaclonal variation have been described ([20] and further references are here reported). In this paper, we report for the first time a technique for plant regeneration starting from flower explants for seven different genotypes of *Sansevieria*. To detect the genetic fidelity of regenerated plants, flow cytometric analysis was performed.

2. Materials and Methods

2.1. Plant Material

Flowers used in this work were collected from seven different genotypes of *Sansevieria*: *S. concinna* N.E.Br. (S-CNR-014; Figure 1(A1,A2)), *S. fasciata* Cornu ex Gérôme & Labroy, × *forskaoliana* (Schult.f.) Hepper & J.R.I. Wood (S-CNR-074; Figure 1(B1,B2)), *S. forskaoliana* (Schult.f.) Hepper & J.R.I.Wood (S-CNR-036; Figure 1(C1,C2)), *S. elliptica* (Chiovenda) Cudofontis (S-CNR-103; Figure 1(D1,D2)), *S. parva* N.E.Br. (S-CNR-054; Figure 1(E1,E2)), *S. pearsonii* N.E.Br. (S-CNR-058; Figure 1(F1,F2)) and *S. caulescens* N.E.Br. (S-CNR-088; Figure 1(G1,G2)). All specimens utilized in the present study come from the former UK National Collection of *Sansevierias* recognized by the Royal Horticultural Society and held by Alan Butler, Brookside Nursery. Other information about area of origin and accession names are reported in Table S1. The plant material was collected from the germplasm repository for perennial plants at the Institute of Biosciences and BioResources of the National Research Council of Italy (CNR-IBBR) located in Collesano Palermo, Italy (38° N, 14° E). Three different floral explants (stigmas/styles (Sti/Sty), anther/filament (Ant/Fil), and ovary (Ov)) used for culture initiation were dissected from flowers harvested 15 days before anthesis. Whole flowers were rinsed with tap water, surface sterilized by immersion for 1 min in 70% ethanol, 7 min in 2% (*w*/*v*) sodium hypochlorite, followed by three 5 min rinses in sterile distilled water. With the aid of a stereo-microscope, the flowers were cut under sterile conditions, and Ov, Ant/Fil and Sti/Sty explants (Figure 2A) were excised from the flowers and plated as single explants under 3 different culture conditions in contact with the medium.

Figure 1. *Sansevieria* plants and inflorescences: (**A1,A2**) *Sansevieria concinna* (S-CNR-014), Bars 8 and 10 cm; (**B1,B2**) *S. fasciata* x *forskaoliana* (S-CNR-074), Bars 15 and 10 cm; (**C1,C2**) *S. forskaoliana* (S-CNR-036), Bars 5 and 3 cm; (**D1,D2**) *S. elliptica* (S-CNR-103), Bars 10 cm; (**E1,E2**) *S. parva* (S-CNR-054), Bars 6 cm; (**F1,F2**) *S. pearsonii* (S-CNR-058), Bars 5 cm; (**G1,G2**) *S. caulescens* (S-CNR-088), Bars 3.5 cm.

Figure 2. Plant regeneration in *Sansevieria* through somatic embryogenesis. (**A**) *Sansevieria* flower collected before opening and cut open under sterile conditions. From left to right: whole flower;

flower with two petals removed; flower with all petals removed and stigma/style (Sti/Sty) and ovary (Ov) emphasized; anther and filament (Ant/Fil) explants dissected from *Sansevieria* flowers collected before opening (bar = 4 mm); (**B**) Creamy-white callus from the Sti/Sty explants (S-CNR-054) 90 days after culture initiation (bar = 1.5 cm); (**C**) Somatic embryos (S-CNR-054) generated after 120 days of culture initiation at the surface of anther explant-derived callus (bar = 2.5 mm); (**D**) Somatic embryos (S-CNR-103) generated after 150 days of culture at the surface of Ov explant-derived callus (bar = 2 mm); (**E**) Somatic embryos (S-CNR-103) generated after 200 days of culture initiation at the surface of Sti/Sty explant-derived callus (bar = 1.5 mm); (**F**) Plant (S-CNR-036) derived from germinated somatic embryos growing in Magenta jars for 150 days on MS medium (bar = 2.5 cm); (**G**,**H**) Somatic embryo-derived plant of S-CNR-074 transferred to sterilized soil and incubated in a growth chamber, respectively 14 and 60 days after acclimatization (bar = 3 cm); (**I**) Regenerated plants after two months of growth under greenhouse conditions (bar = 10 cm).

2.2. Media and Culture Conditions

Explants were cultured on Murashige and Skoog [21] solidified (6 g L^{-1} Plantagar) medium (MS) supplemented with 88 mM sucrose as carbon source under three different plant growth regulator (PGR) combinations: (1) T4 medium, 5 μM N-(2-chloro-4-pyridyl)-N-phenylurea (4-CPPU) + 5 μM 2,4-dichlorophenoxy acetic acid (2,4-D); (2) T5 medium, 20 μM β-naphthoxyacetic acid (NOA) + 4 μM 1-phenyl-3-(1,2,3-thiadiazol-5-yl) urea (TDZ); and (3) T16 medium, 10 μM NOA + 4.4 μM N6-benzylaminopurine (BA). The pH of the media was adjusted to 5.7–5.8 with 1 N NaOH before autoclaving. All chemicals were purchased from Duchefa Biochemie, The Netherlands. Explants were incubated in a climatic chamber at 25 ± 1 °C with a 16 h photoperiod (40 μmol $m^{-2}s^{-1}$ at shelf level, provided by Osram Cool White18 W fluorescent lamps) and subcultured in the same culture medium at 60 d intervals until embryogenic callus was produced (Figure 2B). Usually embryogenic callus was produced after two transfers. Among the explants initially incubated on culture medium, only those showing embryonic response were transferred to basal MS-medium deprived of PGRs, supplemented with 88 mM sucrose and cultured for four more weeks under the same culture conditions described above, to allow embryo proliferation and development (Figure 2C,D). Embryogenic calli were individually transferred to Petri dishes containing 20 mL of basal solid MS medium deprived of PGRs and incubated under the same light and temperature conditions as described above, to allow further growth (Figure 2E).

2.3. Embryo Germination, Plant Development and Acclimatization

Individual somatic embryos were isolated from callus and germination was attempted in Magenta vessels (Figure 2F) containing hormone free MS solidified medium (7 g/L plantagar, Duchefa) supplemented with 88 mM sucrose as carbon source. Magenta vessels were maintained in a climate chamber at 25 ± 1C under the same culture conditions as described above. Embryos were considered as germinated when there was root extension. Plantlets, 2–4 cm tall and with well-developed roots were collected and washed with tap water in order to remove the medium before being transplanted individually into plastic pots 70 mm × 70 mm containing sterile soil. The potted plants, covered with transparent polyethylene bags to maintain temperature and high humidity (Figure 2G), were placed in a climate chamber at 25 ± 1 °C under a 16 h day length and a photosynthetic photon flux of 50 μmol $m^{-2}s^{-1}$ provided by Osram cool-white 18 W fluorescent lamps. After 20–30 days plantlets were exposed to gradual reduction of humidity obtained by gradually perforating the polyethylene bags (Figure 2H) and after 40 days plants were transferred outdoors under natural daylight conditions (Figure 2I). The survival rates were recorded after 2 months.

2.4. Flow Cytometry Analysis

The analyzed plant material consisted of seven different *Sansevieria* genotypes (mother plants: MP) and 16 in vitro regenerated plants (regenerants: R) (Table 1) obtained from different explant types under different culture conditions.

Table 1. The table summarizes the relationship between *Sansevieria*'s genotypes used, the number of the accession, the ratio between the G1 peak of regenerants (R) and corresponding mother plant (MP), the type of explant and the regeneration medium used.

Genotype	Accession ID	G_1R/G_1MP	Explant Type	Regeneration Medium
S. concinna	S-CNR-014/MP		-	-
	S-CNR-014/1	1	Ov	T5
S. fasciata × *forskaoliana*	S-CNR-074/MP		-	-
	S-CNR-074/1	1	Ov	T4
	S-CNR-074/2	1	Ov	T4
	S-CNR-074/3	1	Ov	T4
S. forskaoliana	S-CNR-036/MP		-	-
	S-CNR-036/1	1	Ant/Fil	T5
	S-CNR-036/2	1	Ant/Fil	T5
	S-CNR-036/3	1	Ov	T5
S. elliptica	S-CNR-103/MP		-	-
	S-CNR-103/1	1	Ov	T16
	S-CNR-103/2	1	Sti/Sty	T5
	S-CNR-103/3	1	Sti/Sty	T5
S. parva	S-CNR-054/MP			
	S-CNR-054/1	0,6	Sti/Sty	T16
	S-CNR-054/2	0,6	Ov	T16
S. pearsonii	S-CNR-058/MP			
	S-CNR-058/1	1	Ant/Fil	T16
	S-CNR-058/2	1	Ant/Fil	T16
	S-CNR-058/3	1	Sti/Sty	T16
S. caulescens	S-CNR-088/MP		-	-
	S-CNR-088/1	1	Sti/Sty	T16

Flow Cytometry was carried on *Sansevieria* nuclei in suspension obtained from 2–4 mg leaf tissue, which was homogenized into 200 μL of LB01 extraction buffer [22] and stained with DAPI (4,6-diamidino-2-phenylindole) at the final concentration of 2 μg/mL.

To obtain the nuclei suspensions, mechanical disruption was performed for 12 s at 9500 rpm with a Mini-Turrax T8 with a S10N-5G generator (IKA, Staufen, Germany). Samples were filtered through a 36 μm nylon mesh and their integrity and concentration were preliminarily determined on a Nikon Eclipse TE2000-S epifluorescence microscope (Nikon Corp., Tokyo, Japan).

All analyses were run on a CytoFLEX S flow analyzer (Beckman Coulter Flow Cytometry, Milan, Italy), using as internal reference standard (IRS) 2.5 μm polystyrene microspheres (Alignflow Beads for UV lasers cod. A16502, ThermoFisher, Milan, Italy) to monitor instrument stability and ensuring a true experimental comparison. Genetic stability analysis was estimated as fluorescence DNA emission on 5000 DAPI stained nuclei/sample, excited by a violet laser (ext 405 nm) with a main DAPI fluorescence emission collected at 450/45 nm. In order to prevent staining instability and ensure reliable measurements, all samples were diluted to get a similar concentration of nuclei and a constant number of standard beads (an amount of 10% of beads was added after a preliminary FCM check done sample by sample).

DNA fluorescence histograms were acquired and processed using CytExpert software v. 2.3 (Beckman Coulter Flow Cytometry, Milan, Italy) and the Internal synthetic Reference Standard fluorescence peak (IRS) was carefully set at the same channel value for all acquisitions (Figure 3). The IRS CV was of 3.04 ± 0.05% through all the experiments and its mean fluorescence peak was set at channel 210,436.4 ± 0.25% (on a basis of 16.8×10^6 instrumental channels resolution).

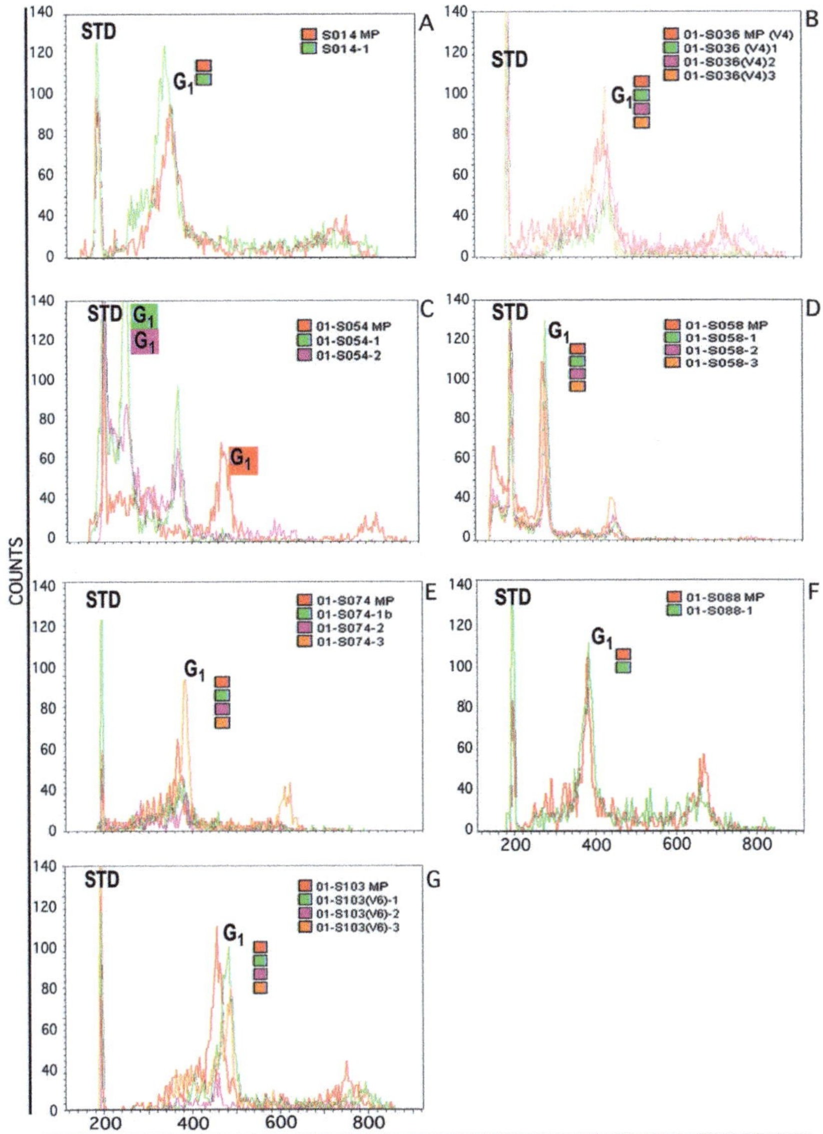

Figure 3. FCM analysis of genetic stability of *Sansevieria* spp. mother plant (MP) and their regenerants. DAPI DNA fluorescence histogram overlapping among MP and respective regenerants are presented. Fluorescence histograms of DAPI stained nuclei in suspension show an IRS (STD) region that contains reference beads to allow histogram comparison among several nuclei isolations from *Sansevieria* individuals. The G_1 region shows basic DNA content for each sample; the use of color coding refers to MP and their Rs per line/species. In the abscissa for DAPI fluorescent emission (arbitrary units) captured by a 450–45 band pass filter after violet laser excitation is shown. The panels show: (**A**) *S. concinna* S-CNR-014, (**B**) *S. forskaoliana* S-CNR-036, (**C**) *S. parva* S-CNR-054, (**D**) *S. pearsonii* S-CNR-058, (**E**) *S. fasciata* × *forskaoliana* S-CNR-074, (**F**). *S. caulescens* S-CNR-088, (**G**) *S. elliptica* S-CNR-103.

The relative stability in DAPI-DNA fluorescence level, which is correlated to the nuclear DNA amount for each mother plant and its regenerants, was calculated on cell cycle histograms and defined in relation to the ratio: G_1 mean peak position of its Rs /G_1 mean peak position of MP (Table 1).

2.5. Data Analyses

All experiments were carried out in a randomized complete block design. Each treatment comprised ten replicates. Five explants each for stigma/style and ovary, and thirty explants for anther/filament were used per Petri plate representing a replicate, making a total of 8400 explants in the experiment. The number of embryogenic explants was recorded after 6 months from culture initiation.

Embryo germination data were collected 2 months from the incubation of somatic embryos on PGR-free medium. Effects of genotype, PGR combination and type of explant on percentages of embryogenic explants, and percentages of embryo germination were tested by ANOVA ($p \leq 0.05$) and the differences among means were tested by Tukey's test. Prior to analysis, percentage data were arcsin-square root transformed. Statistical analysis was performed using SigmaStat 3.5 for Windows.

3. Results

3.1. Plant Regeneration

Callus formation was observed after 60 days from the beginning of the culture under all the PGR combinations tested. Regeneration was obtained from all genotypes with differences according to explant type and medium used. Embryos arose from callus after 4–5 months from culture initiation (Figure 2C,D; Figure 4). Table 2 reports the embryogenic response after 6 months from the culture initiation. The percentages of success varied greatly (0–73.3%) depending on genotypes, explants and PGR combinations. Differences were observed among genotypes in the efficiency of embryogenesis. Percentage of embryogenic explants of the different cultivars calculated across the media and explant types, ranged between 9.7 and 31.1% (Table 3). The capacity of producing new individuals was greater in *S. parva* (31.1%) and in *S. forskaoliana* (28.9%) even if all genotypes produced somatic embryos at remarkable percentages (Table 3). Somatic embryogenesis occurred under all the PGR combinations tested but with significant differences in the response of explants and cultivars according to the PGR combination used. The percentage of embryogenic explants, calculated across the genotypes and explant types ranged between 8.8 and 19.1% (T4 and T16 media, respectively) (Table 4). The best response was obtained with the T16 (19.1%). T5 medium was slightly less effective than T16 while T4 was significantly less effective with the lowest embryogenic potential of 8.8% (Table 4). In our experimental conditions the explant type plays an important role in embryogenic response. When calculated across the cultivars and PGR combinations, the response of different explants tested ranged between 4.08% for style/stigma and 13.75% for anthers/filament, respectively (Table 3). When embryos produced roots, germination was considered to have occurred. About 90% of explants germinated and developed into plantlets when transferred onto MS basal medium. Plantlets reached about 10 cm in height in about 60 days (Figure 2F); these well-developed plantlets were transferred into plastic pots 70 × 70 × 100 mm containing sterile soil and covered with a polythene bag to maintain humidity (Figure 2G). After about 30 days, plantlets were transferred to the greenhouse and exposed to natural daylight conditions at 22/27 °C, night/day. In order to reduce humidity levels holes of increasing size were made in the polyethylene bags. When plantlets were transferred into greenhouse conditions, the percentage of acclimatized plants was about 80%.

Table 2. Embryogenic response of seven *Sansevieria* genotypes using three explant types and three PGR combinations. Data were collected 6 months after culture initiation and each treatment comprised 300 explants for Ant/Fil and 50 explants for Ov and Sti/Sty. Means + SE, values followed by the same letter are not significantly different at $p \leq 0.05$ level (Tukey's test).

		PGR Combinations								
		5 µM CPPU + 5 µM 2,4-D (T4)			20 µM NOA + 4 µM TDZ (T5)			10 µM NOA + 4.4 µM BA (T16)		
		Ov	Sti/Sty	Ant/Fil	Ov	Sti/Sty	Ant/Fil	Ov	Sti/Sty	Ant/Fil
Genotype	**Accessions ID**	Embryogenic Explants (%)								
S. concinna	S-CNR-014	0	0	16 ± 7.5^{d}	40 ± 6.3^{c}	0	0	$12 \pm 4.9^{d,e}$	0	0
S. fasciata × *forskaoliana*	S-CNR-074	$16.8 \pm 7.5^{d,e}$	4 ± 1.9^{e}	$16 \pm 7.5^{d,e}$	100^{a}	60 ± 17^{b}	$8 \pm 4.9^{d,e}$	0	0	0
S. forskaoliana	S-CNR-036	40 ± 6.3^{c}	0	60 ± 14.1^{b}	40 ± 6.3^{c}	0	44 ± 4^{c}	0	0	0
S. elliptica	S-CNR-103	0	0	0	0	60 ± 14^{b}	10 ± 1.1^{d}	20 ± 2.3^{d}	0	$30 \pm 10^{c,d}$
S. parva	S-CNR-054	20 ± 8.9^{d}	0	25 ± 5^{d}	0	24 ± 11.6^{d}	32 ± 18.5^{c}	0	6.7 ± 0.4^{e}	0
S. pearsonii	S-CNR-058	0	0	6.7 ± 1.5^{e}	20 ± 11.5^{d}	33.3 ± 6.7^{c}	0	10 ± 1.8^{e}	70 ± 10^{b}	40 ± 3^{c}
S. caulescens	S-CNR-088	24 ± 4^{d}	30 ± 5.7^{c}	25 ± 5^{d}	73.3 ± 6.7^{b}	40 ± 11.6^{c}	35 ± 5^{c}	20 ± 6.3^{d}	5 ± 1.2^{e}	0

genotypes while it is a disadvantage if the goal is the preservation of elite genotypes and their micropropagation [36]. Among the several methods to study the genetic stability of regenerants during plant tissue culture [37], flow cytometry, more than the long and laborious cytogenetic methodologies, represents a precise and fast method to identify changes in ploidy and DNA content of a species, thus measuring the genetic variability and somaclonal variation in various species [38–40]. Flow cytometry measures fluorescence intensity of nuclei DNA stained with DNA specific fluorochromes, and for DNA ploidy and nuclei content identification the use of a known internal standard is recommended [41]. In the present study the ploidy level of the *Sansevieria* mother plants is not defined; therefore, the analyses tend to evaluate the relative genetic stability in terms of DNA fluorescence variations of the nuclei extracted from plant materials in our particular growth conditions. In particular, our experiments entailed the comparison among MP and their Rs to assess the fluorescence emission, which is related to nuclei DNA amount. The present analysis aims to highlight any variations witnessed in the potential genetic instability of *Sansevieria*'s spp. clones, regenerated from several types of explants, grown in different culture media.

The greater variation in DNA amount has been observed in plants regenerated from S-CNR-054; both regenerants in fact showed a different level of DNA fluorescence compared to MP, probably due to the in vitro culture conditions, while a high degree of uniformity has been observed among other *Sansevieria* lines. In contrast, comparison within mother plants showed differences in relative DNA fluorescence, and this is not surprising since they belong to several species. Few indications are given in the literature regarding the DNA content in *Sansevieria* spp. [42,43]. Additional studies are needed to better characterize the group of regenerants belonging to *S. parva* species (group MP S-CNR-054) by flow cytometric analysis of the absolute DNA content, using the appropriate internal biological reference standard with a known DNA amount. A more precise response regarding inner variability could be achieved also by classic cytogenetic studies on metaphases spreads of regenerants compared to *S. parva*, known to have a chromosome complement of 2n = 40 [44].

The presence of genetic variability in regenerated plants of *S. parva* is an interesting aspect from the perspective of producing new variants for commercial purposes. Nevertheless, the high degree of uniformity in DNA relative content, verified in the other samples, validates the regeneration system starting from floral explants. These results are in agreement with those reported for other species [14,15,35,45] about stability of regenerated plants.

We can conclude that the fast assessment of the flow DNA content stability of in vitro regenerated plants is essential both for valuation of true-to-type clones to be used for production of homogeneous plants and for selection of individuals bearing genetically stable variations, useful in the improvement of ornamental plants.

Supplementary Materials: The following supporting information can be downloaded at: https://www.mdpi.com/article/10.3390/horticulturae9020138/s1, Figure S1: Flow cytometry evaluation of three extraction buffers for *Sansevieria* nuclei isolation in suspension from 2 mg leaf tissue. Figure S2: An example of "clearing up" FCM analysis from debris and unnecessary particles. Table S1: List of seven *Sansevieria* genotypes used for the regeneration of somatic embryos in vitro.

Author Contributions: Conceptualization, C.C., A.C., F.C., M.S., A.B., S.L. and L.A.; methodology, C.C., A.C., D.G., A.F. and L.A.; validation, F.C., M.S. and S.L.; investigation, C.C., A.C., A.M., D.G., A.F. and L.A.; data curation, C.C., A.C., D.G., A.F. and L.A; writing—original draft preparation, C.C., A.C., F.C., D.G., A.F. and L.A.; writing—review and editing, C.C., A.C., F.C., A.M., M.S., A.B., S.L., D.G., A.F. and L.A.; supervision, C.C., A.C., F.C., M.S., S.L. and L.A.; funding acquisition, M.S. All authors have read and agreed to the published version of the manuscript.

Funding: This work was funded by Fondo Finalizzato alla Ricerca di Ateneo dell'Università di Palermo, FFR-D15-005355 to Maurizio Sajeva.

Data Availability Statement: Data are available from the corresponding author upon request.

Acknowledgments: We would like to thank Francesca La Bella for technical support.

Conflicts of Interest: The authors declare no conflict of interest.

References

1. Webb, R.H.; Newton, L.E. *The Genus Sansevieria: A Pictorial Guide to the Species*; Arid Lands Press: Tucson, Arizona, 2022; p. 200.
2. Brown, N.E. *Sansevieria.* A monograph of all the known species. *Bull. Misc. Inform. Kew.* **1915**, *5*, 185–261. [CrossRef]
3. García-Hernández, E.; Loera-Quezada, M.M.; Morán-Velázquez, D.C.; López, M.G.; Chable-Vega, M.A.; Santillán-Fernández, A.; Zavaleta-Mancera, H.A.; Tang, J.Z.; Azadi, P.; Ibarra-Laclette, E.; et al. Indirect organogenesis for high frequency shoot regeneration of two cultivars of *Sansevieria trifasciata* Prain differing in fiber production. *Sci. Rep.* **2022**, *12*, 1–12. [CrossRef] [PubMed]
4. Bandehali, S.; Miri, T.; Onyeaka, H.; Kumar, P. Current state of indoor air phytoremediation using potted plants and green walls. *Atmosphere* **2021**, *12*, 473. [CrossRef]
5. Chatakul, P.; Janpathompong, S. Interior plants: Trends, species, and their benefits. *Build Environ.* **2022**, *222*, 109325. [CrossRef]
6. Suberi, I.; Noor, N.; Darnis, D.; Mukai, Y.; Usup, G. *The Potential of Ornamental Plant, Sansevieria trifasciata to Inhibit the Growth of Harmful Algal Bloom Species*; Isals Publishing: Yazd, Iran, 2014; p. 42.
7. Takawira-Nyenya, R.; Newton, L.E.; Wabuyele, E.; Stedje, B. Ethnobotanical uses of *Sansevieria* Thunb (Asparagaceae) in Coast Province of Kenya. *Ethnobot Res. App.* **2014**, *l12*, 51–69.
8. Yusnita, Y.; Pungkastiani, W.; Hapsoro, D. In vitro organogenesis of two *Sansevieria trifasciata* cultivars on different concentrations of benzyladenine (BA). *AGRIVITA J. Agric. Sci.* **2011**, *33*, 147–153. [CrossRef]
9. Henley, R.W.; Chase, A.R.; Osborne, L.S. *Sansevieria* Production Guide. USA: University of Florida. 1991. Available online: https://mrec.ifas.uf.edu/foliage/folnotes/sansevie.htm (accessed on 10 November 2022).
10. Kaur, J.; Mudgal, G. An efficient and quick protocol for in vitro multiplication of snake plant, *Sansevieria trifasciata* var. Laurentii [Prain]. *Plant Cell Tissue Organ Cult.* **2021**, *147*, 405–411. [CrossRef]
11. Sarmast, M.K.; Salehi, M.; Salehi, H. The potential of different parts of *Sansevieria trifasciata* L. leaf for meristemoids production. *Aust. J. Basic Appl. Sci.* **2009**, *3*, 2506–2509.
12. Agrotan, J.; Inderiati, S. Indirect organogenesis and induction of morphogenic callus for in vitro propagation of *Sansevieria masoniana. Jurnal Agrotan* **2015**, *1*, 1–8.
13. Carra, A.; De Pasquale, F.; Ricci, A.; Carimi, F. Diphenylurea derivatives induce somatic embryogenesis in Citrus. *Plant Cell Tissue Organ Cult.* **2006**, *87*, 41–48. [CrossRef]
14. Carra, A.; Sajeva, M.; Abbate, L.; Siragusa, M.; Pathirana, R.; Carimi, F. Factors affecting somatic embryogenesis in eight Italian grapevine cultivars and the genetic stability of embryo-derived regenerants as assessed by molecular markers. *Sci. Hortic.* **2016**, *204*, 123–127. [CrossRef]
15. Carra, A.; Sajeva, M.; Abbate, L.; Siragusa, M.; Sottile, F.; Carimi, F. In vitro plant regeneration of caper (Capparis spinosa L.) from floral explants and genetic stability of regenerants. *Plant Cell Tissue Organ Cult.* **2012**, *109*, 373–381. [CrossRef]
16. Fiore, M.C.; Carimi, F.; Carra, A.; Sunseri, F. Efficient plant regeneration via somatic embryogenesis in bulbing fennel using immature flower explants. *In Vitro Cell. Dev. Biol. Plant.* **2012**, *48*, 440–445. [CrossRef]
17. Lou, Q.; Liu, H.; Luo, W.; Chen, K.; Liu, Y. Creating a novel petal regeneration system for function identification of colour gene of grape hyacinth. *Plant Methods* **2021**, *17*, 1–10. [CrossRef]
18. de Almeida, N.V.; Rivas, E.B.; Cardoso, J.C. Somatic embryogenesis from flower tepals of *Hippeastrum* aiming regeneration of virus-free plants. *Plant Sci.* **2022**, *317*, 111191. [CrossRef]
19. Botelho, F.B.S.; Rodrigues, C.S.; Bruzi, A.T. Ornamental plant breeding. *Ornam. Hortic.* **2015**, *21*, 9–16. [CrossRef]
20. Datta, S.K. Breeding of ornamentals: Success and technological status. *Nucleus* **2022**, *65*, 107–128. [CrossRef]
21. Murashige, T.; Skoog, F. A revised medium for rapid growth and bioassay with tobacco tissue culture. *Physiol. Plant* **1962**, *15*, 473–497. [CrossRef]
22. Dolezel, J.; Binarova, P.; Lucretti, S. Analysis of nuclear-DNA content in plant-cells by flow-cytometry. *Biol. Plant.* **1989**, *31*, 113–120. [CrossRef]
23. Otto, F. Preparation and staining of cells for high-resolution DNA analysis. In *Flow Cytometry and Cell Sorting*; Radbruch, A., Ed.; Springer: Berlin, Germany, 1992; pp. 101–104. [CrossRef]
24. Catalano, C.; Abbate, L.; Motisi, A.; Crucitti, D.; Cangelosi, V.; Pisciotta, A.; Di Lorenzo, R.; Carimi, F.; Carra, A. Autotetraploid emergence via somatic embryogenesis in Vitis vinifera induces marked morphological changes in shoots, mature leaves, and stomata. *Cells* **2021**, *10*, 1336. [CrossRef]
25. Palma, M.A.; Hall, C.R.; Collart, A.J. Repeat buying behavior for ornamental plants: A consumer profile. *J. Food Distrib. Res.* **2011**, *42*, 67–77. [CrossRef]
26. Carimi, F.; Barizza, E.; Gardiman, M.; Schiavo, F.L. Somatic embryogenesis from stigmas and styles of grapevine. *In Vitro Cell Dev. Biol. Plant* **2005**, *41*, 249–252. [CrossRef]
27. Sajeva, M.; Carra, A.; de Pasquale, F.; Carimi, F. Somatic embryogenesis and plant regeneration from pistil transverse thin cell layers of lemon (Citrus limon). *Plant Biosyst.* **2008**, *142*, 199–203. [CrossRef]
28. Wojciechowicz, M.K. Organogenesis and somatic embryogenesis induced in petal cultures of *Sedum* species. *Acta Biol. Cracov. Bot.* **2009**, *51*, 83–90.

29. Vidal, J.R.; Rama, J.; Taboada, L.; Martin, C.; Ibañez, M.; Segura, A.; González-Benito, M.E. Improved somatic embryogenesis of grapevine (*Vitis vinifera*) with focus on induction parameters and efficient plant regeneration. *Plant Cell Tissue Organ Cult.* **2009**, *96*, 85–94. [CrossRef]
30. Prado, M.J.; Grueiro, M.P.; González, M.V.; Testillano, P.S.; Domínguez, C.; López, M.; Rey, M. Efficient plant regeneration through somatic embryogenesis from anthers and ovaries of six autochthonous grapevine cultivars from Galicia (Spain). *Sci. Hortic* **2010**, *125*, 342–352. [CrossRef]
31. Akin-Idowu, P.E.; Ibitoye, D.O.; Ademoyegun, O.T. Tissue culture as a plant production technique for horticultural crops. *Afr. J. Biotechnol.* **2009**, *8*, 3782–3788.
32. Shumilina, D.; Kornyukhin, D.; Domblides, E.; Soldatenko, A.; Artemyeva, A. Effects of genotype and culture conditions on microspore embryogenesis and plant regeneration in *Brassica rapa* ssp. rapa L. *Plants* **2020**, *9*, 278. [CrossRef]
33. Bednarek, P.T.; Orłowska, R.; Koebner, R.M.D.; Zimny, J. Quantification of the tissue-culture induced variation in barley (*Hordeum vulgare* L.). *BMC Plant Biol.* **2007**, *7*, 10. [CrossRef]
34. Garcia, C.; Furtado de Almeida, A.A.; Costa, M.; Britto, D.; Valle, R.; Royaert, S.; Marelli, J.P. Abnormalities in somatic embryogenesis caused by 2, 4-D: An overview. *Plant Cell Tissue Organ Cult.* **2019**, *137*, 193–212. [CrossRef]
35. De-la-Peña, C.; Nic-Can, G.; Galaz-Ávalos, R.; Avilez-Montalvo, R.; Loyola-Vargas, V. The role of chromatin modifications in somatic embryogenesis in plant. *Front. Plant Sci.* **2015**, *6*, 635. [CrossRef]
36. Krishna, H.; Alizadeh, M.; Singh, D.; Singh, U.; Chauhan, N.; Eftekhari, M.; Sadh, R.K. Somaclonal variations and their applications in horticultural crops improvement. *3 Biotech* **2016**, *6*, 1–18. [CrossRef]
37. Bairu, M.W.; Aremu, A.O.; Van Staden, J. Somaclonal variation in plants: Causes and detection methods. *Plant Growth Regul.* **2011**, *63*, 147–173. [CrossRef]
38. Loureiro, J.; Pinto, G.; Lopes, T.; Doležel, J.; Santos, C. Assessment of ploidy stability of the somatic embryogenesis process in *Quercus suber* L. using flow cytometry. *Planta* **2005**, *221*, 815–822. [CrossRef]
39. Sharma, S.; Bryan, G.; Winfield, M.; Millam, S. Stability of potato (*Solanum tuberosum* L.) plants regenerated via somatic embryos, axillary bud proliferated shoots, microtubers and true potato seeds: A comparative phenotypic, cytogenetic and molecular assessment. *Planta* **2007**, *226*, 1449–1458. [CrossRef]
40. Grosso, V.; Farina, A.; Giorgi, D.; Nardi, L.; Diretto, G.; Lucretti, S. A high-throughput flow cytometry system for early screening of in vitro made polyploids in *Dendrobium* hybrids. *Plant Cell Tissue Organ Cult.* **2018**, *132*, 57–70. [CrossRef]
41. Dolezel, J.; Bartoš, J. Plant DNA flow cytometry and estimation of nuclear genome size. *Ann. Bot.* **2005**, *95*, 99–110. [CrossRef]
42. Zonneveld, B.J.M.; Leitch, I.J.; Bennett, M.D. First nuclear DNA amounts in more than 300 Angiosperms. *Ann. Bot.* **2005**, *96*, 229–244. [CrossRef]
43. Smarda, P.; Bure, P.; Horová, L.; Leitch, I.J.; Mucina, L.; Pacini, E.; Tichý, L.; Grulich, V.; Rotreklová, O. Ecological and evolutionary significance of genomic GC content diversity in monocots. *Proc. Natl. Acad. Sci. USA* **2014**, *111*, E4096–E4102. [CrossRef]
44. Nazeer, M.A.; Khoshoo, T.N. Cytology of Some Species of *Sansevieria* Thunb. *Cytologia* **1984**, *49*, 325–332. [CrossRef]
45. Catalano, C.; Abbate, L.; Fatta Del Bosco, S.; Motisi, A.; Carimi, F.; De Michele, R.; Mercati, F.; D'Onghia, A.M.; Carra, A. Different cell types affect the transition from juvenile to mature phase in *Citrus* plants regenerated through somatic embryogenesis. *Plants* **2022**, *11*, 1811. [CrossRef] [PubMed]

horticulturae

Article

Effects of Brief UV-C Irradiation Treatments on Rooting Performance of *Pelargonium* × *hortorum* (L.H. Bailey) Stem Cuttings

Anastasios I. Darras [1,*], Katerina Grigoropoulou [1], Kallirroi Dimiza [1] and Faisal Zulfiqar [2]

1 Floriculture and Landscape Architecture Laboratory, Department of Agriculture, University of the Peloponnese, 24100 Kalamata, Greece
2 Department of Horticultural Sciences, Faculty of Agriculture and Environment, The Islamia University of Bahawalpur, Bahawalpur 63100, Pakistan
* Correspondence: a.darras@uop.gr; Tel.: +30-272-1104-5199

Abstract: *Pelargonium* × *hortorum* (L.H. Bailey), is a South African native ornamental plant with worldwide commercial recognition used in gardens and terraces. In the present study, we evaluated the effects of low doses of UV-C irradiation on rooting performance of *P.* × *hortorum* stem cuttings. We also tested the hypothesis that UV-C-induced ethylene production directly interacted with rooting process. Over a 40 d evaluation period, the ethylene production of the UV-C-treated stem cuttings was significantly increased. UV-C irradiation positively affected rooting performance. Rooting percentage was increased in the UV-C-irradiated stem cuttings by up to 17%, time to rooting was decreased by 15% (e.g., 5 d) and root weight increased by 17% compared to the nonirradiated controls. UV-C irradiation did not affect net CO_2 assimilation (A_s), but it induced transpiration (E) on the 14, 20, 22 and 24 d of the evaluation period. Positive correlations were found between ethylene production and A_s, E, stomatal conductance (g_s) and root weight, while a negative correlation was recorded between days to rooting and ethylene. UV-C hastened flower production of the cuttings, but it did not affect colour parameters. We suggest that low doses of UV-C may induce endogenous ethylene production, which at low levels, interact with other hormonal mechanisms to activate root development.

Keywords: *Pelargonium*; propagation; stem cuttings; photosynthesis; ornamental plants

Citation: Darras, A.I.; Grigoropoulou, K.; Dimiza, K.; Zulfiqar, F. Effects of Brief UV-C Irradiation Treatments on Rooting Performance of *Pelargonium* × *hortorum* (L.H. Bailey) Stem Cuttings. *Horticulturae* **2022**, *8*, 897. https://doi.org/10.3390/horticulturae8100897

Academic Editors: Chao Ma and Sergio Ruffo Roberto

Received: 22 August 2022
Accepted: 27 September 2022
Published: 29 September 2022

1. Introduction

Pelargonium × *hortorum* (L.H. Bailey), also known as geranium, is an ornamental plant species originating from South Africa with worldwide recognition and production [1]. Geraniums have attracted an increased commercial interest worldwide as they are extensively used in gardens and terraces. *P.* × *hortorum* is propagated by stem cuttings usually prepared by specialised producers around the world. *P.* × *hortorum* stem cuttings are easily rooted in environments of high humidity (>90%), moderate temperatures (18–25 °C) and low light levels, with or without the use of rooting hormones [1]. However, premature leaf senescence, leaf discolouration and disease development (e.g., *Botrytis cinerea*) may occur prior to root formation and result in quality loss of the propagation material. Fast and effective rooting of *P.* × *hortorum* stem cuttings is vitally important for growers and sellers.

The physiological and biochemical responses of plants to preharvest UV-C irradiation exposure have been studied for the past 10 years by scientists around the world [2,3]. These studies have presented the agronomic potential of UV-C irradiation applied to plants, but not to propagation materials. The application of UV-C to ornamental plants induces a cascade of biochemical and biophysical reactions leading to increased disease and herbivore resistance [4,5], faster flowering [6], increased yield [7] and improved postharvest performance. When artificial UV-C lighting is perceived by the epidermal plant cells, various genetic responses are expressed as a result of the up- and down-regulation of certain genes [8]. At molecular level, the presence of UV-C results in a significant induction of antioxidant enzymes such as superoxide dismutase (SOD), catalase (CAT), peroxidase (POD), polyphenol oxidase (PPO) and ascorbate peroxidase (APX) to battle the production of reactive oxygen species (ROS) produced under stress [9–11]. UV-C irradiation induces pathways of the secondary metabolism such as the phenylpropanoid pathway that leads to the production of most phenolic compounds (i.e., lignin, coumarin, flavonoids, tannins, anthocyanins, etc.) associated with plant defence to pathogens, herbivores, biotic and abiotic stressors [6,12]. Recent studies have demonstrated that UV-C irradiation applied postharvest may induce gene expression associated with ethylene production [8,13,14]. In UV-C-treated tomato and strawberry fruits, ethylene production increased by up to 2.0- and 13.3-fold, respectively compared to the nonirradiated fruit [13,14].

The role of ethylene on rooting has been demonstrated in previous studies. The presence of ethylene may induce the rooting response in coleoptiles [15] and in *P.* × *hortorum* cuttings when carbohydrate content is high [16,17]. In transgenic, ethylene-insensitive petunia (*Petunia* × *hybrida*) plants, adventitious root formation was significantly reduced [18]. It was shown that endogenous ethylene sensitivity was necessary for the formation of adventitious roots on vegetative stem cuttings. There is crosstalk and complex interactions between endogenous auxin and ethylene in tissues, which significantly affect root formation and development. For example, endogenous ethylene levels in *Vigna radiata* L. cuttings were higher after treatment with IBA [19]. Moreover, the number of roots per cutting were significantly increased after treatment with 0.1 mM ACC. When 1-MCP was applied to *P.* × *hortorum* cuttings, the ethylene production was blocked, and the rooting percentages were significantly reduced indicating a significant correlation between ethylene and rooting [16]. Apparently, this effect depended on the genotype (i.e., cultivars) and on carbohydrate content. Mutui et al. [17] reported that ethylene markedly increased the rooting percentage of cvs. Greco and Surfing *P.* × *hortorum* cuttings but reduced the total root lengths.

There is limited published research on the effects of UV-C irradiation on the growth and development of plant propagation material. Sukthavornthum et al. [20] showed that long exposures of Persian violet (*Exacum affine* Balf. f. ex Regel) plantlets grown in vitro to UV-C irradiation resulted in a reduced growth, but increased flower development. Phanomchai et al. [21] found that a single low-intensity UV-C exposure of Persian violet microshoots induced the maximum number of roots and the highest root length in vitro without the use of plant growth regulators. The different responses of the propagation material were associated with the UV-C doses used. In most of the cases, UV-C irradiation positively affected root growth and development when used at low dosages.

The aim of the study was to examine, for the first time, the effects of UV-C irradiation treatments on rooting performance of *P.* × *hortorum* stem cuttings. Physiological (i.e., colour, gas exchange patterns, transpiration and stomatal conductance) and morphological changes (i.e., rooting performance, growth and flowering) in response to UV-C treatments were recorded. Ethylene production was measured and the interactions between UV-C irradiation, ethylene production and the rooting performance of *P.* × *hortorum* stem cuttings were studied.

2. Materials and Methods

2.1. Propagation Material and Rooting Environment

Terminal stem cuttings were harvested from mother plantation of *P.* × *hortorum* of cv. Glacis plants cultivated inside a nonheated greenhouse at the University of Peloponnese (Kalamata, Greece, lat. 37°20′20″ N, long. 22°60′51″ E). Sixty stem cuttings in total (thirty for each treatment) were cut with a sterilised blade at the length of 6–8 cm. The cuttings were harvested at midday (11:00 to 13:00). The top two leaves of the axillary shoot on the stock-plant were kept on each cutting. The harvested stem cuttings were immediately planted in individual plastic pots (6 cm height × 4 cm length) filled with peat (PLANTOBALT, Plantaflor, Vechta, Denmark, pH 5.8) and perlite (VIORYP Ltd., Greece, pH 7.0) at 3:1 (v:v). They were held on a bench under mist irrigation running for 1 min for every 3 h. Temperature and relative humidity (R.H.) inside the greenhouse were recorded by data loggers (HOBO, Measurement System Ltd., Berkshire, UK) and ranged between 18 and 36 °C and between 37 and 76%, respectively.

2.2. UV-C Irradiation Treatments

UV-C irradiation was carried out in the greenhouse according to Darras et al. [5]. The UV-C dose rate was chosen from previous studies on *P.* × *hortorum* [4,10] and it was measured at greenhouse temperature using a Multi-Sense optical radiometer fitted with a 254 nm UV-C light sensor (Steril Air, UV-Technologie, Gräfelfing, Germany). The UV-C irradiation dose was calculated in seconds of exposure at a 20 cm distance from the cuttings and was set at 1.0 kJ m^{-2}. The cuttings received three irradiations per week for a total of 6 weeks. Control cuttings were left unirradiated.

2.3. Net CO_2 Assimilation (A_s), Transpiration (E) and Stomatal Conductance (g_s)

CO_2 assimilation (A_s; μmol m^{-2}. s), transpiration (E; mmol m^{-2}. s) and stomatal conductance (g_s; mmol m^{-2}. s) were recorded using a LCpro+ portable photosynthesis system (ADC Bioscientific Ltd., Great Amwell, Hertfordshire, UK). Data were recorded after irradiation twice per week for a total of 6 weeks. Recordings were taken on similarly sized, healthy leaves between 07:00 and 09:00 a.m. Inside the leaf chamber, the photosynthetic photon flux density (PPFD) was set at 1100 μmol m^{-2}. s and temperature at 22 °C. Greenhouse reference CO_2 ranged between 480 and 520 mg L^{-1}.

2.4. Colour Assessments

Colour parameters (a^*, b^*, C^* and L^*) were recorded on designated spots on the surface of fully matured leaves using a Minolta colourimeter (Model CR-300, Minolta Co., Ltd., Osaka, Japan). The instrument was calibrated on a Minolta standard white reflector plate and assessments were carried out by placing the colourimeter sensor (8 mm aperture) on the designated spots. Lightness (L^*), degree of redness to greenness (a^*), degree of yellowness to blueness (b^*), chroma (C^*) and total colour difference (ΔE) were recorded twice per week for a total of 6 weeks.

2.5. Ethylene Production

The ethylene production of *P.* × *hortorum* cuttings was recorded weekly for a total of 6 weeks. Another group of 30 *P.* × *hortorum* cuttings (15 replication samples per treatment, e.g., ± UV-C) were enclosed individually in sealed and gas-tight, 250 mL glass containers for 12 h after UV-C treatment. Past the 12 h, gas samples of 1 mL were withdrawn with a syringe directly from the containers. Ethylene production of stem cuttings was measured using a gas chromatographer (Shimadzu, Model GC-14B; temperature of flame ionisation detector = 200 °C, injector temperature = 180 °C; internal oven temperature = 50 °C; column Porapak P, Duisburg, Germany).

2.6. Rooting Performance, Growth and Development

Days to rooting, rooting percentage (%), fresh root mass (mg), number of roots and root length (cm) were recorded at the end of the experiment (e.g., after 6 weeks). Days to rooting were determined by applying upward force by hand to the cuttings. The time needed for cuttings to show resistance to eradication was recorded as days from the initial plantation. At the end of the 6-week period, the roots were cleared from the substrate under running tap water. Then, the roots were cut from the base of the stem cutting using a razor blade and the number and the root lengths (cm) were recorded. Root fresh weights (mg) were measured using a digital balance (Kern & Sohn GmbH, Balingen, Germany) and the number of flowers and number of leaves of the cuttings were also recorded.

2.7. Experimental Design and Statistical Analysis

Experiments were conducted in a completely randomised design (CRD) with ±UV-C treatment as the experimental factor. Data were subjected to a one-way ANOVA using SPSS v. 21 (SPSS Inc., Chicago, IL, USA). Comparisons between treatment means were carried out using the Duncan's multiple range test at $p = 0.05$. Pearson correlations (2-tailed) were performed in SPSS to highlight the interactions between ethylene production, growth and rooting performance.

3. Results

3.1. Physiological Responses of P. × hortorum Stem Cuttings

Overall, net CO_2 assimilation (A_s), transpiration (E) and stomatal conductance (g_s) were not affected by UV-C irradiation (Table 1).

Table 1. ANOVA outcomes of net CO_2 assimilation (A_s; μmol m^{-2}. s), transpiration (E; mmol m^{-2}. s), stomatal conductance (g_s; mmol m^{-2}. s), a^*, b^*, C^*, L^* and ethylene production (nL L^{-1}) of *P.* × *hortorum* cuttings treated with 1 kJ m^{-2} UV-C for 6 weeks. The ANOVA analysis was performed in SPSS v. 21.

Variables	df	Mean Square	F-Value	Significance (p = 0.05)
A_s	1	6.500	0.272	0.603
E	1	1.755	1.946	0.165
g_s	1	0.007	0.250	0.618
a^*	1	4.273	4.679	0.114
b^*	1	936.084	1.534	0.217
C^*	1	62.657	3.205	0.045
L^*	1	6.260	0.350	0.555
Ethylene	1	1202.311	43.950	0.000

A_s, E and g_s were low for the first 20 d, and then rapidly increased from day 21 to day 36 (Figure 1). The A_s of the UV-C-irradiated stem cuttings maintained at similar levels as those of the controls, although significant differences were recorded on days 8, 30, 34 and 42 (Figure 1A). Similarly, the E of UV-C-irradiated stem cuttings was at similar levels most of times during the 6-week evaluation period, although it was significantly higher on days 14, 20, 22 and 24 (Figure 1B). The stomatal conductance of the UV-C-irradiated stem cuttings was at similar levels as those of the controls, except at day 34, when it was significantly higher (Figure 1C).

UV-C-irradiated stem cuttings produced significantly higher volumes of ethylene compared to the untreated controls (Table 1). Ethylene production followed a clear pattern of gradual increment till day 22 and then a sharp decline till the end of the 6-week evaluation period (Figure 2). Ethylene production of the irradiated cuttings was significantly higher from day 4 to day 34. On day 22, a 47% increase was recorded.

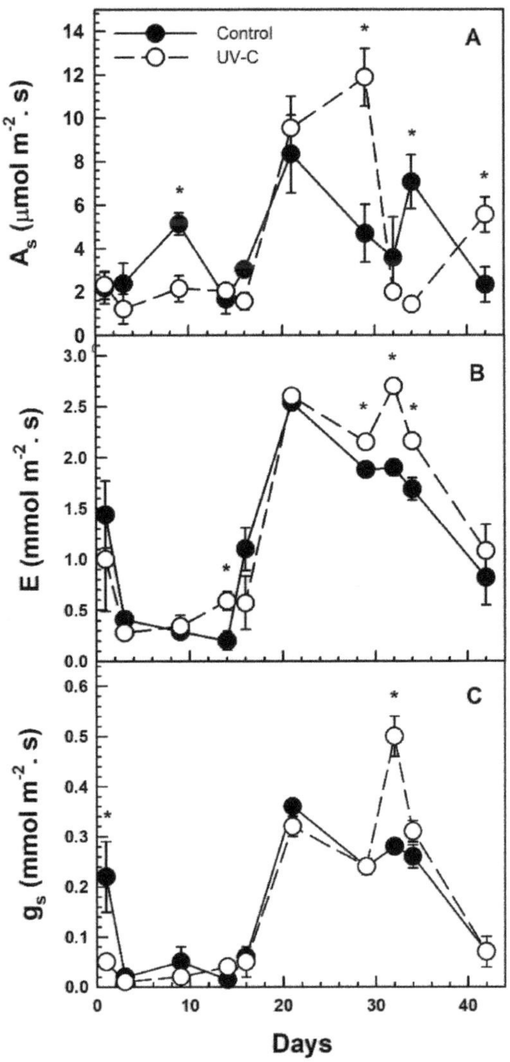

Figure 1. Net CO_2 assimilation (**A**; μmol m^{-2}. s), transpiration (**B**; mmol m^{-2}. s) and stomatal conductance (**C**; mmol m^{-2}. s) of *P.* × *hortorum* stem cuttings irradiated with 1 kJ m^{-2} or left nonirradiated (controls) over a 6-week period. Data (n = 30) are means per treatment per day ± S.E. Asterisks indicate significant differences between treatment means at p = 0.05.

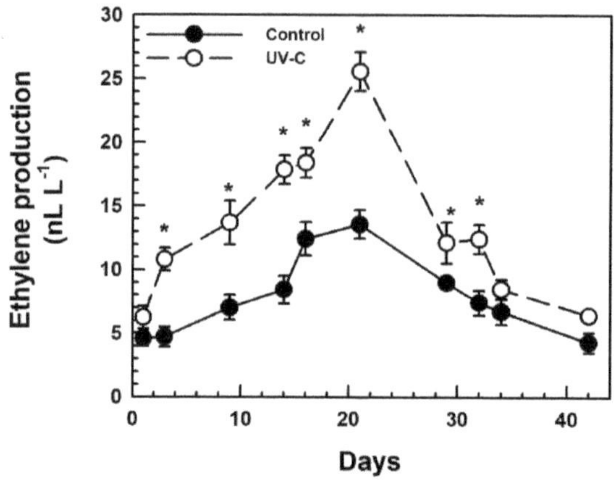

Figure 2. Ethylene production (nL L^{-1}) of *P.* × *hortorum* stem cuttings irradiated with 1 kJ m^{-2} or left nonirradiated (controls) over a 6-week period. Data (n = 15) are means per treatment per day ± S.E. Asterisks indicate significant differences between treatment means at p = 0.05.

3.2. Colour Attributes

The colour attributes (a^*, b^*, C^* and L^*) of UV-C-treated stem cuttings were not different than those of the untreated controls (Table 1 and Figure 3). Both the treated and the untreated stem cuttings showed a moderate level of greenness with a^* values ranging from −12.7 to −14.5 the first 21 d (Figure 3A). Then, the a^* values increased up to −16.3 for the irradiated samples. The b^* values expressing the yellowness of the samples remained positive for both the irradiated and the nonirradiated controls (Figure 3B). Yellowness increased from 21.5 at the beginning of the experiment to 28.9 at day 42. The chroma values (C^*) increased slightly from 23.7 on day 3 to 33.1 on day 42 (Figure 3C).

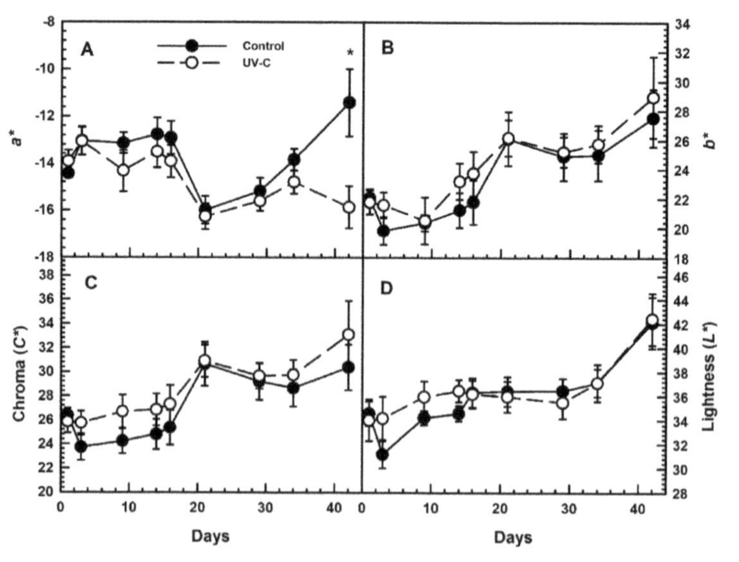

Figure 3. Leaf colour parameters (a^*; **A**), (b^*; **B**), (chroma; **C**) and (lightness; **D**) of *P.* × *hortorum* stem cuttings irradiated with 1 kJ m^{-2} or left nonirradiated (controls) over a 6-week period. Data (n = 30) are means per treatment per day ± S.E. Asterisks indicate significant differences between treatment means at p = 0.05.

No differences in C^* were recorded between the irradiated and the nonirradiated cuttings. Lightness (L^*) increased throughout the 42-day evaluation period for both the UV-C-irradiated and the nonirradiated samples (Figure 3D). L^* values ranged from 31.1 to 42.4, but no significant differences were detected between the UV-C-irradiated and the nonirradiated cuttings. The colour differences (ΔE) ranged considerably from the beginning to the end of the experiment (Figure 4). ΔE values were at 8.5 and 10.6 on day 3 and at 89.4 and 104.0 on day 42, for the control and the UV-C-treated cuttings, respectively. However, these differences were not significant at the $p = 0.05$ level.

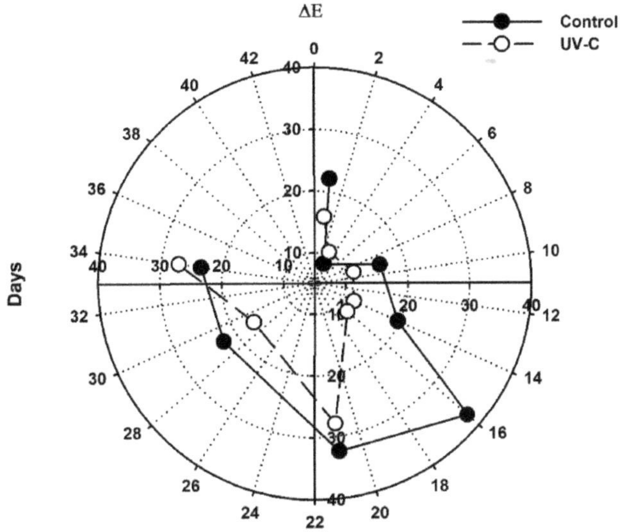

Figure 4. Leaf colour difference (ΔE) of *P.* × *hortorum* stem cuttings irradiated with 1 kJ m^{-2} or left nonirradiated (controls) over a 6-week period. Significant differences between treatment means were calculated at $p = 0.05$.

3.3. Rooting Performance

At the end of the 42-day evaluation period, the growth and rooting performance of the UV-C-treated and the untreated stem cuttings were recorded (Figures 5 and 6). The rooting percentage was increased in the UV-C-irradiated stem cuttings by up to 17% (Figure 5). The number of leaves on *P.* × *hortorum* stem cuttings was not affected by UV-C (Figure 6A). On the contrary, the flowering performance was improved to a significant level by UV-C irradiation (Figure 6B). The number of flowers also increased by up to 62%, compared to the controls.

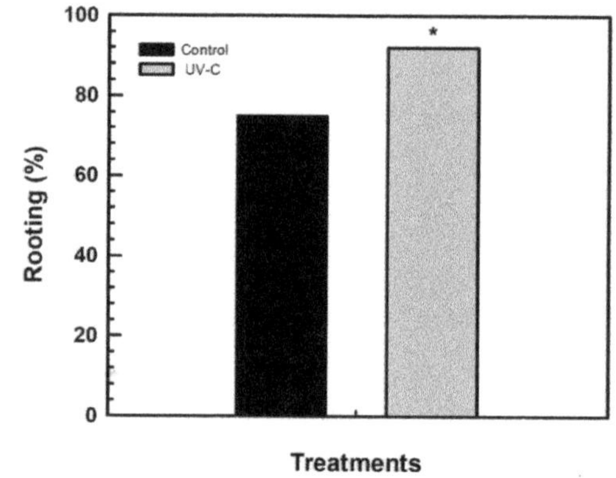

Figure 5. Rooting percentage (%) (n = 30) of *P.* × *hortorum* stem cuttings irradiated with 1 kJ m^{-2} or left nonirradiated (controls) over a 6-week period. Asterisk indicates the significant difference between treatment means at p = 0.05.

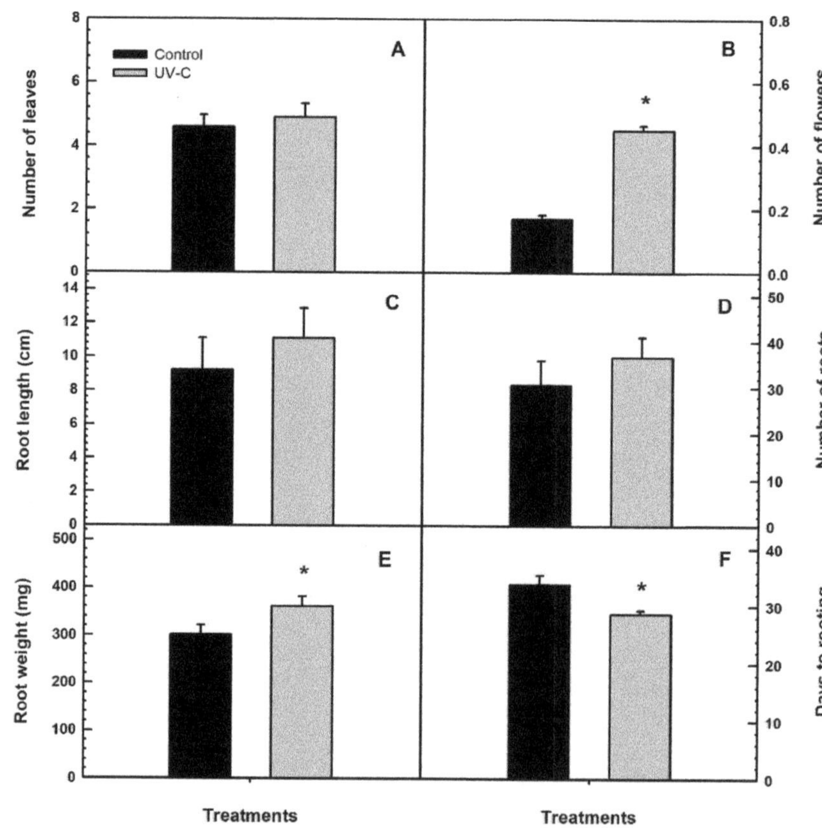

Figure 6. Number of leaves (**A**), number of flowers (**B**), root length (cm; **C**), number of roots (**D**), root weight (**E**) and days to rooting (**F**) of *P.* × *hortorum* stem cuttings irradiated with 1 kJ m^{-2} or left nonirradiated (controls) for 6-week period. Data (n = 30) were collected by the end of the experiment and are means per treatment per day ± S.E. Asterisks indicate significant differences between treatment means at p = 0.05.

In general, UV-C irradiation positively affected the rooting of the cuttings. Although, the root length and the number of roots did not statistically differ between the treated and the untreated controls (Figure 6C,D), the root weight was significantly increased (Figure 6E,F) and days to rooting were decreased from 33.8 to 28.7 d (Figure 6E). The root weight of the UV-C-treated cuttings was 61 mg higher (Figure 6F).

3.4. Correlation Analysis of Ethylene Production and Rooting Performance

Significant correlations were found between ethylene production, physiological responses and rooting performance (Tables 2 and 3). Significant positive correlations were recorded between ethylene production and A_s, E and g_s (Table 2). Ethylene production was not correlated with root number and root length (Table 3). On the contrary, ethylene production was negatively and positively correlated with days to rooting and root weight, respectively (Table 3). The higher the ethylene production the faster the flowering and the higher the root weight.

Table 2. Pearson correlations between net CO_2 assimilation (A_s; μmol m^{-2}. s), transpiration (E; mmol m^{-2}. s), stomatal conductance (g_s; mmol m^{-2}. s) and ethylene as affected by exposure to UV-C irradiation over a period of 6 weeks. Correlation analysis was performed in SPSS v. 21.

	A_s (μmol m^{-2}. s)			E (mmol m^{-2}. s)			g_s (mmol m^{-2}. s)		
	Pearson Correlation	Sum of Squares	Significance	Pearson Correlation	Sum of Squares	Significance	Pearson Correlation	Sum of Squares	Significance
Ethylene	0.336	1188.483	0.000	0.386	298.884	0.000	0.296	39.477	0.000

Table 3. Pearson correlations between root length (cm), root number, days to rooting, root weight (mg) and ethylene production as affected by exposure to UV-C irradiation over a period of 6 weeks. Correlation analysis was performed in SPSS v. 21.

	Root Length (cm)			Root Number			Days to Rooting			Root Weight (mg)		
	Pearson Correlation	Sum of Squares	Significance	Pearson Correlation	Sum of Squares	Significance	Pearson Correlation	Sum of Squares	Significance	Pearson Correlation	Sum of Squares	Significance
Ethylene	0.156	63.837	0.478	0.372	101.784	0.289	−0.618	−148.676	0.005	0.494	2301.144	0.017

4. Discussion

This was the first attempt to improve rooting performance of *P.* × *hortorum* stem cuttings using low doses of UV-C irradiation. Many studies have reported that ethylene is a key regulation factor that affect the rooting responses of stem cuttings [16,19–22]. The connection between UV-C irradiation and ethylene production is well documented in previous research [2,3,8,13]. For example, UV-C-irradiated tomato fruit showed an upregulation of multiple genes' expression such as the EREB (ethylene responsive element binding protein) by 43-fold, the LOC544285 (ethylene forming enzyme) by 39.4-fold, the ETR6 (ethylene receptor-like protein) by 26-fold, the ERF1 (ethylene responsive factor) by 6.8-fold, the ETR4 (Ethylene receptor homolog) by 5.3-fold, the LEJA2 (jasmonic acid 2) by 3.5-fold, the IAA5 (IAA5 protein) by 2.5-fold and many others [7]. UV-C irradiation at 3.7 kJ m^{-2} promoted ethylene production by 2-fold in harvested tomato fruit [13]. Although postharvest UV-C irradiation mediates positive responses in harvested horticultural products [2], it may also increase ethylene production as a result of a mild abiotic stress applied to the produce. In the present study, UV-C irradiation significantly increased ethylene production in *P.* × *hortorum* stem cuttings and affected root formation and development. Strong correlations were found between ethylene production and days to rooting and root weight. Increased ethylene production was negatively correlated with days to rooting, meaning that a higher ethylene production resulted in a quicker rooting response. Generally, the use of UV irradiation may promote root biomass accumulation, but it is a species-depended response. For example, in sunflower seedlings (*Helianthus annuus* L.) wound-induced ethylene production, localised at the lower part of the hypocotyl, promoted rooting [20].

In *Pelargonium* cuttings, ethylene increased rooting percentage, reduced the number of roots and root fresh mass, but it increased dry root mass [23]. Rapaka et al. [24] reported that ethylene may promote root formation in *Pelargonium* cuttings, only when endogenous carbohydrate levels are high. The level of endogenous carbohydrates is always dependent on photosynthetic activity [24] and N_2 fertilisation [25] thus, genetic and environmental conditions during the mother-plant cultivation may also play significant roles. In the study by Mutui et al. [23], exposure of *Pelargonium zonale* cuttings to 1 or 2 μL L^{-1} ethylene significantly reduced the number of roots and the total root length, but not the rooting percentages. We suggest that the endogenous ethylene production in *P.* × *hortorum* stem cuttings used in the current experiments was by up to 10-fold lower compared to that exogenously provided in the Mutui et al. [23] study. Although low ethylene concentrations may promote the initiation of lateral root primordia, higher concentrations strongly inhibit them [26]. The crosstalk between auxin and ethylene in the root formation and elongation has been reported in the past [27–29]. Ethylene effects on root growth are mediated by the regulation of auxin's biosynthesis, transport and local distribution in cells [29]. When IBA hormone was applied to rose (*Rosa hybrida* cv. Royalty) stem cuttings before rooting, ethylene production increased considerably the following days indicating a strong relation between auxin and ethylene during the rooting process [27]. In the present study, UV-C irradiation have elicited endogenous wounding-induced ethylene responses, generated as a result to the abiotic stress. Such responses led to an increase in *P.* × *hortorum*'s rooting percentages and root weights. It also promoted faster rooting response.

Ethylene production was positively correlated with A_s, E and g_s in *P.* × *hortorum* stem cuttings. Increased physiological responses resulted in a higher ethylene production. Although UV-C did not affect A_s, E and g_s, it induced higher transpiration rates on days 26, 30 and 32. In previous studies, treatments with 1 kJ m^{-2} UV-C irradiation positively affected A_s in *P.* × *hortorum* plants but had no effect on *Freesia hybrida* [5] and *Solanum lycopersicum* [7] plants.

UV-C irradiation significantly affected flowering response of the cuttings. Similar findings have been reported for whole plants and propagation materials (e.g., corms and plantlets). For example, the irradiation of freesia (*Freesia hybrida* L.) corms and plants with UV-C hastened flowering, especially when combined with cold treatments [6]. The treatment of *P.* × *hortorum* plants with 1 kJ m^{-2} UV-C hastened the flowering and increased the total flower number [4,11]. The flower size (cm) and flower number developed in Persian violet plantlets was significantly increased after a single, high-intensity UV-C irradiation exposure [22].

5. Conclusions

UV-C irradiation positively affected rooting performance of *P.* × *hortorum* stem cuttings. It induced endogenous ethylene production, which in turn promoted flowering, root growth and development via signalling pathways possibly involving other plant hormones such as auxins.

Author Contributions: Conceptualisation, methodology, writing—original draft, writing—review and editing, supervision and formal analysis by A.I.D.; data curation, experimentation, K.G., K.D., F.Z. and A.I.D. All authors have read and agreed to the published version of the manuscript.

Funding: This research received no external funding.

Informed Consent Statement: Not applicable.

Data Availability Statement: Not applicable.

Conflicts of Interest: The authors declare no conflict of interest.

References

1. Dole, J.M.; Wilkins, H.F. *Floriculture: Principles and Species*; Prentice-Hall Inc.: Hoboken, NJ, USA, 1999.
2. Urban, L.; Charles, F.; de Miranda, M.R.A.; Aarrouf, J. Understanding the physiological effects of UV-C light and exploiting its agronomic potential before and after harvest. *Plant Physiol. Biochem.* **2016**, *105*, 1–11. [CrossRef] [PubMed]
3. Urban, L.; Sari, D.C.; Orsal, B.; Lopes, M.M.D.A.; Miranda, R.; Aarrouf, J. UV-C light and pulsed light as alternatives to chemical and biological elicitors for stimulating plant natural defenses against fungal diseases. *Sci. Hortic.* **2018**, *235*, 452–459. [CrossRef]
4. Darras, A.I.; Bali, I.; Argyropoulou, E. Disease resistance and growth responses in *Pelargonium* × *hortorum* plants to brief pulses of UV-C irradiation. *Sci. Hortic.* **2015**, *181*, 95–101. [CrossRef]
5. Darras, A.I.; Skouras, P.J.; Assimomitis, P.; Labropoulou, C.; Stathas, G.J. Application of UV-C irradiation to *Rosa x hybrida* plants as a tool to minimise *Macrosiphum rosae* populations. *Agronomy* **2021**, *11*, 702. [CrossRef]
6. Darras, A.I.; Vlachodimitropoulou, A.; Dimitriadis, C. Regulation of corm sprouting, growth and flowering of pot *Freesia hybrida* L. plants by cold and UV-C irradiation forcing. *Sci. Hortic.* **2019**, *252*, 110–112. [CrossRef]
7. Darras, A.I.; Tsikaloudakis, G.; Lycoskoufis, I.; Dimitriadis, C.; Karamousantas, D. Low doses of UV-C irradiation affects growth, fruit yield and photosynthetic activity of tomato plants. *Sci. Hortic.* **2020**, *267*, 109357. [CrossRef]
8. Liu, C.; Cai, L.; Han, X.; Ying, T. Temporary effect of postharvest UV-C irradiation on gene expression profile in tomato fruit. *Gene* **2011**, *486*, 56–64. [CrossRef]
9. Tang, K.; Zhan, J.C.; Yang, H.R.; Huang, W.D. Changes of resveratrol and antioxidant enzymes during UV-induced plant defense response in peanut seedlings. *J. Plant Physiol.* **2010**, *167*, 95–102. [CrossRef]
10. Rai, R.; Meena, R.P.; Smita, S.S.; Shukla, A.; Rai, S.K.; Pandey-Rai, S. UV-B and UV-C pre-treatments induce physiological changes and artemisinin biosynthesis in *Artemisia annua* L.–An antimalarial plant. *J. Photochem. Photobiol. B Biol.* **2011**, *105*, 216–225. [CrossRef]
11. Darras, A.I.; Demopoulos, V.; Bali, I.; Tiniakou, C. Photomorphogenic reactions in geranium (*Pelargonium* x *hortorum*) plants stimulated by brief exposures of ultraviolet-C irradiation. *Plant Growth Regul.* **2012**, *68*, 343–350. [CrossRef]
12. Loconsole, D.; Santamaria, P. UV lighting in horticulture: A sustainable tool for improving production quality and food safety. *Horticulturae* **2021**, *7*, 9. [CrossRef]
13. Tiecher, A.; de Paula, L.A.; Chaves, F.C.; Rombaldi, C.V. UV-C effect on ethylene, polyamines and the regulation of tomato fruit ripening. *Postharvest Biol. Technol.* **2013**, *86*, 230–239. [CrossRef]
14. Li, D.; Luo, Z.; Mou, W.; Wang, Y.; Ying, T.; Mao, L. ABA and UV-C effects on quality, antioxidant capacity and anthocyanin contents of strawberry fruit (*Fragaria ananassa* Duch.). *Postharvest Biol. Technol.* **2014**, *90*, 56–62. [CrossRef]
15. Abeles, F.B.; Morgan, P.W.; Saltveit, M.E. Ethylene in plant biology Academic Press San Diego. In *Ethylene in Plant Biology*, 2nd ed.; Academic Press: San Diego, CA, USA, 1992.
16. Rapaka, V.K.; Faust, J.E.; Dole, J.M.; Runkle, E.S. Endogenous carbohydrate status affects postharvest ethylene sensitivity in relation to leaf senescence and adventitious root formation in Pelargonium cuttings. *Postharvest Biol. Technol.* **2008**, *48*, 272–282. [CrossRef]
17. Mutui, T.M.; Mibus, H.; Serek, M. The influence of plant growth regulators and storage on root induction and growth in *Pelargonium zonale* cuttings. *Plant Growth Regul.* **2010**, *61*, 185–193. [CrossRef]
18. Clark, D.G.; Gubrium, E.K.; Barrett, J.E.; Nell, T.A.; Klee, H.J. Root formation in ethylene-insensitive plants. *Plant Physiol.* **1999**, *121*, 53–60. [CrossRef] [PubMed]
19. Riov, J.; Yang, S.F. Ethylene and auxin-ethylene interaction in adventitious root formation in mung bean (*Vigna radiata*) cuttings. *J. Plant Growth Regul.* **1989**, *8*, 131–141. [CrossRef]
20. Liu, J.; Mukherjee, L.; Reid, D.M. Adventitious rooting in hypocotyls of sunflower (*Helianthus annuus*) seedlings. III. The role of ethylene. *Physiol. Plant.* **1990**, *78*, 268–276. [CrossRef]
21. Sukthavornthum, W.; Bodhipadma, K.; Noichinda, S.; Phanomchai, S.; Deelueak, U.; Kachonpadungkitti, Y.; Leung, D.W. UV-C irradiation induced alterations in shoot proliferation and in vitro flowering in plantlets developed from encapsulated and non-encapsulated microshoots of Persian violet. *Sci. Hortic.* **2018**, *233*, 9–13. [CrossRef]
22. Phanomchai, S.; Noichinda, S.; Kachonpadungkitti, Y.; Bodhipadma, K. Differing In vitro rooting and flowering responses of the Persian violet to low and high UV-C irradiation. *Plants* **2021**, *10*, 2671. [CrossRef]
23. Mutui, T.; Mibus, H.; Serek, M. Effects of thidiazuron, ethylene, abscisic acid and dark storage on leaf yellowing and rooting of Pelargonium cuttings. *J. Hortic. Sci. Biotechnol.* **2005**, *80*, 543–550. [CrossRef]
24. Rapaka, V.K.; Bessler, B.; Schreiner, M.; Druege, U. Interplay between initial carbohydrate availability, current photosynthesis, and adventitious root formation in Pelargonium cuttings. *Plant Sci.* **2005**, *168*, 1547–1560. [CrossRef]
25. Druege, U.; Zerche, S.; Kadner, R. Nitrogen-and storage-affected carbohydrate partitioning in high-light-adapted Pelargonium cuttings in relation to survival and adventitious root formation under low light. *Ann. Bot.* **2004**, *94*, 831–842. [CrossRef] [PubMed]
26. Ivanchenko, M.G.; Muday, G.K.; Dubrovsky, J.G. Ethylene–auxin interactions regulate lateral root initiation and emergence in *Arabidopsis thaliana*. *Plant J.* **2008**, *55*, 335–347. [CrossRef]
27. Sun, W.Q.; Bassuk, N.L. Auxin-induced Ethylene Synthesis during Rooting and Inhibition of Budbreak of Royalty' Rose Cuttings. *J. Am. Soc. Hortic. Sci.* **1993**, *118*, 638–643. [CrossRef]

28. Pitts, R.J.; Cernac, A.; Estelle, M. Auxin and ethylene promote root hair elongation in Arabidopsis. *Plant J.* **1998**, *16*, 553–560. [CrossRef]
29. Růžička, K.; Ljung, K.; Vanneste, S.; Podhorská, R.; Beeckman, T.; Friml, J.; Benková, E. Ethylene regulates root growth through effects on auxin biosynthesis and transport-dependent auxin distribution. *Plant Cell* **2007**, *19*, 2197–2212. [CrossRef]

horticulturae

Article

Mycorrhization Enhances Vegetative Growth, Leaf Gas Exchange, and Root Development of Micropropagated *Philodendron bipinnatifidum* Schott ex Endl. Plantlets during Acclimatization

Yaser Hassan Dewir [1,*], Muhammad M. Habib [1], AbdulAziz A. AlQarawi [1], Thobayet S. Alshahrani [1], Ahmed Ali Alaizari [1], Jahangir A. Malik [1], Mona S. Alwahibi [2] and Hosakatte Niranjana Murthy [3,4]

1 Plant Production Department, College of Food and Agriculture Sciences, King Saud University, Riyadh 11451, Saudi Arabia
2 Department of Botany and Microbiology, College of Science, King Saud University, Riyadh 11495, Saudi Arabia
3 Department of Horticultural Science, Chungbuk National University, Cheongju 28644, Republic of Korea
4 Department of Botany, Karnatak University, Dharwad 580003, India
* Correspondence: ydewir@ksu.edu.sa

Abstract: *Philodendron bipinnatifidum* Schott ex Endl. is a popular ornamental plant that is normally propagated by tissue culture methods. However, the growth and acclimatization of micropropagated plants are tarrying processes. Therefore, in the present study we examined the effect of arbuscular mycorrhizal fungi (AMF) *Gigaspora albida* and *G. marginata* on the success in the establishment, growth, and development of *P. bipinnatifidum* plantlets during the acclimatization phase. AMF plants had significantly more leaves (10.67 per plant), leaf area (75.63 cm^2), plant height (14.17 cm), shoot fresh weight (3.30 g) and shoot dry weight (0.31 g), according to an analysis of growth characteristics. In comparison, non-AMF plants had lower values for these metrics. In addition, AMF plants had significantly longer main roots (23 cm), total length roots per plantlet (485.73 cm), average root diameter (4.58 mm) per plantlet, number of root tips (236) per plant, total root surface area (697.76 cm^2), total root volume (79.98 cm^3), roots fresh weight (1.51 g), roots dry weight (0.16 g) than non-AMF plants. AMF-treated plants showed better performance in leaf gas exchange, chlorophyll, and carotenoid content. These results emphasize the need for mycorrhization of micropropagated plants to promote vegetative growth, especially during the acclimatization stage.

Keywords: acclimatization; in vitro regeneration; micropropagation; *Philodendron bipinnatifidum*; plantlets; ornamental plant

Citation: Dewir, Y.H.; Habib, M.M.; AlQarawi, A.A.; Alshahrani, T.S.; Alaizari, A.A.; Malik, J.A.; Alwahibi, M.S.; Murthy, H.N. Mycorrhization Enhances Vegetative Growth, Leaf Gas Exchange, and Root Development of Micropropagated *Philodendron bipinnatifidum* Schott ex Endl. Plantlets during Acclimatization. *Horticulturae* **2023**, *9*, 276. https://doi.org/10.3390/horticulturae9020276

Academic Editors: Konstantinos Bertsouklis, Epaminondas Kartsonas and Angela Carra

Received: 14 January 2023
Revised: 13 February 2023
Accepted: 14 February 2023
Published: 17 February 2023

1. Introduction

An important phase in the micropropagation of plant species is the acclimatization of in vitro regenerated plants during ex vitro transplantation. Since they have been transplanted from controlled conditions of light, temperature, and greater relative humidity to high light, temperature, and lower humidity conditions, micropropagated plants typically experience transplantation stress. Because of inadequate stomatal control and abnormally increased water loss through cuticles, which cause wilting and necrosis of the leaves and ultimately cause the death of the plants, the transpiration rate in micropropagated plants is significantly higher than in field-grown plants [1,2]. Some characteristics of in vitro-generated plants include low net carbon dioxide uptake and fixation, poor morphological differentiation of leaves, increased stomatal frequency, defective stomatal structure and movement, and inadequate deposition of protective epicuticular wax on the leaves [3,4]. Even plants that are growing in waterlogged conditions also demonstrate variability in stomatal density and stomatal morphological characteristics. However, normal stomatal features and density are useful to plants for the proper uptake of carbon dioxide from the environment [5].

Researchers have adopted photoautotrophic micropropagation methods to overcome the morphological, anatomical, and physiological variations with in vitro regenerated plants, such as growing the plantlets under enhanced light, and carbon dioxide conditions, lowering sucrose concentration in the medium and using ventilated culture vessels that facilitate the availability of abundant resources for the plantlets to develop their photosynthetic ability and involve in photosynthesis [6,7]. Alternately, after ex vitro transplantation, colonization of micropropagated plantlets with arbuscular mycorrhiza fungi (AMF) can also improve the plantlets' acclimatization, survival, and performance. Mycorrhization of in vitro regenerated plants with AMF after ex vitro transplantation has enhanced survival and plantlet performance. Field crops [2,8,9], horticultural plants [10,11], medicinal plants [12–14], and ornamentals [15] have all shown that mycorrhization in vitro grown plants during acclimatization is a beneficial technique for their involvement in successful growth and development. AMF has been shown to boost early development, improve nutrient uptake, boost disease resistance, and boost tolerance to environmental challenges [2,16–18]. Recent pieces of evidence have suggested that mycorrhization of in vitro-raised plants has promoted the accumulation of enzymes and metabolites and depicted higher antioxidant activities compared to non-inoculated plants [14].

A large genus of flowering plants in the Araceae family, *Philodendron* is a lovely indoor ornamental plant. The popular indoor plants cultivated all over the world, philodendrons have a wide range of growth habits including epiphytic, hemi-epiphytic, and terrestrial. There are countless hybrids with leaves in the colors green, red, yellow, and orange [19]. The self-heading species *Philodendron bipinnatifidum* Schott ex Endl. also referred to as the 'Lacy tree philodendron', may grow to heights of 4 to 5 m and have deeply cut, green to dark green leaves that can extend up to 1 m in length [20]. To meet the need of the horticulture industry micropropagation is typically used for the propagation of these plants.

In commercial micropropagation systems plant losses of 10 to 40% have been reported [18,21]. Some micropropagated plants may develop functional roots during the tissue culture or development deficient vascular connections between the root and shoot system [22]. The slow growth of micropropagated *Philodendron* plants was also on record [23]. Therefore, inoculants containing AMF are a viable biotechnological tool, especially when plants are in the acclimatization stage.

The hypothesis to be tested in this study is that inoculation with compatible AMF species is essential for the survival and development of *P. bipinnatifidum* in the acclimatization phase. John [24] and Burndrett [25] have reported the mycorrhizal association of natural philodendron plants. Consequently, in the current study we inoculated the in vitro regenerated plants of *P. bipinnatifidum* with two AMF strains, namely *Gigospora albida* and *G. marginata*. We analyzed the morphological (plant height, number of leaves, leaf area, shoot fresh and dry weight, total root length, number of root tips, average diameter, total root surface area, total root volume, root fresh and dry weight, number of stomata, stomatal length, pore/aperture length and width), and physiological responses (chlorophyll and carotenoid concentration, net CO_2 assimilation, stomatal conductance, and transpiration rate) of AMF treated and without AMF-treatment plants. The results are useful for the in vitro regeneration of *P. bipinnatifidum* and successful transplantation to ex vitro conditions.

2. Materials and Methods

2.1. Plant Materials and Arbuscular Mycorrhizal Fungi (AMF) Treatment

The King Saud University College of Food and Agricultural Science's plant tissue culture laboratory served as the site for this investigation. The Alawaadh et al. [20] and Dewir et al. [23] procedure was used to multiply and root the axillary shoots of *P. bipinnatifidum* in vitro. The shoots were multiplied using Murashige and Skoog's medium (MS) [26] supplemented with a 6-benzyl amino purine (BA; 1 mg·L^{-1}) and indole-3-butyric acid IBA (0.5 mg·L^{-1}). The axillary shoots (3.5–4.0 cm) were divided up and put individually into MS media supplemented with 5.4 µM -naphthaleneacetic acid (NAA) for their rooting. Agar–agar (Duchefa, Haarlem, The Netherlands) was used to solidify all media that contained sucrose (30 g·L^{-1}), and the pH of the medium was adjusted to 5.8 before autoclaving at 121 °C and 118 kPa pressure for 15 min. The cultures were maintained at a temperature of 25 ± 2 °C with a cool white fluorescent tube-provided photosynthetic photon flux density (PPFD) of 25 mol m^{-2}·s^{-1} under a 16:8 h (light: dark) photoperiod.

After six weeks, the plantlets were carefully removed from the gelled medium, rinsed with tap water, and then placed into plastic pots (21.5 cm in length, 4 cm in height, and 1.7 cm in width) filled with sterile sand and soil mixture (1:1). The potted plants were cultivated in a growth room for the first two weeks at 25 ± 2 °C, 50%–60% RH, and 100 µmol m^{-2}·s^{-1} PPFD (16:8 h photoperiod under white fluorescent lamps). Regular irrigation with Hoagland nutrition solution devoid of phosphorus was given to the plantings. Eight weeks after being moved into the growth chamber, the vegetative growth of the plantlets was assessed. Each treatment had 30 repetitions, and each replicate was represented by a container with a single plantlet.

2.2. Collection of Soil Samples and Identification of Arbuscular Mycorrhizal Fungi (AMF)

Soil samples at a depth of 10–20 cm were collected from four different sites (20 m away) of King Saud University Botanic Garden, Riyadh, Saudi Arabia (24°44′31.79″ N 46°51′25.19″ E) in January 2022. Three replicates per sample were collected, sealed in plastic bags, and kept at 4 °C until usage. A trap culture with Maize (*Zea mays* L.) as a target mycotrophic plant was used to propagate AMF by the pooled soil samples. AMF spores were scrupulously extracted using the wet sieving and decanting method after 12 weeks of maize cultivation in the soil [27]. The AMF spores were identified on the basis of morphology (i.e., shape, surface ornamentation, color, contents, and wall structure) [28,29] and compared to the morphological descriptions of species presented in the *International Culture Collection of Vesicular-Arbuscular Mycorrhizal Fungi (INVAM)* and other literature [30–32]. The AMF species found in the soil were identified (Figure 1, Spores Plater).

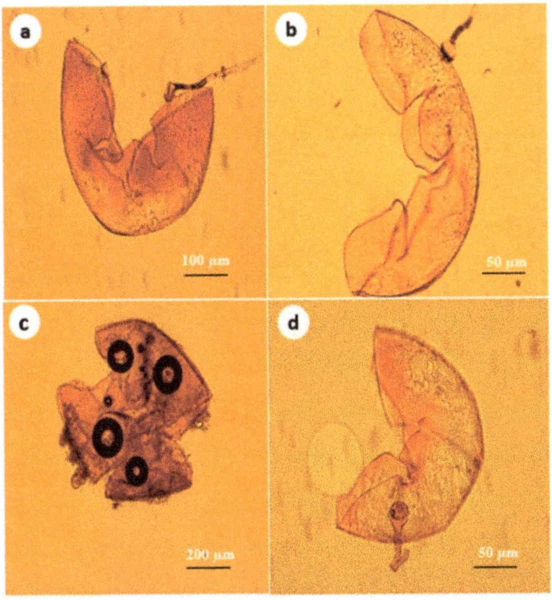

Figure 1. Arbuscular mycorrhizal fungi spores collected from the trap culture: crushed spore spores of *Gigaspora albida* (**a**,**b**); crushed spores of *Gigaspora margarita* (**c**,**d**).

2.3. *AMF Colonization with Transplanted Philodendron Plants*

The AMF inoculum (*Gigaspora albida* and *Gigaspora margarita*) contained growth substrate, spores (density of approximately 100 per 10 dry substrates), mycelium, and infected roots of maize. Each pot was inoculated with 50 g each of inoculum for mycorrhizal treatment. Non-mycorrhizal treatment received 100 g sterilized AM inoculum (160 °C, 3 h).

2.4. *Estimation of Symbiotic Development and Spore Count*

Philodendron bipinnatifidum roots were separated and cleansed with distilled water. The roots were subsequently processed for 30 min at 80 degrees Celsius in 10% potassium hydroxide (KOH), rinsed once more, and then exposed for 3 min to 3% hydrogen peroxide (H_2O_2) before being acidified for 10 min with 1% HCl. They were then stained with Trypan blue for an additional 20 min at 80 °C [33]. In a lactoglycerol solution, the dyed root segments were put on glass slides. An optical microscope was used to examine various features in the root segment (at 400×). To evaluate intraradical colonization, at least 50 root segments from each sample of *P. bipinnatifidum* were examined. It was noted that there was mycelium, vesicles, and arbuscules present. Calculations were made about the proportion and intensity of intraradical mycorrhizal colonization (mycelium, vesicles, and arbuscular development) within the roots [34,35]. The spores were extracted from the substrate of each treatment using the techniques detailed in Section 2 previously [27]. Dry soil (100 g) was used to calculate the total spore population in each treatment.

2.5. *Leaf Gas Exchange Parameters*

Net CO_2 assimilation, stomatal conductance, and transpiration were measured at 8 weeks of acclimatization for both non-AMF and AMF-treated plantlets of *P. bipinnatifidum*, as described by Dewir et al. [36]. An LI-6400 portable photosynthesis system (Li-Cor, Inc., Lincoln, NE, USA) outfitted with a typical 2 × 3 cm leaf cuvette and a Li-Cor LI-6400-02B light source was used to collect the data. The leaf temperature was 23 °C, and the photosynthetic parameters were assessed in inflow air with a 350 µmol CO_2 concentration and relative humidity of 60%. Ten randomly chosen plants from each treatment were used for the measurements, which were made in triplicate.

2.6. *Measurements of Vegetative Parameters*

After 8 weeks of cultivation, growth responses were measured in terms of shoot fresh and dry weight (g), plant height (cm), and the number of leaves and leaf area (cm^2) per plantlet. Using a portable area meter, the leaf area was calculated (CI-202; CID, Inc., Vancouver, WA, USA). Ten randomly selected plantlets served as the source of all the measurements, which were made in triplicate.

2.7. *Measurements Leaf Pigments*

Philodendron bipinnatifidum plantlets with and without AMF treatment had their chlorophyll and carotenoid concentrations measured. Cold acetone (80%) was used to remove three replicates of young leaves (weighing 0.5 g each) from each treatment for two days. The absorbance was measured at the wavelengths of 663.2, 646.8, and 470.0 nm with calculations made following the method described by Lichtenthaler et al. [37].

2.8. *Microscopic Observations of Stomata*

Strips from the cuticle of the leaves of non-AMF and AMF-treated *Philodendron* plantlets were prepared following the method described by Cotton [38]. The dry leaves were soaked for 24 h and the transparent thin layer of the surface cells of the epidermal layer of the leaf was carefully removed using pointed forceps and placed on a glass slide. They were then stained with a mixture solution of 0.1 g triaryl methane dye and 2 mL of glacial acetic acid in 100 mL distilled water (a light-green dye) for several seconds and covered with a slide cover. The glass slides were examined to identify stomata types, stomata size (measured with an ocular ruler), and stomatal density (number of stomata per unit area) using an optical microscope with a SwiftCam 20 Megapixel camera for microscopes (DeltaPix, Smørum, Denmark). The microscopic images of the leaf surfaces were captured at 40× magnification. The type of stomata, stomatal density, and aperture length and width were determined in the microscopic view field. A total of 30 measurements were conducted for the analysis of stomatal characteristics from randomly selected plants with different leaves (10 × 3 = 30).

2.9. Measurement of the Root Growth Parameters

The roots of *P. bipinnatifidum* plantlets, both non-AMF and AMF-treated plantlets, were removed from the pots and washed with tap water to establish three root replicates for three plants from each treatment. Before scanning, the roots were toluidine red dyed for around 8 h. A flatbed scanner (Cannon unit 101, Green Island, NY, USA) was used for scanning, and WinRHIZO software was used to analyze the images (V5.0, Regent Instruments, Quebec, QC, Canada). We measured a few root system characteristics, including root fresh weight, total root length, root diameter, root volume, and root surface area.

2.10. Experimental Design and Data Analysis

With 30 replicates for each treatment, the studies were properly randomized. One plantlet in a pot served as the representation of each replicate. Utilizing ANOVA and the unpaired *t*-Test, the treatment effects were statistically evaluated.

3. Results

3.1. Mycorrhizal Colonization

After eight weeks of cultivation, the roots of AMF-treated plants were harvested and examined for colonization with the host plants. The microscopic observation of the mycorrhizal status of *P. bipinnatifidum* plantlets indicated the presence of all predicted AMF structures (mycelium, vesicles, arbuscules, and spores) in the roots (Figure 2). The analysis of the mycorrhizal colonization showed the colonization percentage as mycelium 66.66%, vesicles 11.11%, and arbuscules 51.11%. The total spore count was also recorded as (169/100 g soil).

3.2. Plant Growth

Overall, plant growth was initially low. However, growth increased as the plantlets were adjusted to the ex vitro environment. Analysis of growth parameters such as the number of leaves per plant, leaf area, plant height, and shoot fresh and dry weight revealed that AMF plants had a significantly greater number of leaves (10.67 per plant), leaf area (75.63 cm^2), plant height (14.17 cm), shoot fresh weight (3.30 g) and dry weight (0.315 g). These parameters were comparatively lower in non-AMF plants (Table 1). Figure 3a shows the promotive effect of vegetative growth AMF plants when compared to non-AMF plants. Root growth characteristics of AMF-treated plants were compared with non-AMF-treated plants and presented in Figures 3b and 4. The length of the main root (23 cm), total root length/plantlet (485.73 cm), number of root tips/plantlet (236), average root diameter/plantlet (4.58 mm), total root surface area (697.76 cm^2), total root volume/plantlet (79.98 cm^3), root fresh weight (1.51 g) and root dry weight (0.16 g) were all significantly higher when compared to non-AMF plants (Figure 4).

Table 1. Stoma density, leaf gas exchange and pigments, and vegetative growth characteristics of *Philodendron bipinnatifidum* in response to arbuscular mycorrhizal fungi after 8 weeks acclimatization.

Growth Parameters	Non-AMF	AMF Treated
Number of stomata (mm^2)	78.93 ± 9.874	136.68 ± 5.775 **
Aperture length (μm)	11.00 ± 1.354	11.5 ± 0.2886 NS
Aperture width (μm)	8.25 ± 0.6292	9.5 ± 0.2887 NS
Net CO_2 assimilation (μmol $CO_2 \cdot m^{-2} \cdot s^{-1}$)	7.1905 ± 0.3865	9.2673 ± 0.4032 ***

Table 1. *Cont.*

Growth Parameters	Non-AMF	AMF Treated
Stomatal conductance (mol $H_2O \cdot m^{-2} \cdot s^{-1}$)	0.0239 ± 0.0007	0.0492 ± 0.0029 ***
Transpiration rate (mol $H_2O \cdot m^{-2} \cdot s^{-1}$)	0.8784 ± 0.0234	1.6324 ± 0.0901 ***
Chlorophyll a (mg·g^{-1} FW)	1.379 ± 0.0024	1.512 ± 0.0130 **
Chlorophyll b (mg·g^{-1} FW)	0.571 ± 0.0088	0.598 ± 0.0096 NS
Chlorophyll a+b (mg·g^{-1} FW)	1.950 ± 0.0112	2.110 ± 0.0226 **
Chlorophyll a/b ratio	2.414 ± 0.2751	2.529 ± 0.0188 NS
Carotenoids (mg·g^{-1} FW)	0.423 ± 0.0004	0.533 ± 0.0070 ***
Number of leaves/plants	7.00 ± 0.408	10.67 ± 0.624 **
Leaf area/plant (cm^2)	31.11 ± 3.058	75.63 ± 1.207 ***
Plant height (cm)	12.13 ± 0.409	14.17 ± 0.118 **
Shoot fresh weight/plant (g)	1.776 ± 0.081	3.307 ± 0.047 ***
Shoot dry weight/plant (g)	0.148 ± 0.013	0.315 ± 0.004 ***

NS = not significant, ** and *** = significant at $p \leq 0.01$, and $p \leq 0.001$, respectively, according to unpaired *t*-test.

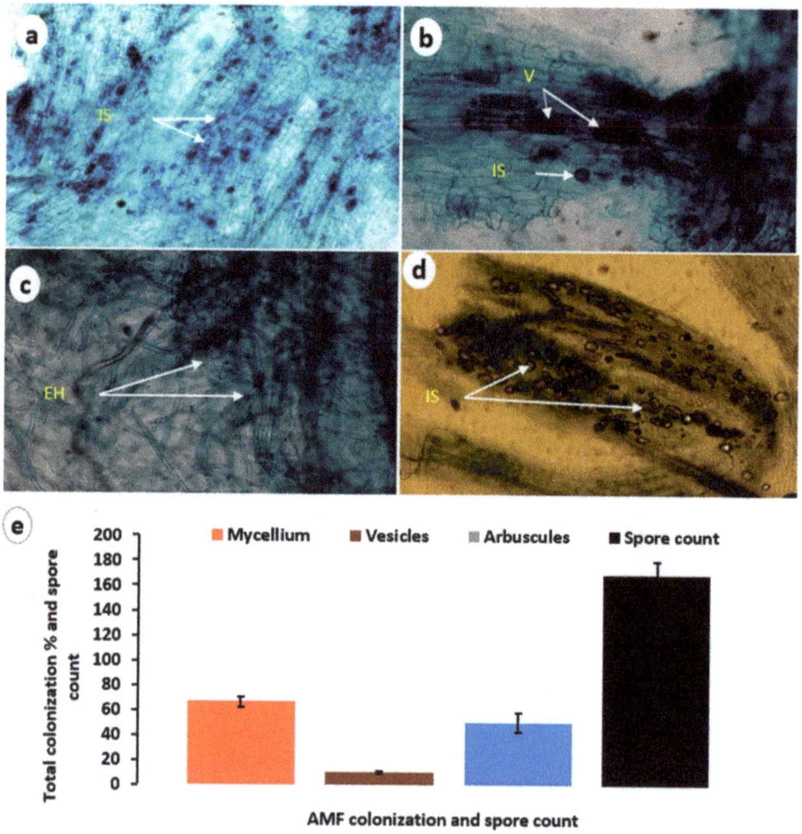

Figure 2. Photomicrographs indicating the root colonization structures and colonization status in the roots of *Philodendron bipinnatifidum* in response to arbuscular mycorrhizal fungi after 8 weeks acclimatization. ((**a–d**) 400× magnification). The presence of extraradical hyphea (EH) indicated initiation of AMF colonization, which later propagated and developed various structural forms such as intraradical intact spores (IS); Vesicles (V). (**e**) AMF root colonization (Mycelium, Vesicles, and Arbuscules) and spore count of *Philodendron* plants.

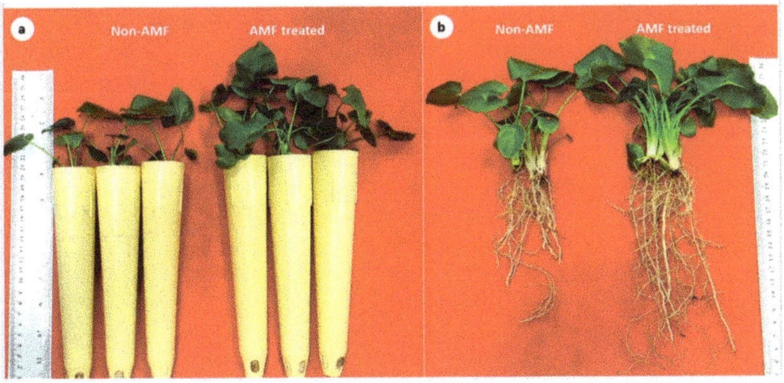

Figure 3. Photograph showing vegetative growth (**a**) and root growth (**b**) in non-AMF and AMF-treated *Philodendron bipinnatifidum* plants after 8 weeks of acclimatization.

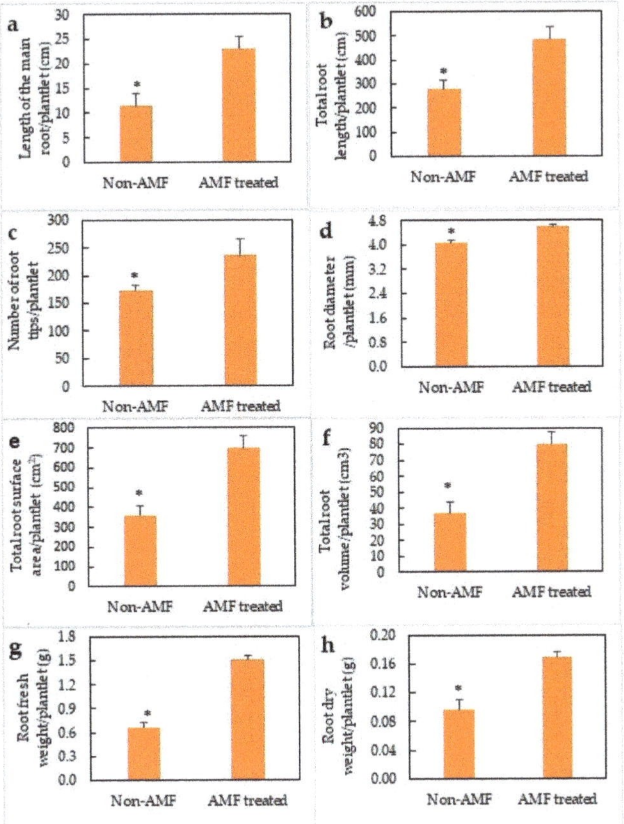

Figure 4. Root growth characteristics in non-AMF and AMF-treated *Philodendron bipinnatifidum* plants after 8 weeks of acclimatization. (**a**) Length of the main root per plantlet, (**b**) Total root length per plantlet, (**c**) Number of root tips per plantlet, (**d**) Root diameter per plantlet, (**e**) Total root surface area per plantlet, (**f**) Total root volume per plantlet, (**g**) Root fresh weight per plantlet and (**h**) Root dry weight per plantlet. * = Significant at $p \leq 0.05$ according to unpaired *t*-test.

3.3. Stomatal Frequency, Stomatal Conductance, Leaf Gas Exchange, and Transpiration Rate

In the current study, the plants which were treated with AMF showed higher leaf gas exchange, net-CO_2 assimilation, high transpiration rate, and stomatal conductance (Table 1) compared to non-AMF plants. The number of stomata per unit area was higher in the AMF-treated plants (136.68 mm^2), whereas the aperture length and aperture width of the stomatal apparatus were almost similar in both AMF and non-AMF treated plants (Table 1; Figure 5). The net CO_2 assimilation was 9.26 $\mu mol\ CO_2 \cdot m^{-2} \cdot s^{-1}$ in AMF plants, whereas CO_2 assimilation was 7.10 $\mu mol\ CO_2 \cdot m^{-2} \cdot s^{-1}$ in non-AMF plants. Stomatal conductance was 0.049 $mol\ H_2O \cdot m^{-2} \cdot s^{-1}$ and 0.023 $mol\ H_2O \cdot m^{-2} \cdot s^{-1}$, respectively, in AMF and non-AMF plants. The transpiration rate was 1.6324 $mol\ H_2O \cdot m^{-2} \cdot s^{-1}$ in AMF-treated plants and it was 0.8784 $mol\ H_2O \cdot m^{-2} \cdot s^{-1}$ in non-AMF-treated plants (Table 1).

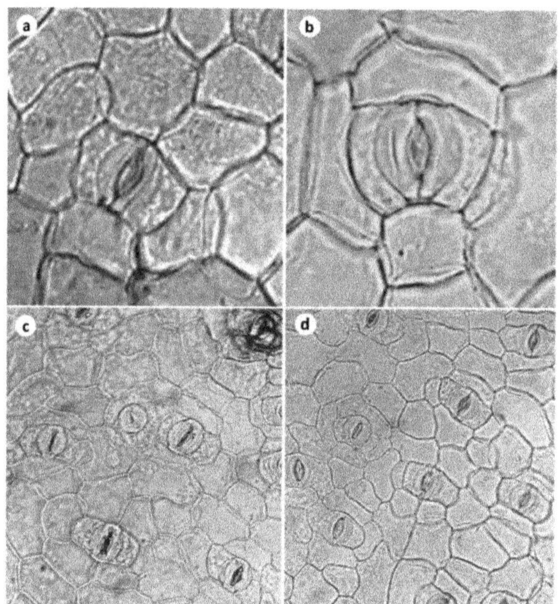

Figure 5. Stomatal density in non-AMF (**a**,**c**) and AMF-treated (**b**,**d**) *Philodendron* after 8 weeks of acclimatization (40× magnification).

3.4. Chlorophyll and Carotenoid Content

The leaf chlorophyll and carotenoid content were estimated with AMF-treated and non-treated plants, and results are presented in Table 1. The chlorophyll a (1.512 $mg \cdot g^{-1}$ FW), chlorophyll b (0.598 $mg \cdot g^{-1}$ FW), chlorophyll a+b (2.529 $mg \cdot g^{-1}$ FW), chlorophyll a/b ratio (2.529) and carotenoid (0.533 $mg \cdot g^{-1}$ FW) were higher in AMF-treated plants compared to non-treated plants (Table 1).

4. Discussion

The acclimatization phase is one of the pivotal steps during the micropropagation of plants and it is the transition phase wherein plants are entering into autotrophic conditions (ex vitro) from heterotopic conditions (in vitro). Plants that have adjusted/habituated to heterotrophic nutrition and physical conditions (such as light, temperature, and humidity) may fail to overcome the unfavorable related ex vitro cultivation. During this transition, pivotal physiological processes for the plant's survival must be enhanced. The acclimatization phase may affect the plant's photosynthetic efficiency, reduce its defense against pathogens and hinder the proper development of the root and sap conditions systems [39].

Many researchers have used AMF at this crucial stage of the plant's transition stage, as they provide several benefits to the plants with which they are associated. AMF provides protection against biotic and abiotic stresses as well as the improvement in plant growth and development [40]. AMF are obligatory biotrophs and the symbiosis basically involving of nutrients, in which the plant provides carbon through products of photosynthesis, and the AMF transfer nutrients from the soil to the plants [41]. In general, the AMF develops their hyphae rapidly through the soil for longer distances and adsorb nutrients efficiently to the plants. These features are useful to plants at the pivotal time of acclimatization. Due to these advantages, mycorrhization/AMF inoculation is widely recommended for the ex vitro establishment of horticultural plants [39,42]. In the present study we isolated two AMFs, namely *Gigospora albida* and *G. marginata*, and inoculated them to the plantlets in equal amounts of *P. bipinnatifidum* after transplantation to potting medium. After eight weeks of cultivation, the roots of treated plants were harvested and examined for colonization with the host plants. The microscopic observation of the mycorrhizal status of *P. bipinnatifidum* plantlets indicated the presence of all predicted AMF structures (mycelium, vesicles, arbuscules, and spores) in the roots (Figure 2). The analysis of the mycorrhizal colonization showed the colonization percentage as mycelium 66.66%, vesicles 11.11%, and arbuscules 51.11%. Similarly, *Etilingera elatior* microporpagated plants showed very good colonization of AMF fungi *Gigospora albida* and *Claroideogloums etunicatum* after the treatment with these fungi [10].

Varied reports suggest that colonization of micropropagated plants with AMF after ex vitro transplantation can enhance growth and plant performance. For example, AMF-inoculated plants have demonstrated an increase in plant height, shoot biomass, leaf number, and leaf area in *Gerbera* and *Heliconia* [43], *Gloriosa superba* [44], *Musa acuminata* [45], and *Paeonia* spp. [46]. The results of the present study also support this view and AMF plants had a significantly greater number of leaves (10.67 per plant), leaf area (75.63 cm^2), plant height (14.17 cm), shoot fresh weight (3.30 g), and shoot dry weight (0.314 g) than control plants, according to an analysis of growth characteristics. In comparison, non-AMF plants had lower values for these metrics (Table 1; Figure 3a). It has been demonstrated very well that reduction in plant morphological characteristics, i.e., reduced stem length, number of leaves, leaf area (length and width), and overall vegetative growth of plants causes a severe reduction in photosynthesis, leaf water potential, and sap movement [47,48]. About these features, the use of beneficial AMF during acclimatization will help the plants to involve in proper growth and achieve higher biomass (stem length, number of leas, and leaf area) so that plants in the process of acclimatization will involve proper physiological functions including photosynthesis.

Figures 3b and 4 provide data comparing the root growth traits of plants treated with AMF and without AMF. In comparison to non-AMF plants, the main root's length (23 cm), total root length per plantlet (485.73 cm), number of root tips per plantlet (236), average root diameter per plantlet (4.58 mm), total root surface area (697.76 cm^2), total root volume (79.98 cm^3), root fresh weight (1.51 g), and root dry weight (0.16 g) were all significantly higher in AMF plants. A similar increase in root growth and biomass accumulation was reported in AMF-treated plants of *Alpinia purpurata* [15], *Citrus limon* [8], and *Scutellaria integrifolia* [12].

Tissue culture plants upon transplantation showed very low chlorophyll content and reduced net-CO_2 assimilation, high transpiration rate, and stomatal conductance [7]. However, plants subjected to AMF treatments have restored these physiological functions several days after transplantation [2]. Furthermore, stomatal density and stomatal morphology are important characteristics that enable plants to involve net physiological functions such as CO_2 assimilation, stomatal conductance, and transpiration [49]. Moreover, higher stomatal density observed in AMF plants is known to have a direct positive influence on stomatal conductance and enhanced photosynthetic rate [50]. The plants treated with AMF in the current study had a higher stomatal density and showed increased stomatal conductance, net CO_2 assimilation, and leaf gas exchange, when compared to non-AMF plants (Table 1).

In contrast to non-AMF plants, AMF plants had higher leaf stomatal densities, with more stomata per unit area (136.68 mm^{-2}) in the AMF-treated plants (Table 1; Figure 5), while the aperture length and aperture width of the stomatal apparatus were nearly the same in AMF-treated plants and non-AMF-treated plants. The chlorophyll A (1.512 $mg \cdot g^{-1}$ FW) and carotenoid (0.533 $mg \cdot g^{-1}$ FW) were higher in AMF-treated plants compared to non-treated plants (Table 1). However, the quantity of chlorophyll b and chlorophyll a/b ratio were similar in AMF-treated and control plants (Table 1). An increment in chlorophyll was also reported in AMF-treated plants of *Capsicum annuum* [2], *Glycyrrhiza glabra* [13], and *Vitis vinifera* [47]. The higher stomatal frequency and stomatal conductance in AMF plants compared to non-AMF plants were attributed to the higher chlorophyll content, higher leaf area, and a greater number of leaves. Previous studies have also suggested a positive correlation between chlorophyll content and net photosynthetic rate [48–51]. Overall, the results of current experiments demonstrate the beneficial effects of AMF with the micropropagated plants enabling their physiological adjustments during acclimatization.

5. Conclusions

The current study offers new knowledge about the development and physiological response of control and AMF-treated *P. bipinnatifidum* plants grown in tissue culture. Plantlets of the *P. bipinnatifidum* benefited from the establishment, growth, and development of the arbuscular mycorrhizal fungi (*Gigaspora albida* and *G. marginata*). In terms of shoot and root growth, leaf gas exchange, chlorophyll content, and carotenoid content, the AMF-treated plants performed better. In addition, AMF-treated plants had more stomata, stomatal conductance, net CO_2 assimilation, and transpiration rate as compared with non-AMF plants. Therefore, we conclude that AMF's interaction with acclimatizing plants was the cause of superior growth, development, and physiological characteristics. Further, tissue culture plant mycorrhizal treatments are advantageous for the establishment during the acclimatization period. The use of additional AMF fungi and establishing consortium AMF fungi will be investigated further to aid in the acclimatization of tissue-cultured plants.

Author Contributions: Conceptualization, Y.H.D.; methodology, M.M.H., A.A.A. (AbdulAziz A. AlQarawi), T.S.A., A.A.A. (Ahmed Ali Alaizari) and J.A.M.; software, Y.H.D. and T.S.A.; validation, M.S.A. and H.N.M.; formal analysis Y.H.D., M.M.H., A.A.A. (AbdulAziz A. AlQarawi), T.S.A., J.A.M.; investigation, Y.H.D., M.M.H., A.A.A. (AbdulAziz A. AlQarawi), T.S.A., A.A.A. (Ahmed Ali Alaizari), J.A.M. and M.S.A.; resources, Y.H.D. and A.A.A. (AbdulAziz A. AlQarawi); data curation, Y.H.D., M.M.H., A.A.A. (AbdulAziz A. AlQarawi), T.S.A., J.A.M. and M.S.A.; writing—original draft preparation, Y.H.D., M.M.H. and H.N.M.; writing—review and editing, Y.H.D., M.M.H., A.A.A. (AbdulAziz A. AlQarawi), M.S.A. and H.N.M.; visualization, M.S.A. and H.N.M. All authors have read and agreed to the published version of the manuscript.

Funding: The authors acknowledge Researchers Supporting Project number (RSP2023R375), King Saud University, Riyadh, Saudi Arabia.

Data Availability Statement: All data are presented in the article.

Acknowledgments: The authors acknowledge the Researchers Supporting Project number (RSP2023R375), King Saud University, Riyadh, Saudi Arabia. Hosakatte Niranjana Murthy is thankful to the 'Brain Pool (BP)' program of the National Research Foundation of Korea (Grant No. 2022H1D3A2A02056665).

Conflicts of Interest: The authors declare no conflict of interest.

References

1. Diaz-Perez, J.C.; Shackel, K.A.; Sutter, E.G. Effect of in vitro formed roots and acclimatization on water status and gas exchange of tissue-cultured apple shoots. *J. Am. Soc. Hortic. Sci.* **1995**, *120*, 435–440. [CrossRef]
2. Estrada-Luna, A.; Davies, F.T., Jr. Arbuscular mycorrhizal fungi influence water relations, gas exchange, abscisic acid, and growth of micropropgated chile ancho pepper (*Capsicum annuum*) plantlets during acclimatization and post-acclimatization. *J. Plant Physiol.* **2003**, *160*, 1073–1083. [CrossRef] [PubMed]
3. Wetzstein, H.Y.; Sommer, H.E. Leaf anatomy of tissue cultured *Liquidambar styraciflua* (Hemamelidacae) during acclimatization. *Am. J. Bot.* **1982**, *69*, 1579–1586. [CrossRef]

4. Ziv, M.; Schwarz, A.; Fleminger, D. Malfunctioning stomata in vitreous leaves of carnation (*Dianthus caryophyllus*) plants propagated in vitro; implications for hardening. *Plant Sci.* **1987**, *52*, 127–134. [CrossRef]
5. Bhusal, N.; Kim, H.S.; Han, S.G.; Yoon, T.M. Photosynthetic traits and plant-water relations of apple cultivars grown as bi-leader trees under long-term waterlogging conditions. *Environ. Exp. Bot.* **2020**, *176*, 104111. [CrossRef]
6. Kozai, T. Acclimatization of micropropagated plants. In *Biotechnology in Agriculture and Forestry, High-Tech and Micropropagation I*; Bajaj, Y.P.S., Ed.; Springer: Berlin/Heidelberg, Germany, 1991; Volume 17, pp. 127–141.
7. Preece, J.E.; Sutter, E.G. Acclimatization of micropropagated plants to the greenhouse and field. In *Micropropagation Technology and Applications*; Debergh, P.C., Zimmerman, R.H., Eds.; Kluwer Academic Publishers: Dordrecht, The Netherlands, 1991; pp. 71–93.
8. Quatrini, P.; Gentile, M.; Carimi, F.; Pasquale, F.D.; Puglia, M. Effect of native arbuscular mycorrhizal fungi and *Glomus mosseae* on acclimatization and development of micropropgated *Citrus limon* (L.) Burm. *J. Hortic. Sci. Biotechnol.* **2003**, *78*, 39–45. [CrossRef]
9. Hajiboland, R.; Joudmand, A.; Aliasgharzad, N.; Tolra, R.; Poschenrieder, C. Arbuscular mycorrhizal fungi alleviate low-temperature stress and increase freezing resistance as a substitute for acclimation treatment in barley. *Crop Pasture Sci.* **2019**, *70*, 218–233. [CrossRef]
10. Yano-Melo, A.M.; Saggin, O.J., Jr.; Lima-Filho, J.M.; Melo, N.F.; Maia, L.C. Effect of arbuscular mycorrhizal fungi on the acclimatization of micropropagated banana plantlets. *Mycorrhiza* **1999**, *9*, 119–123. [CrossRef]
11. Da Silva, A.R.; de Melo, N.F.; Yano-Melo, A.M. Acclimatization of micropropagated plants *Etlingera elatior* (Jack) R.M. Sm. Inoculated with arbuscular mycorrhizal fungi. *S. Afr. J. Bot.* **2017**, *113*, 164–169. [CrossRef]
12. Joshee, N.; Mentreddy, S.R.; Yadav, A.K. Mycorrhizal fungi and growth and development of micropropagated *Scutellaria integrifolia* plants. *Ind. Crop. Prod.* **2007**, *25*, 169–177. [CrossRef]
13. Yadav, K.; Aggarwal, A.; Singh, N. Arbuscular mycorrhizal fungi induced acclimatization and growth enhancement of *Glycyrrhiza glabra* L.: A potential medicinal plant. *Agric. Res.* **2013**, *2*, 43–47. [CrossRef]
14. Ferrari, M.P.D.S.; Queiroz, M.D.S.; Andrade, M.M.D.; Alberton, O.; Goncalves, J.; Gazim, Z.C.; Magalhaes, H.M. Substrate-associated mycorrhizal fungi promote changes in terpene composition, antioxidant activity, and enzymes in *Curcuma longa* L. acclimatized plants. *Rhizosphere* **2020**, *13*, 100191. [CrossRef]
15. Campos, M.A.D.S.; Silva, F.S.B.; Yano-Melo, A.M.; Melo, N.F.D.; Maia, L.C. Application of arbuscular mycorrhizal fungi during the acclimatization of *Alpinia purpurata* to induce tolerance to *Meloidogyne arenaria. Plant Pathol. J.* **2017**, *33*, 329–336. [CrossRef]
16. Fortuna, P.; Citernesi, A.S.; Morini, S.; Viaglano, C.; Giovannetti, M. Influence of arbuscular mycorrhizae and phosphate fertilization on shoot apical growth of micropropagated apple and plum rootstocks. *Tree Physiol.* **1996**, *16*, 757–763. [CrossRef]
17. Guillemin, J.P.; Gianinazzi, S.; Gianinazzi-Pearson, V.; Marchal, J. Contribution of arbuscular mycorrhizas of biological protection of micropropagated pineapple (*Ananas comosus* (L.) Merr) against *Phytophthora cinnamomi* Rands. *Agric. Food Sci.* **1994**, *3*, 241–251. [CrossRef]
18. Varma, A.; Schuepp, H. Mycorrhization of the commercially important micropropagated plants. *Crit. Rev. Biotechnol.* **1995**, *15*, 313–328. [CrossRef]
19. Chen, F.C.; Wang, C.Y.; Fang, J.Y. Micropropagation of self-heading *Philodendron* via direct shoot regeneration. *Sci. Hortic.* **2012**, *141*, 23–29. [CrossRef]
20. Alawaadh, A.A.; Dewir, Y.H.; Alwihibi, M.S.; Aldubai, A.A.; El-Hendawy, S.; Naidoo, Y. Micropropagation of lacy tree philodendron (*Philodendron bipinnatifidum* Schott ex Endl.). *HortScience* **2020**, *55*, 294–299. [CrossRef]
21. Grout, B.W.W.; Aston, M.J. Transplanting of cauliflower plants regenerated from meristem culture. II. Carbon dioxide fixation and development of photosynthetic ability. *Hortic. Res.* **1978**, *17*, 65–71.
22. Grout, B.W.W.; Milan, S. Photosynthetic development of micropropagated strawberry plantlets following transplanting. *Ann. Bot.* **1985**, *55*, 129–131. [CrossRef]
23. Dewir, Y.H.; Ibrahim, A.; Murthy, H.N.; Aldubai, A.A.; Al-Ali, A.M.; Al-Aizari, A.A.; Migdadi, H.; Alsadon, A.; Al-Suhaibani, N.A. Integration of salicylic acid in in vitro rooting medium improves photosynthesis and growth during acclimatization of *Philodendron bipinnatifidum* Schott ex endl. Plantlets. *Propag. Ornam. Plants* **2020**, *20*, 103–108.
24. John, T.S. A list of naturally infected Brazilian tropical plant species with vesicular-arbuscular mycorrhiza. *Acta Amaz.* **1980**, *10*, 229–234. [CrossRef]
25. Brundrett, M.C. Mycorrhizal associations and other means of nutrition of vascular plants: Understanding the global diversity of host plants by resolving conflicting information and developing reliable means of diagnosis. *Plant Soil* **2009**, *320*, 37–77. [CrossRef]
26. Murashige, T.; Skoog, F. A revised medium for rapid growth and bioassays with tobacco tissue cultures. *Physiol. Plant.* **1962**, *15*, 473–497. [CrossRef]
27. Gerdemann, J.; Nicolson, T.H. Spores of mycorrhizal *Endogone* species extracted from soil by wet sieving and decanting. *Trans. Br. Mycol. Soc.* **1963**, *46*, 235–244. [CrossRef]
28. Schenck, N.C.; Perez, Y. *Manual for Identification of VA Mycorrhizal Fungi*; Synergistic Publication: Gainesville, FL, USA, 1990.
29. Walker, C. *Spore Extraction by Centrifugation-Sugar Flotation*; Biological Research and Imaging Laboratory: Hampshire, UK, 1997.
30. International Culture of Vesicular-Arbuscular Mycorrhizal Fungi (INAVAM). Available online: http://invam.caf.wvu.edu (accessed on 1 July 2022).
31. Schüssler, A.; Walker, C. The Glomeromycota: A Species List with New Families and New Gener. In *The Glomeromycota*; CreateSpace Independent Publishing Platform: Scotts Valley, CA, USA, 2010.

32. Redecker, D.; Schüßler, A.; Stockinger, H.; Stürmer, S.L.; Morton, J.B.; Walker, C. An evidence-based consensus for the classification of arbuscular mycorrhizal fungi (Glomeromycota). *Mycorrhiza* **2013**, *23*, 515–531. [CrossRef]
33. Phillips, J.M.; Hayman, D. Improved procedures for clearing roots and staining parasitic and vesicular-arbuscular mycorrhizal fungi for rapid assessment of infection. *Trans. Br. Mycol. Soc.* **1970**, *55*, 158–161. [CrossRef]
34. Trouvelot, A.; Kough, J.; Gianinazzi-Pearson, V. Evaluation of VA infection levels in root systems. Research for estimation methods having a functional significance. *Physiol. Genet. Asp. Mycorrhizae* **1986**, *1*, 217–221.
35. Al-Qarawi, A.; Mridha, M.; Alghamdi, O. Diversity of structural colonization and spore population of arbuscular mycorrhizal fungi in some plants from Riyadh, Saudi Arabia. *J. Pure Appl. Microbiol.* **2012**, *6*, 1119–1125.
36. Dewir, Y.H.; Alsadon, A.; Ibrahim, A.; El-Mahrouk, M. Effects of growing substrate, mode of nutrient supply, and Saffron corm size on flowering, growth, photosynthetic competence, and cormlet formation in hydroponics. *HortTechnology* **2022**, *32*, 234–240. [CrossRef]
37. Lichtenethaler, H.K. Biosynthesis, accumulation and emission of carotenoids, α-tocopherol, plastoquinone and isoprene in leaves under high photosynthetic irradiance. *Photosynth. Res.* **2007**, *92*, 163–179. [CrossRef]
38. Cotton, R. *Cytotaxonomy of the Genus Vulpia*; The University of Manchester: Manchester, UK, 1974.
39. Kapoor, R.; Sharma, D.; Bhatnagar, A.K. Arbuscular mycorrhizae in micropropagation systems and their potential applications. *Sci. Hortic.* **2008**, *116*, 227–239. [CrossRef]
40. Smith, S.E.; Read, D.J. *Mycorrhizal Symbiosis. Mycorrhizal Symbiosis*, 3rd ed.; Academic Press: New York, NY, USA, 2008.
41. Smith, F.A.; Smith, S.E. What is the significance of the arbuscular mycorrhizal colonization of many economically important crop plants? *Plant Soil* **2011**, *348*, 63–79. [CrossRef]
42. Azcon-Aguilar, C.; Barea, J.M. Applying mycorrhiza biotechnology to horticulture: Significance and potentials. *Sci. Hortic.* **1997**, *68*, 1–24. [CrossRef]
43. Sato, A.Y.; Nannetti, D.D.C.; Pinto, J.E.B.P.; Siqueira, J.O.; Blank, M.D.F.A. Application of arbuscular mycorrhiza to micropropagated Heliconia and Gerbera plants during acclimatization period. *Hortic. Bras.* **1999**, *17*, 25–28. [CrossRef]
44. Yadav, K.; Aggarwal, A.; Singh, N. Arbuscular mycorrhizal fungi (AMF) induced acclimatization, growth enhancement and colchicine content of micropropagated *Gloriosa superba* L. plantlets. *Ind. Crop. Prod.* **2013**, *45*, 88–93. [CrossRef]
45. Koffi, M.C.; Declerck, S. In vitro mycorrhization of banana (*Musa acuminata*) plantlets improves their growth during acclimatization. *Vitr. Cell. Dev. Biol.-Plant* **2015**, *51*, 265–273. [CrossRef]
46. Wen, S.S.; Cheng, F.Y.; Zhong, Y.; Wang, X.; Li, L.Z.; Zhang, Y.X.; Qiu, J.M. Efficient protocols for the micropropagation of tree peony (*Paeonia suffruticosa* "Jin Pao Hong", *P. suffruticosa* "Wu Long Peng Sheng", and *P.* × lemoninei "High Noon') and application of arbuscular mycorrhizal fungi to improve plantlet establishment. *Sci. Hortic.* **2016**, *201*, 10–17. [CrossRef]
47. Bavaresco, L.; Fogher, C. Lime-induced chlorosis of grape-vine as affected by rootstock and root infection with arbuscular mycorrhiza and *Pseudomonas fluorescence*. *Vitis* **1996**, *35*, 119–123.
48. Bhusal, N.; Bhusal, S.J.; Yoon, T.M. Comparisons of physiological and anatomical characteristics between two cultivars in bi-leader apple trees (*Malus* × *domestica* Borkh.). *Sci. Hortic.* **2018**, *231*, 73–81. [CrossRef]
49. Salomon, Y.; Dietrich, L.; Sevanto, S.; Holtta, T.; Dannoura, M.; Epron, D. Drought impacts on tree phloem: From cell-level responses to ecological significance. *Tree Physiol.* **2019**, *39*, 173–191. [CrossRef] [PubMed]
50. Pardiso, R.; Arena, C.; De Micco, V.; Giordano, M.; Aronne, G.; De Pascale, S. Changes in leaf anatomical traits enhanced photosynthetic activity of soybean grown in hydroponics with plant growth promoting microorganisms. *Front. Plant Sci.* **2017**, *8*, 674. [CrossRef] [PubMed]
51. Massonnet, C.; Costes, E.; Rambal, S.; Dreyer, E.; Regnard, J.L. Stomatal regulation of photosynthesis in apple leaves: Evidence for different water-use strategies between two cultivars. *Ann. Bot.* **2007**, *100*, 1347–1356. [CrossRef] [PubMed]

horticulturae

MDPI

Article

Effects of Different Media and Their Strengths in In Vitro Culture of Three Different *Cistus creticus* L. Populations and Their Genetic Assessment Using Simple Sequence Repeat Molecular Markers

Kostas Ioannidis [1,*] and Polyxeni Koropouli [2]

[1] Laboratory of Forest Genetics and Biotechnology, Institute of Mediterranean and Forest Ecosystems, Hellenic Agricultural Organization "Demeter", Ilissia, 11528 Athens, Greece
[2] Ministry of Education, 15180 Athens, Greece; koropouli.x@gmail.com
* Correspondence: ioko@fria.gr; Tel.: +30-210-7783-750

Abstract: *Cistus creticus* L. (rockrose), a species of ecological and medicinal significance, constitutes a valuable component of the Mediterranean ecosystem. The present study investigated the effect of the inorganic salt concentration of Murashige and Skoog medium (MS), woody plant medium (WPM), and Driver and Kuniyaki Walnut medium (DKW) at several strengths (1/8×, 1/4×, 1/2×, 1×, and 2×) on the in vitro growth and organogenesis of rockrose. Significant interactions were observed throughout the experiments between pairs of plant origins, medium types, and strengths, and we also examined the extent to which they affected the studied traits was examined. The types of nutrient medium affected all studied traits except shoot and root percentages. The maximum growth percentage (143.49%) was gained using full-strength WPM. The best performance in shoot percentage was obtained using MS (100%) at several strengths along with 1× WPM (100%). The topmost rooting percentage values (98.61%) were obtained using 1× WPM and 1/2× DKW. The highest number of shoots and roots were observed using full-strength MS (9.39) and half-strength WPM (6.49), respectively. The maximum values for shoot and root length were achieved using 1/2× MS (0.78 cm) and 1/8× WPM (1.55 cm), respectively. The origin of the plant material did not influence any studied trait. Moreover, the genetic relations among the populations used in the in vitro culture were assessed using simple sequence repeats (SSR) markers. Twenty-eight alleles were identified across all five STR loci. The different and effective alleles per locus were 5.60 and 4.72, respectively. The average observed and expected heterozygosity was estimated at 0.52 and 0.72, respectively. Shannon's information index and the inbreeding coefficient (F) were assessed at 1.48 and 0.30, respectively, revealing a narrow genetic base and high genetic similarity among origins, suggesting that they belong to the same population.

Keywords: rockrose; MS; WPM; DKW; in vitro micropropagation; shoot induction; root induction; microsatellites; population genetics

Citation: Ioannidis, K.; Koropouli, P. Effects of Different Media and Their Strengths in In Vitro Culture of Three Different *Cistus creticus* L. Populations and Their Genetic Assessment Using Simple Sequence Repeat Molecular Markers. *Horticulturae* **2024**, *10*, 104. https://doi.org/10.3390/horticulturae10010104

Academic Editors: Konstantinos Bertsouklis, Epaminondas Kartsonas and Angela Carra

Received: 22 December 2023
Revised: 17 January 2024
Accepted: 18 January 2024
Published: 22 January 2024

1. Introduction

Cistus creticus L. (rockrose) and other species of Cistaceae, a medium-sized taxon of shrubs and commonly herbs, are prominent characteristics of the Mediterranean flora, covering broad dry and sun-exposed areas [1]. Evolutionarily, they developed survival adaptations to overcome the disturbances of the frequent fires that Mediterranean ecosystems experience [2]. The *Cistus* species' ecological value has been acknowledged since they are considered typical pyrophytes that spread through seeds and creat pure stands after fire [3,4]. Cistaceae distribution is extensive; i.e., several species are distributed over most of Europe [5], North Africa, and even North and South America [6], although they are present mainly in the Mediterranean region, with some species distributed almost exclusively in the Mediterranean rim [7,8].

Cistus species are major components of Greek phryganic ecosystems [9]. In Greece, the species have been known since antiquity [10,11] because of their resin properties, called ladanon [12], which are used in Mediterranean folk medicine for the healing of several diseases [10]. The resin is a source of pharmaceutical and aromatic properties, is secreted from the glandular hairs of leaves and stems, and contains labdane diterpenes, which have antimicrobial [13], anti-inflammatory [14], and cytotoxic activity [15–17].

The economic interest of several *Cistus* spp. is due to their use as ornamental and melliferous flora [18]. Also, ladanon resin is traditionally used in medicine and perfumery [19–22]. A quite impressive use is to inoculate them with the mycorrhizae of *Tuber nigrum* Bull (a black truffle) and then plant them in the primary stage of truffle forest repopulation [23].

In a previous study [1], an efficient in vitro propagation protocol was developed for the large-scale production of *C. creticus* L. Prior to our propagation protocol, there were many successful efforts in in vitro propagation for other *Cistus* species [18,24] and only one in rockrose [25]. To complement our initial research as well as that of other authors [20,26,27], in the present work, we investigated the effects of three different agar-solidified nutrient mediums, i.e., MS, WPM, and DKW, at several strengths, i.e., 1/8×, 1/4×, 1/2×, 1×, and 2×, on in vitro growth and organogenesis. The studied traits concerned the shoot proliferation and root induction of three populations of *C. creticus* plantlets for the large-scale clonal propagation of selected rockrose clones. Moreover, the current research aimed to assess the genetic population structure of the three populations of *C. creticus* used in an in vitro propagation study. To the best of our knowledge, this research is the first attempt at a population genetic analysis of *C. creticus* L. based on SSR markers. Our genetic analysis of the three populations used in the in vitro culture experiments was based on prior studies performed by Astuti et al. [28] for *Cistus laurifolius* L. and by Bertolasi et al. [29] for *Cistus albidus* L.

2. Materials and Methods

2.1. In Vitro Culture

2.1.1. Plant Material

Three large, healthy mature *C. creticus* L. individuals with attractive purple flowers and green leaves, growing in three natural environments of Attica (Greece), i.e., Mt. Parnitha, Mt. Pateras, and Mt Pendeli, were selected as explant sources (Supplementary Materials, Figure S1). The lateral shoots of mother plants (explant donors) were collected in April. These were derived from actively growing stems during the most recent vegetative growth season, i.e., they were 1–2 months old. They were stored in a moist cotton cloth at 4 °C until subsequent handling. The following day, explants that were 1.0–1.5 cm long, i.e., nodal segments and apical shoot tips, were excised from the explant donors.

2.1.2. Explant Surface Sterilization

Explants were effectively disinfected through consecutive immersions in two distinct aqueous solutions. Initially, the explants were immersed in a solution of 70% (v/v) ethanol with continuous stirring for 1 min. Subsequently, they were treated with a second aqueous solution of sodium hypochlorite (10% w/v NaOCl, Merck KGaA, Darmstadt, Germany) at a concentration of 1.5% (v/v), supplemented with 0.05% (v/v) Tween 20 (Fisher Bioreagents, Pittsburgh, PA, USA), under continuous stirring for 15 min. After each immersion, a sequence of three rinses with sterile deionized water, each lasting three minutes, was conducted.

2.1.3. Initial Culture Establishment

Single node explants were placed in culture vessels containing basal medium. Three media were employed to initiate the in vitro culture: the Murashige and Skoog (MS) [30] (Duchefa Biochemie, Haarlem, The Netherlands), the Lloyd and McCown wood plant medium (WPM) [31] (Duchefa Biochemie, Haarlem, The Netherlands), and the Driver and Kuniyaki Walnut (DKW) medium [32] (Duchefa Biochemie, Haarlem, The Netherlands).

Each medium contained 3% (*w*/*v*) sucrose (Duchefa Biochemie, Haarlem, The Netherlands) and was solidified with 2.4 g/L agar (Phytagel, Sigma, Burlington, MA, USA). Before the addition of agar, each medium was adjusted to pH 5.8. All media were autoclaved at 121 °C and 122 kPa for 20 min. The cultures were grown in a growth chamber at 22 ± 1 °C under a 16 h light/8 h dark photoperiod. The photosynthetic photon flux density at culture level was maintained at 100 µmol m^{-2} s^{-1}, which was supplied by cool-white fluorescent lamps.

2.1.4. Initial Culture Shoot Multiplication

Contaminated-free explants derived from those three *C. creticus* clones were subcultured in each WPM, MS and DKW nutrient media. The media had no growth regulators, and multiple shoot induction was performed to achieve an adequate number of shoots for subsequent experiments, i.e., shoot regeneration and root induction. The cultures were maintained under the same conditions as previously described.

2.1.5. Cultures for the Evaluation of Growth, Shooting and Rooting Performance

Nodal segments, from the previous initial shoot multiplication cultures, were placed onto five different strengths (1/8×, 1/4×, 1/2×, 1× and 2×) in three different nutrient media (WPM, MS and DKW), containing no growth regulators. To avoid the effect of different salt concentration deriving from different media, explants were subcultured in the same medium. The media were prepared as described above. The cultures were maintained for 4 weeks in the same conditions as mentioned. The effect of the three solid nutrient media of different strengths had on growth percentage (%, (initial height-terminal height)/initial height), shoot and root formation percentage (%), number and length of shoots, and roots was assessed after a 4-week period of culture. The height was determined as the distance from the cut to the top of the explant. Three replications of eight explants per replication were used for each medium and its strength. Each experiment was arranged in a growth chamber in a completely randomized design.

2.1.6. Statistical Analysis

Analysis was based on the individual values of average percentage of growth, shoot and root formation, the average number of shoots and roots per explant, and the average length of shoots and roots per explant per treatment. The following linear model was used in the analysis, at a significance level of $a = 0.05$, to determine the influence of the clone, the nutrient medium, the strength of the nutrient medium and the interactions among clone–nutrient medium–medium strength, and finally clone–nutrient medium–medium strength:

$$y_{ijkl} = \mu + c_i + m_j + s_k + c_i{}^*m_j + c_i{}^*s_k + c_i{}^*m_j{}^*s_k + e_{ijkl} \quad (1)$$

where y_{ijkl} is the phenotypic measurement for a trait of the l_{th} explant, the jth nutrient medium, the kth strength of jth nutrient medium and the ith explant clone, as dependent variables; μ is the fixed population mean of all explants; c_i is the fixed effect of the ith clone; m_j is the random effect of the jth nutrient medium, s_k is the random effect of the kth strength of the nutrient medium; $c_i{}^*m_j$ is the interaction of the ith clone with the jth nutrient medium; $c_i{}^*s_k$ is the interaction of the ith clone with the kth strength of jth nutrient medium; $c_i{}^*m_j{}^*s_k$ is the interaction of the ith clone with the jth nutrient medium being at the kth strength; and e_{ijkl} is the random residual error of the lth explant, the kth strength, the jth nutrient medium and the ith clone. The restricted maximum likelihood (REML) method was performed to assess the variance components. Analysis of variance (ANOVA) and Duncan's multiple range test (MRT) at $a = 0.05$ were performed on the mean shoot and root number, the mean shoot and root length and the shooting and rooting percentage per treatment. The data, initially expressed as percentages, underwent appropriate log or arcsine transformation for statistical analysis. Subsequently, the transformed data were reverted to percentages presented in the relevant tables and graphs. All statistical analyses were performed using SPSS v.20 software for Windows (IBM SPSS Statistics 2011, IBM Corp., Armonk, NY, USA).

2.2. Genetic Population Analysis

2.2.1. Plant Material—DNA Isolation and Quantification

Twelve plants from the three regions (populations) were used in the genetic population analysis. Genomic DNA extraction was performed on 100 mg fresh leaf samples, using the Higher Purity Plant DNA Purification Kit (Canvax, Córdoba, Spain) following the manufacturer's instructions. Before storing at −20 °C, the purified total DNA of the samples underwent quantification, and its quality was documented using the UV-Vis spectrophotometer Q5000 UV-Vis (Quawell, San Jose, CA, USA).

2.2.2. Microsatellite Loci

Five microsatellite loci developed by Astuti et al. [28] for *Cistus laurifolius* L. and had been successfully tested for their ability to amplify *C. albidus* DNA samples according to Bertolasi et al. [29] were used to analyze the genetic population structure of the three *C. creticus* L. populations. The sequences and traits of the SSR markers are presented in Table 1.

Table 1. Documentation on the 5 microsatellite markers used in the genetic analysis of *C. creticus* L.

Locus	Repeat Motif	Primer Sequences 5′→3′		Expected Allele Size Range (bp *)
		Forward	Reverse	
cislau1	(TC)5	TCGATCGGGTGAAAACAAAT	TTCCTTCCAGAGGCTTCTCA	227–255
cislau7	(AT)7	TCAAAAGCTTCTTCCCCTCT	GATGTATGATGAAGGGCAGG	158–166
cislau11	(TC)5	GCGAGATCCCGAAACACT	AAAAACCCTAGAAGTCCTCGA	163–167
cislau12	(AT)7	TAATTGTCGCTTTGCTGTGC	TCATGCACAAGTTGAATCAAGA	202–232
cislau14	(AAAG)4	GACAACTCACACGACTCTAAACG	AAATTGGGCATGGACCAAG	180–184

* bp = base pair.

2.2.3. PCR Reaction Mix and Amplification

The PCR reaction was performed in a total volume of 20 μL comprising 30 ng genomic DNA. The final concentrations of reaction mix components were 1×PCR buffer (10×) (Nippon Genetics, Tokyo, Japan), 2 mM of $MgCl_2$ (50 mM) (Nippon Genetics, Tokyo, Japan), including the amount of $MgCl_2$ contained in PCR buffer, 1 U Fast Gene Taq DNA polymerase (5 U μL^{-1}, Nippon Genetics, Tokyo, Japan), 100 μM each of dNTPs (10 mM) (Nippon Genetics, Tokyo, Japan) and 0.8 μM of each forward and reverse primer (Eurofins Genomics, Ebersberg, Germany).

The amplifications reactions were performed twice in the 96-well thermal cycler Bio Rad C1000 Touch (Bio-Rad, Hercules, CA, USA) as follows: 95 °C for 3 min, 30 cycles at 95 °C for 45 s, 60 °C for 90 s and 72 °C for 1 min and a final extension of 8 min at 72 °C. The ultimate holding temperature was set at 4 °C. Each reaction comprised both a negative and a positive control.

2.2.4. Capillary Electrophoresis, Genotyping and Statistical Data

DNA fragment capillary electrophoresis was performed in a 3730 Genetic Analyzer (Applied Biosystems, Thermo Fisher Scientific Co., Waltham, MA, USA) using LIZ500 (Applied Biosystems, Thermo Fisher Scientific Co., Waltham, MA, USA) as the molecular weight standard. Geneious Prime v. 2022.1.1 software (Dotmatics, Boston, MA, USA) was employed for genotyping. The analytical threshold was set at 150 relative fluorescence units (RFUs).

Genetic diversity, i.e., the number of observed alleles (Na), effective number of alleles (Ne), Shannon's information index (I), percentage of polymorphic loci (PPL), Nei genetic distance (Nei D) and Nei Genetic Identity (Nei I), was estimated using the GenAlEx package [33]. The PIC (Polymorphism Information Content) of each SSR marker was calculated using Cervus 3.0.7 [34,35]. Hardy–Weinberg tests were not conducted due to small within-population sample sizes.

3. Results

3.1. *In Vitro Culture*

Table 2 presents the source of variation retrieved from the ANOVA table of the in vitro culture of rockrose explants for the studied traits, according to GLM: $y_{ijkl} = \mu + c_i + m_j + s_k + c_i{}^*m_j + c_i{}^*s_k + c_i{}^*m_j{}^*s_k + e_{ijkl}$.

Table 2. Source of variation retrieved from ANOVA table of in vitro culture of *C. creticus* explants for the studied traits, according to GLM: $y_{ijkl} = \mu + c_i + m_j + s_k + c_i{}^*m_j + c_i{}^*s_k + c_i{}^*m_j{}^*s_k + e_{ijkl}$.

Source of Variation	df	Growth Percentage (%)	Shoot Percentage (%)	Root Percentage (%)	Shoot Number	Shoot Length (cm)	Root Number	Root Length (cm)
Clone	2	ns	ns	ns	ns	ns	ns	ns
Medium	2	***	ns	ns	***	**	***	*
Strength	4	***	*	***	***	***	***	***
Clone * Medium	4	**	ns	ns	ns	ns	*	ns
Clone * Strength	8	*	ns	*	ns	ns	ns	ns
Clone * Medium * Strength	24	***	*	***	***	***	***	***

*** $p \leq 0.001$; ** $p \leq 0.01$, * $p \leq 0.05$, ns = non-significant.

Shoot formation was achieved after two weeks of culture and root initiation of in vitro-propagated explants commenced during the third week, both depending on the applied treatment (Figure 1). Blastogenesis was not affected by the origin of *C. creticus* explants (Table 2) but from the nutrient medium and its strength. The origin of plant material did not influence any studied traits. The medium type had a significant effect on the growth percentage ($p \leq 0.001$), number ($p \leq 0.001$) and length ($p \leq 0.01$) of shoots per explant. Conversely, the medium type had no effect on shooting and rooting percentage. Moreover, statistically significant differences in the mean number ($p \leq 0.001$) and length ($p \leq 0.05$) of roots per explant among different nutrient mediums were observed. The strength of the medium had an impact on all the studied traits. The medium strength had a greater effect on the growth and rooting percentage ($p \leq 0.001$) compared to the shooting percentage ($p \leq 0.05$) as well as on the number and length of shoots and roots ($p \leq 0.001$) (Figure 2, Table 2).

(a) (b)

Figure 1. Media and strength selection have affected the number and length of *C. creticus* L. explants' shoots after 4 weeks of culture: (**a**) explants in full strength MS; (**b**) explants in half-strength DKW. Bar = 25 mm.

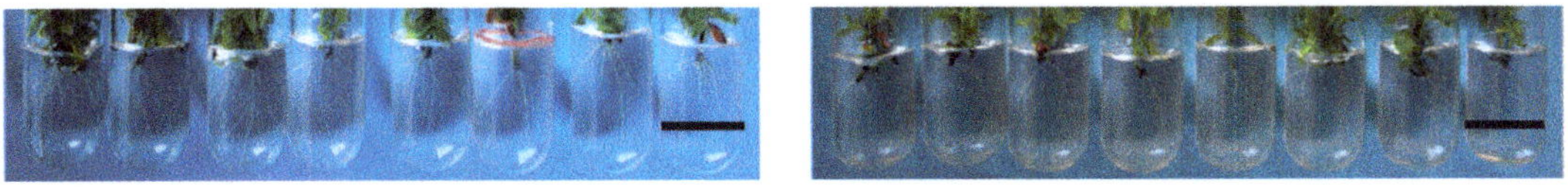

(a) (b)

Figure 2. Media and strength selection have affected the number and length of *C. creticus* L. explant roots after 4 weeks of culture: (**a**) explants in half-strength MS; (**b**) explants in half-strength WPM. Bar = 25 mm.

Significant interactions were observed among clone–nutrient medium, clone–medium strength, and nutrient medium–medium strength as well as among clone–nutrient medium–medium strength in relation to studied traits (Table 2). Significant interactions between explant origins and medium types were noticed only for explant growth ($p \leq 0.01$) and root number ($p \leq 0.05$). The interaction between explant origin and medium strength was significant only for growth ($p \leq 0.05$) and rooting ($p \leq 0.05$) percentage. The interaction among plant origin, medium type and its strength had a greater effect on growth ($p \leq 0.001$) and rooting percentage ($p \leq 0.001$) compared to shooting percentage ($p \leq 0.05$), as well as on the number and length of shoots and roots ($p \leq 0.001$).

The influence of various media on the overall mean growth, blastogenesis and root induction percentage (%), mean number of shoots and roots per explant as well as on the mean length of shoots and roots of *C. creticus* L. explants are presented in Figures 3 and 4 and in the Supplementary Material Table S1.

The applied treatment, i.e., the type of nutrient medium, significantly affected the average growth percentage of explant, blastogenesis and root induction (Figure 3, Supplementary Material Table S1). Explants cultured in DKW exhibited superior growth (103.32%) compared to those cultured in WPM and MS. Conversely, MS showed the lowest value (67.52%), which exhibited a statistically significant difference compared to DKW and WPM (93.30%). No differences in shoot percentage were observed among the treatments using full-strength mediums, i.e., MS (97.78%), DKW (96.67%) and WPM (95.28%). Likewise, no differences in rooting percentage were observed among the treatments using full-strength mediums, i.e., WPM (89.72%), DKW (87.78%) and MS (84.72%), too.

The MS medium presented the highest mean number of shoots per explant (7.00), which was statistically significant different ($p \leq 0.001$) compared to WPM (5.56) and DKW (5.66). The last two treatments were not statistically different from each other. A similar trend was presented concerning the average length of shoots per explant, in which the best results were obtained using MS (0.62 cm), being statistically significant different ($p \leq 0.01$) compared to WPM (0.55 cm) and DKW (0.58 cm). The last two treatments presented non-significant differences.

The DKW and WPM did not exhibit statistical differences in the number of roots per explant with the first medium showing the highest value (4.87) and the second showing a similar value (4.71). The number of roots using MS was the lowest (3.64) and statistically different compared to the previous treatments ($p \leq 0.001$). Statistically significant differences ($p \leq 0.05$) were observed between WPM and MS in the mean length of roots per explant with WPM presenting the best performance (1.07 cm) followed by the DKW (0.99 cm) and MS (0.96 cm). The last two treatments showed non-significant differences.

Additionally, the effect of different media on the examined parameters in relation to the rockrose origins used in in vitro culture are presented in Supplementary Material Figures S2–S4.

The impact of the combination of different nutrient media and their strengths on the overall mean growth, blastogenesis and root induction percentage (%), mean number of shoots and roots per explant as well as on the mean length of shoots and roots of *C. creticus* L. explants are presented in Figures 5–7 and in Supplementary Material Tables S2–S4.

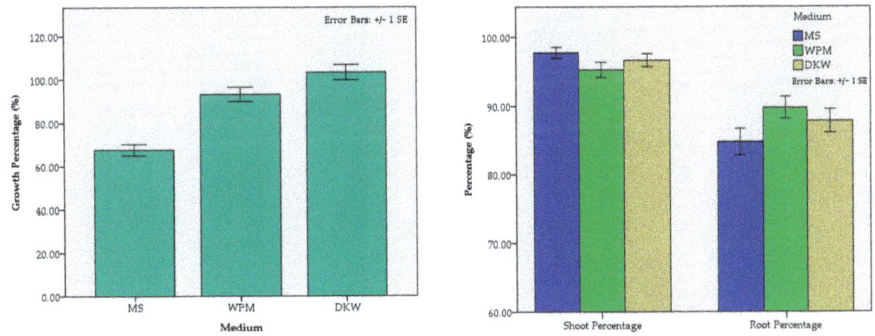

(a) (b)

Figure 3. Effect of the medium on the mean growth percentage of shoots (%) (**a**) and on the mean shoot and root percentage (%) (**b**) of *C. creticus* explants.

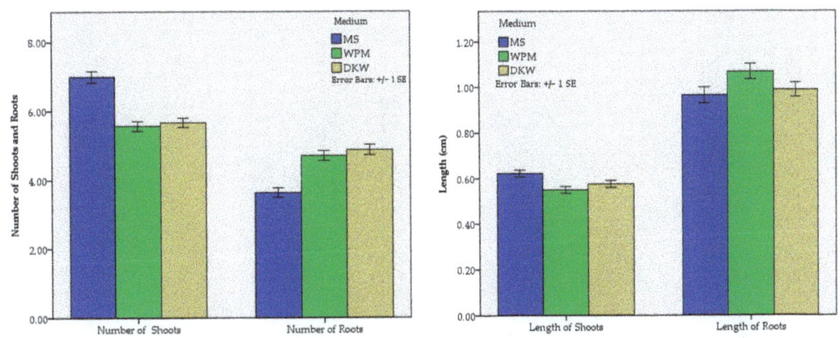

(a) (b)

Figure 4. Effect of the medium on the mean shoot and root number (**a**) and on the mean shoot and root length (**b**) per *C. creticus* explants.

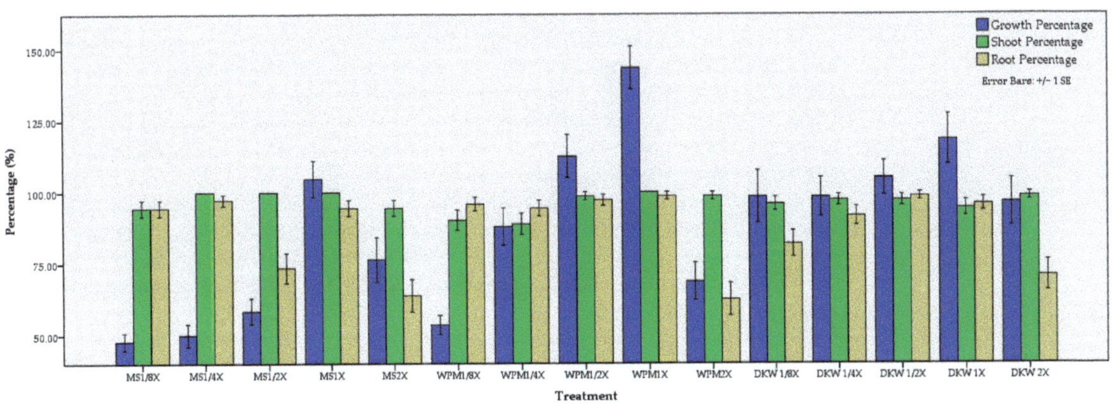

Figure 5. Effect of medium and its strength on the average growth, shoot and root percentage (%) of *C. creticus* explants.

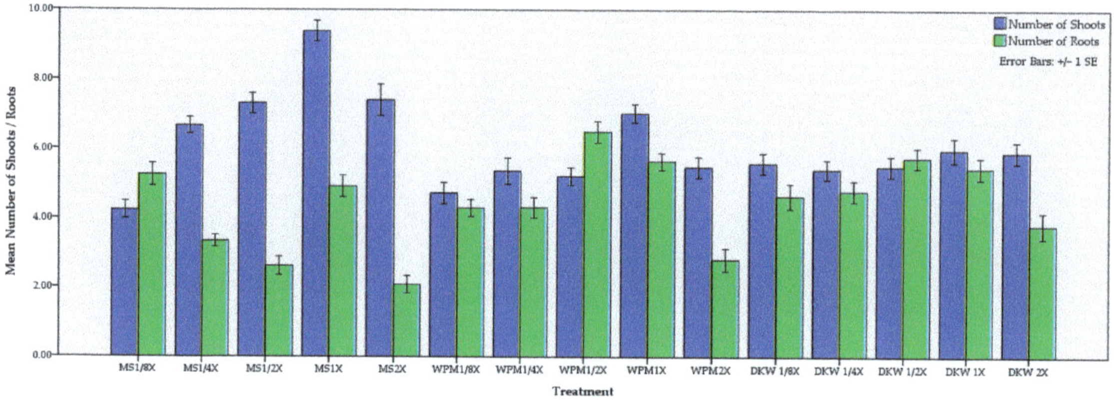

Figure 6. Effect of medium and its strength on the average shoot and root number of *C. creticus* explants.

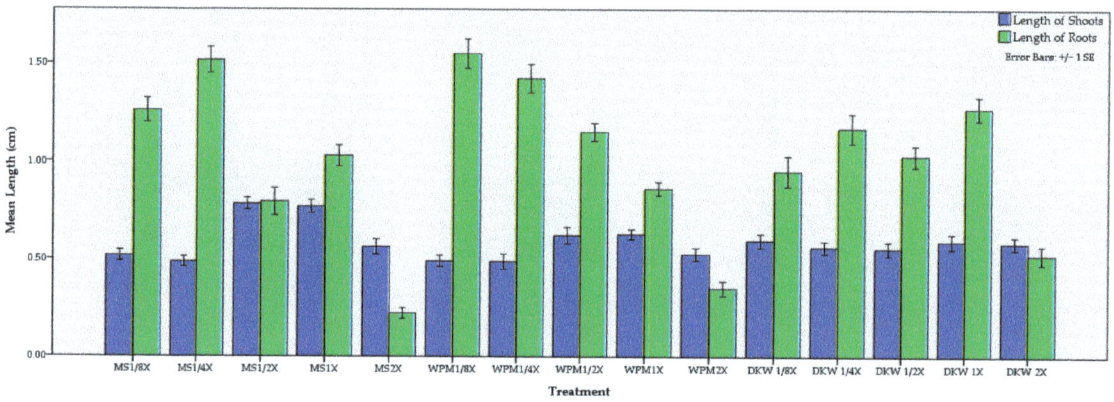

Figure 7. Effect of medium and its strength on the average shoot and root length of *C. creticus* explants.

Regarding the growth percentage, the maximum (143.49%) was obtained in the full-strength WPM medium followed by the full-strength DKW (118.45%) and half-strength WPM (112.66%) (Figure 5). Full-strength MS also showed good results (104.75%). In general, when the media were diluted, the growth decreased. This was obvious for MS and WPM media but not for DKW. The best performance in shooting percentage was obtained in MS (100%) at several strengths (1/4×, 1/2× and 1×) as well as in full-strength WPM (100%). Overall, the shooting percentage was adequate in all treatments presenting non-significant differences. To some extent, the same pattern was observed for rooting percentage with the highest values (98.61%) obtained in full-strength WPM and in half-strength DKW. Dilution of media lead to sufficient rooting percentage (97.22%) in some treatments, e.g., 1/4× MS and 1/2× WPM (Figure 5).

Concerning the number of shoots, highest values were observed in all media when these were at full strength, i.e., MS (9.39), WPM (7.03) and DKW (5.96) (Figure 6). The best medium proved to be MS except when diluted to 1/8 of its strength. Normally, a shift from full strength resulted in a decrease in shoot number values that were often statistically significant different. To a limited degree, the above motif was applied for number of roots with reference to WPM and DKW media. Both half-strength WPM and DKW presented the highest root number values, 6.49 and 5.74, respectively, while the best MS score (5.25) was exhibited in the 1/8 dilution. On the contrary, the lowest values were obtained in 2× MS (2.08), 1/2× MS (2.61), and 2× WPM (2.81) (Figure 6).

No differences were observed in shoot length between 1/2× MS (0.78 cm) and 1× MS (0.77 cm). These two values were significantly different from those of the other treatments that also involved the same nutrient medium (Figure 7). The same pattern was noticed concerning WPM, where no statistical differences were observed between 1× WPM (0.63 cm) and 1/2× WPM (0.62 cm). Whether diluting or doubling the DKW, the mean shoot length per explant, which was close to 0.57 cm, was not actually affected. In general, when those media were diluted, the mean length of roots was increased. This was obvious for MS and WPM media but did not apply to DKW. The best performance in root length was achieved in 1/8× WPM (1.55 cm) and in 1/4× MS (1.52 cm) followed by the 1/4× WPM (1.42). Regarding the DKW medium, the maximum mean root length (1.27 cm) was attained in full strength (Figure 7).

3.2. Genetic Population Analysis

All SSR primers generated amplicons in all three populations. The number of alleles and their range for each locus and *C. creticus* population are presented in Table 3.

Table 3. Characteristics of PCR amplicons for studied STR loci and plant material of *C. creticus* populations.

STR [1] Locus	Mt. Pateras		Mt. Parnitha		Mt. Pendeli		Overall	
	Alleles Number	Allele Range (bp [2])	Alleles Number	Allele Range (bp [2])	Alleles Number	Allele Range (bp [2])	Alleles Number	Allele Range (bp [2])
cislau1-1	10	231–251	9	231–251	9	231–249	10	231–251
cislau7-1	5	159–166	5	159–166	5	159–166	5	159–166
cislau11-1	3	163–167	3	163–167	3	163–167	3	163–167
cislau12-1	8	208–230	8	208–230	8	208–230	8	208–230
cislau14-1	2	180–184	2	180–184	2	180–184	2	180–184

[1] Short tandem repeat, [2] bp = base pairs.

In total, 28 alleles were identified across all five STR loci for the 36 *C. creticus* L. samples derived from the three populations of Attica. The average allele per locus was 5.60 ± 1.23 (mean ± standard error). The mean of different alleles per locus (Na) was estimated at 5.60 ± 1.50, while the effective alleles (Ne) were estimated at 4.72 ± 1.14 (Table 4) respectively. The mean observed (Ho) and expected (He) heterozygosity were evaluated at 0.52 ± 0.08 and 0.72 ± 0.07, respectively. The inbreeding coefficient (fixation index *F*) and Shannon's information index (I) were calculated at 0.30 ± 0.06 and 1.48 ± 0.28, respectively. The percentage of polymorphic loci (PPL) in all populations was estimated at 100% (Table 4). The F_{ST} indices between the pairs of populations, i.e., Mt. Pateras and Mt. Parnitha, Mt. Pateras and Mt. Pendeli and Mt. Parnitha and Mt. Pendeli were 0.009, 0.004 and 0.007, respectively. The overall mean F_{ST} value was estimated at 0.009 ± 0.002. Nei genetic distances (Nei D) for the same pairs of populations were calculated at 0.046, 0.021 and 0.040, respectively. The genetic identities (Nei I) were evaluated at 0.955, 0.979 and 0.961, respectively (Table 5). The minimum polymorphism information content (PIC) values were observed in two loci, i.e., cislau14-1 (0.374) and cislau11-1 (0.535) (Table 6). The most informative marker was cislau1-1 (PIC = 0.853) followed by cislau12-1 (PIC = 0.845) (Table 6).

Table 4. Genetic informative parameters of *C. creticus* populations.

n = 36	Na	Ne	I	Ho	He	F	Percentage of Polymorphic Loci
Mean	5.60 ± 1.50	4.72 ± 1.14	1.48 ± 0.28	0.52 ± 0.08	0.72 ± 0.07	0.30 ± 0.06	100.00%

n = number of samples, Na = number of different alleles, Ne = number of effective alleles, I = Shannon's information index, Ho = observed heterozygosity, He = expected heterozygosity, F = fixation index.

Table 5. Pairwise population matrices of genetic distance, genetic identity, and inbreeding coefficient of *C. creticus* populations.

	Nei Genetic Distance (Nei D)		Nei Genetic Identity (Nei I)		F_{st} * Values		Mean F_{st} * Value
	Mt. Pateras	Mt. Parnitha	Mt. Pateras	Mt. Parnitha	Mt. Pateras	Mt. Parnitha	
Mt. Parnitha	0.046		0.955		0.009		0.009 ± 0.002
Mt. Pendeli	0.021	0.040	0.979	0.961	0.004	0.007	

* F_{ST} = inbreeding coefficient.

Table 6. Characteristics of studied STR locus of *C. creticus* L. populations.

Locus	Overall		
	Ho [1]	He [2]	PIC [3]
cislau1-1	0.639	0.879	0.853
cislau7-1	0.444	0.779	0.731
cislau11-1	0.500	0.624	0.535
cislau12-1	0.722	0.873	0.845
cislau14-1	0.278	0.505	0.374

[1] Ho: observed heterozygosity, [2] He: expected heterozygosity, [3] PIC: polymorphism information content.

4. Discussion

4.1. In Vitro Culture

Successful shoot organogenesis was observed across all types of nutrient media. All cultures exhibited satisfactory shooting percentages, and the newly formed shoots exhibited significant elongation. Our most favorable results in terms of growth were obtained using the DKW medium, which has an intermediate nitrogen concentration compared to the other two media. In contrast to MS, WPM also demonstrated superior results, as it was the medium with the lowest nitrogen availability. The success of in vitro blastogenesis relies on the composition and concentration of basal salts, growth regulators, and organic components [36]. Specifically, the nitrogen content in the nutrient medium appears to affect the formation of shoots in explants [37].

Maximum growth percentage was achieved in full- or half-strength WPM and DKW. The same results were those of Hatzilazarou et al. [38] in *Nerium oleander* L. They reported higher shoot elongation on WPM or MS, regardless of their strengths, and stressed that the cultures on DKW or B5 nutrient media produced shorter shoots. Conversely, Rezali et al. [39] reported that an increase in MS strength resulted in a decrease in shoot height concerning *Typhonium flagelliforme*.

The shoot percentage was higher on MS, but there was no significant difference in the other two media. Our results were contrary to the findings of Bell et al. [40], where DKW presented superiority to WPM and MS for shoot proliferation and shoot number per explant of pear cultivars. The dilution of MS has a significant negative effect on shoot formation but does not impact the mean shoot length. In line with the findings of this study were those of Hatzilazarou et al. [38] in *Nerium oleander* L., where higher shoot formation was achieved on full-strength WPM or MS media, with no statistically significant differences between them, as opposed to DKW or B5. The shoot number obtained in full-strength medium was higher compared to the diluted media as reported by Wan Nurul Hidayah et al. [41] in *Pogostemon cablin* (patchouli). They reported that the number of shoots obtained in full-strength medium was greater compared to the diluted media. In addition, varieties of *Cannabis sativa* L. displayed better response in the full-strength MS compared to the 1/2× MS medium [42]. Kumar et al. [43] also reported similar findings in *Litchi chinensis*.

Full-strength MS exhibited higher shoot regeneration rates in *Harpagophytum procumbens* with no significant differences observed when half-strength MS medium was used [44].

The results of this study are consistent with the previously mentioned findings, especially concerning goji berry (*Lycium barbarum* L.). The highest shoot multiplication was recorded on MS and DKW media compared to WPM [45]. Similar results were found by Parris et al. [46] in a *Magnolia* cultivar. The DKW medium exhibited the highest shoot multiplication rate in *Betula pendula* L [47]. In contrast to our results, Fadel et al. [48], in *Mentha spicata* L., observed the maximum number of shoots on half-strength MS medium. Tetsumura et al. [49] in *Vaccinium corymbosum* and *V. virgatum* observed that a reduction in the strength of MS medium resulted in the increase in in vitro shoot formation. In similar experiments, Villamor [50] in *Zingiber officinale*, Rezali et al. [39] in *Typhonium flagelliforme* as well as Taheri et al. [51] in *Ziziphora persica* observed a decrease in both shoot number and length with the reduction in MS strength. Grigoriadou et al. [52] reported conflicting results regarding the dilution of nutrient media for two pear cultivars. Moreover, Bertsouklis et al. [53] reported that using DKW increased *Juniperus phoenicea* L. (Phoenicean juniper) shoot formation, which were findings that differed from ours. The type of nutrient medium had a significant impact on the average number of shoots formed in *Juniperus oxycedrus* L. [54,55]. Jain et al. [44] found that MS medium produced more shoots compared to WPM, which exhibited a significantly lower number of shoots in *Harpagophytum procumbens* cultures.

Our results, regarding shoot length, were similar to those of Hatzilazarou et al. [38] in *Nerium oleander* L. The maximum shoot length was achieved on full-strength WPM or MS, showing a significant difference compared to DKW or B5 media. Similar findings were those of Sokolov et al. [56], where MS was found to be superior to DKW in terms of shoot length in *Magnolia* sp. However, DKW performed better in shoot number. In contrast, Halstead et al. [57] reported a better shoot height performance in DKW compared to MS. DKW was proved to be the best medium for shoot length in a *Magnolia* cultivar [46] and in *Juglans regia* L. [58]. DKW was also identified as the most effective medium for shoot length in Brazilian ginseng (*Pfaffia glomerata* (Spreng.) Pedersen) [59]. The results of Villamor [50] indicated that MS dilution decreased shoot length in *Zingiber officinale* Rosc., too. According to Jain et al. [44], reducing the salt concentration in the media resulted in a decrease in shoot elongation capacity. On half- and quarter-strength WPM, shoot elongation was significantly reduced compared to the MS medium in *Harpagophytum procumbens*. Conversely, Fadel et al. [48] observed the maximum shoot length on half-strength MS medium in *Mentha spicata* L.. Our results did not coincide with those of Loureiro et al. [60], who observed that *Juniperus phoenicea* L. explants growing in DKW exhibited significantly better results compared to those growing in WPM or MS. Several studies in *Juniperus* species [61–67] have showed that optimal results in blastogenesis were achieved in media with a lower nitrogen content, i.e., diluted media. However, Bertsouklis et al. [53] encountered challenges using the MS medium with low nitrogen content in an in vitro propagation of *Juniperus phoenicea*.

Nodal explants exhibited satisfactory rooting in all types and strengths of nutrient media. All cultures demonstrated a satisfactory rooting percentage with notable root elongation. The optimal rooting percentages and the highest root number per explant were observed in half- and full-strength DKW and WPM. High rooting percentage was also observed in quarter-strength MS. Diluting WPM and MS to 1/8 and 1/4 resulted in achieving the maximum root length. Our findings aligned with those of Halstead et al. [57] who reported the best root percentage in full-strength DKW compared to MS. Many researchers have also documented the advantageous impact of reducing the medium strength on root initiation [68,69]. The strength dilution of MS basal medium increased root induction in *Rosa* spp. [68], *Cannabis sativa* L. [42,65], *Typhonium flagelliforme* (G. Lodd.) Blume [39], *Camellia sinensis* L. [70] *Mentha spicata* L. [48], *Zingiber officinale* Roscoe [50] and *Syzygium alternifolium* (Wight) Walp. [71]. Reducing the strength of MS medium by half resulted in an increase in the rooting of *Mentha arvensis* L. explants, too [72]. Tetsumura et al. also observed that lowering the strength of the MS medium resulted in the increase in in vitro root formation in *Vaccinium corymbosum* and *V. virgatum*. Decreasing the strength of the medium was advantageous for the in vitro root induction of *Cynara scolymus* L. [73–75]. Moreover, Li and Eaton [76], in grapevine, reported that rooting in half-strength MS salts was superior compared to full strength. The beneficial effect of reducing the concentration of the MS basal medium on in vitro rooting ability has also been demonstrated in trees such as *Quercus sobur* L. [77] and *Wrightia tomentosa* Roem. & Schult. [78].

In the in vitro culture of *Magnolia* cultivar, WPM outperformed DKW and MS in terms of root number [46]. Villamor [50] in *Zingiber officinale* reported that the root number was significantly increased in half-strength MS basal medium, which are results that align to ours. Patel and Shah [69] mentioned that the root number and root length of *Stevia rebaudiana* explants were influenced by the strength of the MS medium, although these results were not statistically significant, with the best results achieved in quarter-strength MS medium.

DKW proved to be the most successful medium for Brazilian ginseng (*Pfaffia glomerata* (Spreng.) Pedersen) in terms of root length [59]. Controversial results were reported by Fadel et al. [48], in *Mentha spicata* L., where the highest root length was observed on full-strength MS, while no roots were induced on 1/4 MS. In *Prunus* sp., MS failed to induce roots, too [79]. Rezali et al. [39] recorded an increment in the number of roots in *Typhonium flagelliforme* by decreasing the MS strength. Bidarigh and Azarpour [80] also reported that the highest root length and root number in tea explants (*Camellia sinensis* L.) were obtained by reducing the medium strength. Reducing the strength of the MS medium by half resulted in enhanced rooting characteristics of *Mentha spicata* L. [48] and *Mentha arvensis* L. [72]. The root number in *Zingiber officinale* Roscoe [50] was increased with the dilution of the MS basal medium. The highest rooting percentage, root number and root length in *Syzygium alternifolium* were obtained in quarter-strength and half-strength MS medium [71].

The variations in the effects of medium strength are likely associated with the specific components of the culture medium [48] and may vary among origins, depending on the type and physiological condition of the explants [38]. Even minor alterations in the concentration of trace elements can dramatically affect in vitro plant organogenesis [48]. In the in vitro culture of *Populus alba* L., an increase in zinc concentration in the medium resulted in a significant reduction in the number and length of induced roots [81]. In addition, the maximum in vitro growth of the epiphyte bromeliads *Vriesea friburguensis* Mez, *V. hieroglyphica* E. Morren and *V. unilateralis* Mez were obtained on the medium which was full of Ca [82]. The omission of having omitted KNO_3 in both full and half-strength MS media was proven to be significantly deleterious to root formation [50], thus having a detrimental effect on the growth of all the genotypes tested according to Jean and Cappadocia [83]. In other cases, the observed deleterious effect could not be solely attributed to potassium deficiency. Instead, this result might be attributed to a high ratio of NO_3:NH_4 in the media, as observed in *Dioscorea opposita* Thunb [84]. Bell et al. [40] also highlighted the differential responses of explants in in vitro cultures due to various salt contents.

4.2. Genetic Population Analysis

Allele peaks were detected within the size range reported in the literature [28,29]. The most diverse loci were cislau1-1 followed by cislau12-1, presenting 10 and 8 alleles, respectively, which were findings that were comparable to the results of Astuti et al. [28] and Bertolasi et al. [29]. This specific SSR combination may detect the variation and the structure among different *Cistus* populations. In general, the system could allow the discrimination among populations, sub-populations and possibly identifying divergent individuals within different geographical regions [28,29].

Our results provided, to a certain extent, allele frequency estimations for the three *C. creticus* L. populations. However, they did not deliver a substantial genetic population survey due to the limited geographic area included in the analysis. The genetic base was relatively narrow, which can be explained by the fact that only twenty-eight alleles were identified. The genetic diversity, though, between the three origins was very low due to minimal, close to zero, F_{ST} values [85,86]. This conclusion is confirmed by both values—the very low value of Nei D (<0.05) and the especially high value of Nei I (>0.95) [87–89]—thus suggesting the genetic similarity of the three populations and indicating that they belong to the same population. Furthermore, as the F value was not close to zero, it could be assumed that the plants coming from all three origins are not undergoing random mating [90]. The results of Paolini et al. (REF), who studied the differences between two subspecies of *C. creticus* using ISSR markers, revealed significant divergence between the two groups alongside low genetic diversity within them. In addition, the results of Astuti et al. [28] suggested low genetic diversity within the population of *Cistus laurifolius* L. Another study on *Cistus ladanifer* L. also reported low levels of variability [91]. In contrast, Bertolasi et al. [29] found a good reservoir of genetic variability in a population of *Cistus albidus* L. using SSR markers.

5. Conclusions

In vitro culture of *Cistus creticus* L. occurred effortlessly. An absence of complication was observed in both blastogenesis and root induction. The results showed that in addition to the type of nutrient medium and its strength, their combination had a significant effect on in vitro growth, shoot and root formation of rockrose explants in vitro culture. DKW demonstrated the highest values for explant growth and rooting percentage along with a parallel increase in root number. On the other hand, MS medium presented the best results in shooting percentage and mean number and length of shoots per explant, while WPM mainly affected root length. The origin of plant material did not influence any of the studied in vitro culture traits. Correspondingly, specific types of nutrient media combined with certain strengths could be employed at particular stages of in vitro culture, i.e., culture establishment, shoot multiplication or root formation. Moreover, the SSR markers used to assess the genetic relations of the three populations of *C. creticus* L. plants, which were involved in in vitro culture, indicated that the genetic base was relatively narrow. The very low estimated genetic diversity was attributed to the limited geographic sampling area included in the analysis. Low genetic diversity, along with the high evaluated genetic similarity, suggested that the three origins belonged to the same population.

Supplementary Materials: The following supporting information can be downloaded at https://www.mdpi.com/article/10.3390/horticulturae10010104/s1, Figure S1. The sampling sites of the *C. creticus* L. populations in Attica (Greece). Figure S2. Effect of the medium on the mean shoot growth percentage (%) (a), on the mean shoot percentage (%) (b) and on the mean root percentage (%) (c) in relation to clone origin of *C. creticus* L. explants. Figure S3. Effect of the medium on the mean shoot number (a) and mean shoot length (b) in relation to clone origin of *C. creticus* L. explants. Figure S4. Effect of the medium on the mean root number (a) and mean root length (b) in relation to clone origin of *C. creticus* L. explants. Table S1: Effect of the medium type on the studied traits of *C. creticus* L. explants. Means followed by the same letter do not differ statistically at $p \leq 0.05$ according to the Duncan test; Table S2: Effect of medium and its strength on the average growth, shoot and root percentage (%) of *C. creticus* L. explants. (Means followed by the same letter do not differ statistically at $p \leq 0.05$ according to the Duncan test.) MS: Murashige and Skoog medium, WPM: wood plant medium, DKW: Driver and Kuniyaki Walnut medium; Table S3: Effect of medium and its strength on the average number of shoots and length per *C. creticus* L. explants. (Means followed by the same letter do not differ statistically at $p \leq 0.05$ according to the Duncan test.) MS: Murashige and Skoog medium, WPM: wood plant medium, DKW: Driver and Kuniyaki Walnut medium; Table S4: Effect of medium and its strength on the average number of roots and length per *C. creticus* L. > explants. (Means followed by the same letter do not differ statistically at $p \leq 0.05$ according to the Duncan test.) MS: Murashige and Skoog medium, WPM: wood plant medium, DKW: Driver and Kuniyaki Walnut medium.

Author Contributions: Conceptualization, K.I.; methodology, K.I. and P.K.; validation, K.I. and P.K.; formal analysis, K.I. and P.K.; investigation, K.I. and P.K.; resources, K.I. and P.K.; data curation, K.I. and P.K.; writing—original draft preparation, K.I.; writing—review and editing, K.I. and P.K.; visualization, K.I.; supervision, K.I.; project administration K.I. All authors have read and agreed to the published version of the manuscript.

Funding: This research received no funding.

Data Availability Statement: Data are contained within the article and Supplementary Materials.

Acknowledgments: The authors want to express their sincere thanks to the librarian of the Forest Research Institute of Athens, D. Panayiotopoulou (MSc Library and Information Science), for information seeking and retrieving processes as well as her additional proofreading services.

Conflicts of Interest: The authors declare no conflicts of interest.

References

1. Zygomala, A.M.; Ioannidis, C.; Koropouli, X. In Vitro Propagation of *Cistus creticus* L. *Acta Hortic.* **2003**, *616*, 391–396. [CrossRef]
2. Trabaud, L. Fire and Survival Traits of Plants. In *The Role of Fire in Ecological Systems*; Trabaud, L., Ed.; SPBAcademic Publishing: The Hague, The Netherlands, 1987; pp. 65–89.
3. Le Houérou, H.N. Fire and Vegetation in the Mediterranean Basin. In Proceedings of the 13th Annual Tall Timbers Fire Ecology Conference, Talahassee, FL, USA, 22–23 March 1973.
4. Proctor, M. Cistaceae. In *Flowering Plants of the World*; Heywood, V., Ed.; Oxford University Press: Oxford, UK, 1978; pp. 100–108.
5. Tutin, T.G.; Heywood, V.H.; Burges, N.A.; Valentine, D.H.; Walters, S.M.; Webb, D.A. *Flora Europaea: Cistaceae*; Cambridge University Press: Cambridge, UK, 1968; Volume 2.
6. Meusel, H.-J.; Rauschert, E.; Weinert, S.E. *Vergleichende Chorologie Der Zentraleuropäischen Flora. Band II*; VEB Gustav Fischer Verlag: Jena, Germany, 1978.
7. Dansereau, P.M. *Monographie du Genre* Cistus *L.*; Boissiera 4; Conservatoire de Botanique: Nord, France, 1939; pp. 1–90.
8. Greuter, W.; Burdet, H.; Long, G. Cistaceae. In *Med-Checklist*; Med-Checklist Trust of OPTIMA; Greuter, W., Burdet, H.M., Long, G., Eds.; Conservatoire et Jardin Botaniques: Geneve, Switzerland, 1984; Volume 1, pp. 315–330.
9. Margaris, N.S. Structure and Dynamics in a Phryganic (East Mediterranean) Ecosystem. *J. Biogeogr.* **1976**, *3*, 249–259. [CrossRef]
10. Pedanius, D.; Osbaldeston, T.A.; Wood, R.P. *De Materia Medica: Being an Herbal with Many Other Medicinal Materials: Written in Greek in the First Century of the Common Era: A New Indexed Version in Modern English*; IBIDIS: Johannesburg, South Africa, 2000; ISBN 978-0-620-23435-1.
11. Hort, F. *Theophrastus: De Signis Vol. II in the Enquire into PlantsHeinemann: London, UK*; Harvard University Press: Cambridge, UK, 1926.
12. Rosén, H.B. *Herodoti Historiae. Vol. I: Libros I–IV*; Springer: Berlin, Germany, 1987; ISBN 3-322-00359-0.
13. Moujir, L.; Gutiérrez-Navarro, A.M.; San Andrés, L.; Luis, J.G. Structure—Antimicrobial Activity Relationships of Abietane Diterpenes from Salvia Species. *Phytochemistry* **1993**, *34*, 1493–1495. [CrossRef]
14. Demetzos, C.; Dimas, K.; Hatziantoniou, S.; Anastasaki, T.; Angelopoulou, D. Cytotoxic and Anti-Inflammatory Activity of Labdane and Cis-Clerodane Type Diterpenes. *Planta Med.* **2001**, *67*, 614–618. [CrossRef] [PubMed]
15. Chinou, I.; Demetzos, C.; Harvala, C.; Roussakis, C.; Verbist, J. Cytotoxic and Antibacterial Labdane-Type Diterpenes from the Aerial Parts of *Cistus incanus* subsp. *Creticus. Planta Med.* **1994**, *60*, 34–36. [CrossRef]
16. Demetzos, C.; Mitaku, S.; Couladis, M.; Harvala, C.; Kokkinopoulos, D. Natural Metabolites of Ent-13-Epi-Manoyl Oxide and Other Cytotoxic Diterpenes from the Resin "LADANO" of *Cistus creticus. Planta Med.* **1994**, *60*, 590–591. [CrossRef]
17. Dimas, K.; Demetzos, C.; Marsellos, M.; Sotiriadou, R.; Malamas, M.; Kokkinopoulos, D. Cytotoxic Activity of Labdane Type Diterpenes against Human Leukemic Cell Lines in Vitro. *Planta Med.* **1998**, *64*, 208–211. [CrossRef]
18. Iriondo, J.M.; Moreno, C.; Pérez, C. Micropropagation of Six Rockrose (*Cistus*) Species. *HortScience* **1995**, *30*, 1080–1081. [CrossRef]
19. Polunin, O.; Huxley, A. *Flowers of the Mediterranean*; Chatto and Windus: London, UK, 1972.
20. López González, G. *La Guía de Incafo de Los Árboles y Arbustos de La Península Ibérica*; Guias verdes de INFACO, 4; Infaco: Madrid, Spain, 1982; ISBN 9788485389346.
21. Brickell, C. *Royal Horticultural Society Gardeners' Encyclopedia of Plants and Flowers*; Dorling Kindersley: London, UK, 1989; ISBN 0-86318-386-7.
22. Heywood, V.H. *Flowering Plants of the World*; Dorling Kindersley Publishers Ltd.: London, UK, 1993; ISBN 0-7134-7422-X.
23. Pacioni, G. *El Cultivo Moderno y Rentable de La Trufa*; De Vecchi: Milan, Italy, 1992; ISBN 84-315-0516-8.
24. M'Kada, J.; Dorion, N.; Bigot, C. In Vitro Propagation of *Cistus × purpureus* Lam. *Sci. Hortic.* **1991**, *46*, 155–160. [CrossRef]
25. Pela, Z.; Pentcheva, M.; Gerasopoulos, D.; Maloupa, E. In Vitro Induction of Adventitious Roots and Proliferation of *Cistus creticus creticus* L. Plants. *Acta Hortic.* **2000**, 317–322. [CrossRef]
26. Ruta, C.; Morone-Fortunato, I. In Vitro Propagation of *Cistus clusii* Dunal, an Endangered Plant in Italy. *Vitr. Cell. Dev. Biol. Plant* **2010**, *46*, 172–179. [CrossRef]

27. Airò, M.; Farruggia, G.; Giardina, G.; Giovino, A. Micropropagation protocol of a threatened species: *Cistus crispus* L. In Proceedings of the Acta Horticulturae; International Society for Horticultural Science (ISHS), Leuven, Belgium, 26 May 2015; pp. 549–552.
28. Astuti, G.; Roma-Marzio, F.; D'Antraccoli, M.; Bedini, G.; Carta, A.; Sebastiani, F.; Bruschi, P.; Peruzzi, L. Conservation Biology of the Last Italian Population of *Cistus laurifolius* (Cistaceae): Demographic Structure, Reproductive Success and Population Genetics. *Nat. Conserv.* **2017**, *22*, 169–190. [CrossRef]
29. Bertolasi, B.; Zago, L.; Gui, L.; Sitzia, T.; Vanetti, I.; Binelli, G.; Puppi, G.; Buldrini, F.; Pezzi, G. Phenological and Genetic Characterization of Mediterranean Plants at the Peripheral Range: The Case of *Cistus albidus* near Lake Garda. *Flora* **2019**, *252*, 26–35. [CrossRef]
30. Murashige, T.; Skoog, F. A Revised Medium for Rapid Growth and Bio Assays with Tobacco Tissue Cultures. *Physiol. Plant.* **1962**, *15*, 473–497. [CrossRef]
31. Lloyd, G.; McCown, B. Woody Plant Medium: A Mineral Nutrient Formulation for Microculture of Woody Plant Species. *HortScience* **1981**, *16*, 453.
32. Driver, J.A.; Kuniyuki, A.H. In Vitro Propagation of Paradox Walnut Rootstock. *HortScience* **1984**, *19*, 507–509. [CrossRef]
33. Peakall, R.; Smouse, P.E. GENALEX 6: Genetic Analysis in Excel. Population Genetic Software for Teaching and Research. *Mol. Ecol. Notes* **2006**, *6*, 288–295. [CrossRef]
34. Botstein, D.; White, R.L.; Skolnick, M.; Davis, R.W. Construction of a Genetic Linkage Map in Man Using Restriction Fragment Length Polymorphisms. *Am. J. Hum. Genet.* **1980**, *32*, 314.
35. Kalinowski, S.T.; Taper, M.L.; Marshall, T.C. Revising How the Computer Program Cervus Accommodates Genotyping Error Increases Success in Paternity Assignment. *Mol. Ecol.* **2007**, *16*, 1099–1106. [CrossRef]
36. Ge, X.; Chu, Z.; Lin, Y.; Wang, S. A Tissue Culture System for Different Germplasms of Indica Rice. *Plant Cell Rep.* **2006**, *25*, 392–402. [CrossRef]
37. Momeni, M.; Ganji-Moghadam, E.; Kazemzadeh-Beneh, H.; Asgharzadeh, A. Direct Organogenesis from Shoot Tip Explants of *Juniperus polycarpos* L.: Optimizing Basal Media and Plant Growth Regulators on Proliferation and Root Formation. *Plant Cell Biotechnol. Mol. Biol.* **2018**, *19*, 40–50.
38. Hatzilazarou, S.; Kostas, S.; Economou, A.; Scaltsoyiannes, A. Efficient propagation of *Nerium oleander* L. through tissue culture. *Propag. Ornam. Plants* **2017**, *17*, 12.
39. Rezali, N.I.; Jaafar Sidik, N.; Saleh, A.; Osman, N.I.; Mohd Adam, N.A. The Effects of Different Strength of MS Media in Solid and Liquid Media on in Vitro Growth of *Typhonium flagelliforme*. *Asian Pac. J. Trop. Biomed.* **2017**, *7*, 151–156. [CrossRef]
40. Bell, R.L.; Srinivasan, C.; Lomberk, D. Effect of Nutrient Media on Axillary Shoot Proliferation and Preconditioning for Adventitious Shoot Regeneration of Pears. *Vitr. Cell. Dev. Biol. Plant* **2009**, *45*, 708–714. [CrossRef]
41. Wan Nurul Hidayah, W.A.; Norrizah, J.S.; Sharifah Aminah, S.M.; Sharipah Ruzaina, S.A.; Faezah, P. Effect of Medium Strength and Hormones Concentration on Regeneration of Pogostemon Cablin Using Nodes Explant. *Asian J. Biotechnol.* **2012**, *4*, 46–52. [CrossRef]
42. Ioannidis, K.; Dadiotis, E.; Mitsis, V.; Melliou, E.; Magiatis, P. Biotechnological Approaches on Two High CBD and CBG *Cannabis sativa* L. (Cannabaceae) Varieties: In Vitro Regeneration and Phytochemical Consistency Evaluation of Micropropagated Plants Using Quantitative 1H-NMR. *Molecules* **2020**, *25*, 5928. [CrossRef]
43. Kumar, M.; Shiva Prakash, N.; Prasad, U.S.; Bhalla-Sarin, N. A Novel Approach of Regeneration from Nodal Explants of Field-Grown Litchi (*Litchi chinensis* Sonn.) Fruit Trees. *J. Plant Sci.* **2006**, *1*, 240–246. [CrossRef]
44. Jain, N.; Bairu, M.W.; Stirk, W.A.; Van Staden, J. The Effect of Medium, Carbon Source and Explant on Regeneration and Control of Shoot-Tip Necrosis in *Harpagophytum procumbens*. *S. Afr. J. Bot.* **2009**, *75*, 117–121. [CrossRef]
45. Silvestri, C.; Sabbatini, G.; Marangelli, F.; Rugini, E.; Cristofori, V. Micropropagation and Ex Vitro Rooting of Wolfberry. *HortScience* **2018**, *53*, 1494–1499. [CrossRef]
46. Parris, J.K.; Touchell, D.H.; Ranney, T.G.; Adelberg, J. Basal Salt Composition, Cytokinins, and Phenolic Binding Agents Influence In Vitro Growth and Ex Vitro Establishment of Magnolia 'Ann'. *HortScience* **2012**, *47*, 1625–1629. [CrossRef]
47. Rathwell, R.; Shukla, M.R.; Jones, A.M.P.; Saxena, P.K. In Vitro Propagation of Cherry Birch (*Betula lenta* L.). *Can. J. Plant Sci.* **2016**, *96*, 571–578. [CrossRef]
48. Fadel, D.; Kintzios, S.; Economou, A.S.; Moschopoulou, G.; Constantinidou, H.-I.A. Effect of Different Strength of Medium on Organogenesis, Phenolic Accumulation and Antioxidant Activity of Spearmint (*Mentha spicata* L.). *Open Hortic. J.* **2010**, *3*, 31–35. [CrossRef]
49. Tetsumura, T.; Matsumoto, Y.; Sato, M.; Honsho, C.; Yamashita, K.; Komatsu, H.; Sugimoto, Y.; Kunitake, H. Evaluation of Basal Media for Micropropagation of Four Highbush Blueberry Cultivars. *Sci. Hortic.* **2008**, *119*, 72–74. [CrossRef]
50. Villamor, C.C. Influence of Media Strength and Sources of Nitrogen on Micropropagation of Ginger, *Zingiber officinale* Rosc. *Int. Sci. Res. J.* **2010**, *2*, 6.
51. Taheri, A.; Kosari-Nasab, M.; Movafeghi, A. Effects of Different Strengths of Medium on Production of Phenolic and Flavonoid Compounds in Regenerated Shoots of *Ziziphora persica*. *Russ. Agric. Sci.* **2015**, *41*, 225–229. [CrossRef]
52. Grigoriadou, K.; Leventakis, N.; Vasilakakis, M. Effect of Various Culture Conditions on Proliferation and Shoot Tip Necrosis in the Pear Cultivars 'William's' and 'Highland' Grown In Vitro. *Acta Hortic.* **2000**, *520*, 103–108. [CrossRef]

53. Bertsouklis, K.; Paraskevopoulou, A.; Zarkadoula, N. In Vitro Propagation of *Juniperus phoenicea* L. In Proceedings of the International Symposium on Botanical Gardens and Landscapes, Bangkok, Thailand, 2–4 December 2019; pp. 331–334.
54. Gómez, M.P.; Segura, J. Axillary Shoot Proliferation in Cultures of Explants from Mature *Juniperus oxycedrus* Trees. *Tree Physiol.* **1995**, *15*, 625–628. [CrossRef]
55. Gómez, M.P.; Segura, J. Factors Controlling Adventitious Bud Induction and Plant Regeneration in mature *Juniperus oxycedrus* Leaves Cultured in Vitro. *Vitr. Plant* **1994**, *30*, 210–218. [CrossRef]
56. Sokolov, R.S.; Atanassova, B.Y.; Iakimova, E.T. Physiological Response of Cultured. to Nutrient Medium Composition. *J. Hortic. Res.* **2014**, *22*, 49–61. [CrossRef]
57. Halstead, M.A.; Garfinkel, A.R.; Marcus, T.C.; Hayes, P.M.; Carrijo, D.R. Hemp Growth in Vitro and in Vivo: A Comparison of Growing Media and Growing Environments across 10 Accessions. *HortScience* **2022**, *57*, 1041–1047. [CrossRef]
58. Saadat, Y.A.; Hennerty, M.J. Factors Affecting the Shoot Multiplication of Persian Walnut (*Juglans regia* L.). *Sci. Hortic.* **2002**, *95*, 251–260. [CrossRef]
59. Silva, T.D.; Chagas, K.; Batista, D.S.; Felipe, S.H.S.; Louback, E.; Machado, L.T.; Fernandes, A.M.; Buttrós, V.H.T.; Koehler, A.D.; Farias, L.M.; et al. Morphophysiological in Vitro Performance of Brazilian Ginseng (*Pfaffia Glomerata* (Spreng.) Pedersen) Based on Culture Medium Formulations. *Vitr. Cell. Dev. Biol. Plant* **2019**, *55*, 454–467. [CrossRef]
60. Loureiro, J.; Capelo, A.; Brito, G.; Rodriguez, E.; Silva, S.; Pinto, G.; Santos, C. Micropropagation of *Juniperus phoenicea* from Adult Plant Explants and Analysis of Ploidy Stability Using Flow Cytometry. *Biol. Plant.* **2007**, *51*, 7–14. [CrossRef]
61. Khater, N.; Benbouza, H. Preservation of *Juniperus thurifera* L.: A Rare Endangered Species in Algeria through In Vitro Regeneration. *J. For. Res.* **2019**, *30*, 77–86. [CrossRef]
62. Salih, A.M.; Al-Qurainy, F.; Khan, S.; Tarroum, M.; Nadeem, M.; Shaikhaldein, H.O.; Alabdallah, N.M.; Alansi, S.; Alshameri, A. Mass Propagation of *Juniperus procera* Hoechst. Ex Endl. From Seedling and Screening of Bioactive Compounds in Shoot and Callus Extract. *BMC Plant Biol.* **2021**, *21*, 192. [CrossRef]
63. Thorpe, T.A.; Harry, I.S.; Kumar, P.P. Application of Micropropagation to Forestry. In *Micropropagation: Technology and Application*; Debergh, P.C., Zimmerman, R.H., Eds.; Springer: Dordrecht, The Netherlands, 1991; pp. 311–336. ISBN 978-94-009-2075-0.
64. Al-Ramamneh, E.A.; Dura, S.; Daradkeh, N. Propagation Physiology of *Juniperus phoenicea* L. from Jordan Using Seeds and in Vitro Culture Techniques: Baseline Information for a Conservation Perspective. *Afr. J. Biotechnol.* **2012**, *11*, 7684–7692. [CrossRef]
65. Ślusarkiewicz-Jarzina, A.S.; Ponitka, A.; Kaczmarek, Y. Influence of Cultivar, Explant Source and Plant Growth Regulator on Callus Induction and Plant Regeneration of *Cannabis sativa* L. *Acta Biol. Crac. Ser. Bot* **2005**, *47*, 145–151.
66. Ioannidis, K.; Tomprou, I.; Panayiotopoulou, D.; Boutsios, S.; Daskalakou, E.N. Potential and Constraints on In Vitro Micropropagation of *Juniperus drupacea* Labill. *Forests* **2023**, *14*, 142. [CrossRef]
67. Al-Ramamneh, E.A.-D.; Daradkeh, N.; Rababah, T.; Pacurar, D.; Al-Qudah, M. Effects of Explant, Media and Growth Regulators on in Vitro regeneration and Antioxidant Activity of *Juniperus phoenicea*. *Aust. J. Crop Sci.* **2017**, *11*, 828–837. [CrossRef]
68. Sauer, A.; Walther, F.; Preil, W. Different Suitability for in Vitro Propagation of Rose Cultivars. *Gartenbauwissenschaft* **1985**, *3*, 133–138.
69. Patel, R.M.; Shah, R.R. Regeneration of Stevia Plant through Callus Culture. *Indian J. Pharm. Sci.* **2009**, *71*, 46–50. [CrossRef]
70. Bidarigh, S.; Hatamzadeh, A.; Azarpour, E. The Study Effect of IBA Hormone Levels on Rooting in Micro Cuttings of Tea (*Camellia sinensis* L.). *World Appl. Sci. J.* **2012**, *20*, 1051–1054.
71. Sha Valli Khan, P.S.; Hausman, J.F.; Rao, K.R. Effect of Agar, MS Medium Strength, Sucrose and Polyamines on in Vitro Rooting of *Syzygium alternifolium*. *Biol. Plant.* **1999**, *42*, 333–340. [CrossRef]
72. Phatak, S.V.; Heble, M.R. Organogenesis and Terpenoid Synthesis in *Mentha arvensis*. *Fitoterapia* **2002**, *73*, 32–39. [CrossRef]
73. Ancora, G. Globe Artichoke (*Cynara scolymus* L.). In *Crops I*; Bajaj, Y.P.S., Ed.; Springer: Berlin/Heidelberg, Germany, 1986; pp. 471–519. ISBN 978-3-642-61625-9.
74. Iapichino, G. Micropropagation of Globe Artichoke (*Cynara scolymus* L.) from Underground Dormant Buds ("Ovoli"). *Vitr. Cell. Dev. Biol. Plant* **1996**, *32*, 249–252. [CrossRef]
75. Lauzer, D.; Vieth, J. Micropropagation of Seed-Derived Plants of *Cynara scolymus* L., Cv. 'Green Globe'. *Plant Cell Tissue Organ Cult.* **1990**, *21*, 237–244. [CrossRef]
76. Li, J.-R.; Eaton, G. Growth and Rooting of Grape Shoot Apices In Vitro. *HortScience* **1984**, *19*, 64–66. [CrossRef]
77. Manzanera, J.A.; Pardos, J.A. Micropropagation of Juvenile and Adult *Quercus suber* L. *Plant Cell Tissue Organ Cult.* **1990**, *21*, 1–8. [CrossRef]
78. Purohit, S.D.; Kukda, G.; Sharma, P.; Tak, K. In Vitro Propagation of an Adult Tree Wrightia Tomentosa through Enhanced Axillary Branching. *Plant Sci.* **1994**, *103*, 67–72. [CrossRef]
79. Fallahpour, M.; Miri, S.M.; Bouzari, N. Effects of Media Cultures and Plant Growth Regulators on Micropropagation of CAB-6P Cherry Semi-Dwarf Rootstock. *Iran. J. Hortic. Sci.* **2019**, *50*, 187–196.
80. Bidarigh, S.; Azarpour, E. Evaluation of the Effect of MS Medium Levels on Rooting in Micro Cuttings of Tea (*Camellia sinensis* L.) under In-Vitro Culture Condition. *ARPN J. Agric. Biol. Sci.* **2013**, *8*, 5.
81. Castiglione, S.; Franchin, C.; Fossati, T.; Lingua, G.; Torrigiani, P.; Biondi, S. High Zinc Concentrations Reduce Rooting Capacity and Alter Metallothionein Gene Expression in White Poplar (*Populus alba* L. Cv. Villafranca). *Chemosphere* **2007**, *67*, 1117–1126. [CrossRef]

82. Aranda-Peres, A.N.; Peres, L.E.P.; Higashi, E.N.; Martinelli, A.P. Adjustment of Mineral Elements in the Culture Medium for the Micropropagation of Three Vriesea Bromeliads from the Brazilian Atlantic Forest: The Importance of Calcium. *HortScience* **2009**, *44*, 106–112. [CrossRef]
83. Jean, M.; Cappadocia, M. In Vitro Tuberization in *Dioscorea alata* L. 'Brazo Fuerte' and 'Florido' and *D. abyssinica* Hoch. *Plant Cell Tissue Organ Cult.* **1991**, *26*, 147–152. [CrossRef]
84. Asahira, T.; Yazawa, S. Bulbil Formation of *Dioscorea opposita* Cultured In Vitro. *Mem. Coll. Agric. Kyoto Univ.* **1979**, *113*, 39–51.
85. Nei, M. Genetic Distance between Populations. *Am. Nat.* **1972**, *106*, 283–292. [CrossRef]
86. Nei, M. Estimation of Average Heterozygosity and Genetic Distance from a Small Number of Individuals. *Genetics* **1978**, *89*, 583–590. [CrossRef]
87. Wright, S. The Interpretation of Population Structure by F-Statistics with Special Regard to Systems of Mating. *Evolution* **1965**, *19*, 395–420. [CrossRef]
88. Wright, S. *Evolution and the Genetics of Populations. Volume 4. Variability within and among Natural Populations*; University of Chicago Press: Chicago, IL, USA, 1978.
89. Bairu, M.W.; Aremu, A.O.; Van Staden, J. Somaclonal Variation in Plants: Causes and Detection Methods. *Plant Growth Regul.* **2011**, *63*, 147–173. [CrossRef]
90. Larkin, P.J.; Scowcroft, W.R. Somaclonal Variation—A Novel Source of Variability from Cell Cultures for Plant Improvement. *Theor. Appl. Genet.* **1981**, *60*, 197–214. [CrossRef] [PubMed]
91. Quintela-Sabarís, C.; Ribeiro, M.M.; Poncet, B.; Costa, R.; Castro-Fernández, D.; Fraga, M.I. AFLP analysis of the pseudometallophyte *Cistus ladanifer*: Comparison with cpSSRs and exploratory genome scan to investigate loci associated to soil variables. *Plant Soil* **2012**, *359*, 397–413. [CrossRef]

horticulturae

MDPI

Article

The Dormancy Types and Germination Characteristics of the Seeds of *Berberis koreana* Palibin, an Endemic Species of Korea

Do-Hyun Kim [1,2], Sang-Geun Kim [1], Hayan Lee [1], Chae-Sun Na [1,*] and Do-Hyung Lee [2,*]

1 Forest Biological Resource Department, Baekdudaegan National Arboretum, BongHwa 36209, Gyeongsangbuk-do, Republic of Korea; kdh88@koagi.or.kr (D.-H.K.)
2 College of Natural Resources, Forest Resources, Yeungnam University, 280 Daehak-ro, Gyeongsan-si 38541, Gyeongsangbuk-do, Republic of Korea
* Correspondence: chaesun.na@koagi.or.kr (C.-S.N.); dhlee@yu.ac.kr (D.-H.L.)

Abstract: *Berberis koreana* Palibin is an endemic plant native to Korea. In this study, we aimed to study the seed germination of this species using a water imbibition experiment, gibberellic acid (GA_3) treatment (0, 10, 100, or 1000 mg·L^{-1}), cold stratification (0, 2, 4, 8, or 12 weeks at 4 °C), move-along experiment, and phenology studies. In the water imbibition experiment, the weight of the seeds increased by more than 120% in 24 h. An analysis of the internal and external morphological characteristics of the seed revealed that the embryo was already fully grown from the fruit and did not grow thereafter. The final germination percentages for the cold stratification at 0, 2, 4, 8, and 12 weeks at 4 °C were 12 ± 3.65, 32 ± 9.09, 59 ± 1.00, 59 ± 9.59, and 71 ± 1.91%, respectively. In the move-along experiment and phenology studies, a longer low-temperature treatment period resulted in a higher germination percentage. However, the GA_3 treatment had little effect on the seed germination. Our results indicate that *B. koreana* exhibits an intermediate physiological seed dormancy.

Keywords: cold stratification; gibberellic acid; seed dormancy; barberry; endemic species; germination

Citation: Kim, D.-H.; Kim, S.-G.; Lee, H.; Na, C.-S.; Lee, D.-H. The Dormancy Types and Germination Characteristics of the Seeds of *Berberis koreana* Palibin, an Endemic Species of Korea. *Horticulturae* **2023**, *9*, 547. https://doi.org/10.3390/horticulturae9050547

Academic Editors: Konstantinos Bertsouklis, Epaminondas Kartsonas and Angela Carra

Received: 17 April 2023
Revised: 27 April 2023
Accepted: 28 April 2023
Published: 1 May 2023

1. Introduction

Plant propagation methods include various methods such as sowing, stem cutting, tissue cultures, and layering. The sowing method has the following advantages: (1) a large number of plants can be secured at once, (2) it reduces labor, and (3) it does not require a relatively large number of auxiliary facilities, in comparison to those required by a tissue culture facility. (4) In addition to these, there are various other advantages. However, some species must have the conditions required for seed germination. "Seed dormancy", an innate seed property, defines the environmental conditions that must be met before a seed can germinate [1]. Thus, to propagate plants using the sowing method, a suitable dormancy-breaking technique specific to their seeds must first be determined [2].

Various types of seed dormancy exist depending on the life cycle of the plants, the ambient environmental conditions, and their geographical distribution; seed dormancy has been studied and classified according to plant species and genera [3]. Lang [4] classified seed dormancy into three types: eco-dormancy, para-dormancy, and endo-dormancy. Baskin and Baskin [3] classified seed dormancy into five types by comprehensively considering physiological and morphological factors: physiological dormancy (PD), where inhibitory compounds inside and outside the seeds prevent germination; morphological dormancy (MD), where the seeds contain underdeveloped or immature embryos; morphophysiological dormancy (MPD), which is a combination of PD and MD; physical dormancy (PY), which involves the suppression of water absorption by the seeds; and combinational dormancy (PY + PD), which is a combination of PY and PD.

Berberis koreana Palibin, commonly known as Korean barberry, is a deciduous shrub of Berberidaceae, which are endemic to Korea and used in oriental medicine for their anti-inflammatory, analgesic, anti-cancer, anti-conjunctivitis, and antibacterial properties [5,6], and as a source of functional foods [7]. *Berberis* spp. is also a spinous shrub that is highly ornamentally valued for its evergreen or multicolored leaves, brilliant flowers, and often showy fruit, and is often used as garden trees or hedges [8].

Berberis seeds are known for their physiologically non-deep dormancy [9]. Baskin and Baskin [3] reported that the seeds of five *Berberis* species, namely, *B. aristata*, *B. dictrophylla*, *B. dubia*, *B. kansuensis*, and *B. vernae*, displayed PD. Thakur et al. [10] reported that the dormancy of *B. aristata* seeds was broken under light conditions during the growth phase at 20 °C. Wang et al. [11,12] reported that *B. dictrophylla* and *B. kansuensis* seeds were subjected to a low-temperature wet treatment for 80 days to break their dormancy. The dormancy of *B. dubia* and *B. vernae* seeds was broken after culturing at 20/15 °C in the growth phase after 168 days of a low-temperature wet treatment [3]. Deb at al. [13] proved that the effect of chilling and light can break the seed dormancy of *B. manipurana* seeds. The seeds of most plants in the *Berberis* genus have PD and reports have suggested that that their dormancy is broken under cold stratification treatments and temperature conditions of 20 °C [14–17].

Some studies have reported on the inter- and intraspecific variability in the fruit and/or seed traits in the *Berberis* genus [18–22]. These studies have suggested that the differences between the populations for all the analyzed fruit and seed traits depend on environmental factors and species-related differences. Therefore, in this study, we present the external morphological measurements of *B. koreana* seeds.

Therefore, we hypothesize as follows: (a) it is expected that *B. koreana* seeds will also have PD, (b) *B. koreana* seeds will absorb water, (c) a low-temperature wet treatment will break the dormancy of *B. koreana* seeds, and (d) a hormone treatment (GA_3) will break the dormancy of *B. koreana* seeds. The results of this research can be used as reference data for further research on the morphological characteristics of barberry plants in Korea.

2. Materials and Methods

2.1. Experimental Materials

The seeds of *B. koreana* used in this study were obtained from wild plants growing near Samneung in Paju (37°44′35″ N, 126°49′27″ E), Republic of Korea, on 17 October 2019. The harvested fruit was removed using the repair method and the selected seeds were shade-dried for 7 days in a well-ventilated space. The seeds were then stored under refrigeration (4–5 °C) until further use.

2.2. Investigation of Internal and External Characteristics of Seeds

To investigate the external morphological characteristics of the seeds, images were acquired using a scanning electron microscope (SEM; CX-200, COXEM, Daejeon, Republic of Korea). The weights of the dried seeds were measured per 1000 dried seeds in triplicate.

To investigate the internal morphological characteristics, the seeds were cut in half using a double-edged razor (stainless steel blade, Dorco, Seoul, Republic of Korea) and photographed using a digital microscope (DVM6, Leica, Land Hessen, Germany). Changes in the embryos and endosperms were observed before and after the seed germination.

2.3. Seed Disinfection and Setting

Before the start of the experiment, the seeds were surface-sterilized via soaking in 1000 $mg \cdot L^{-1}$ of benomyl (FarmHannong, Seoul, Republic of Korea) for 24 h; the seeds were then washed three times with distilled water. The surface-sterilized seeds were placed over two sheets of filter paper (Whatman No. 1; GE Healthcare, Buckinghamshire, UK) in a Petri dish containing 5 mL of distilled water.

After the GA_3 and cold stratification treatments (described below), the germination percentage was investigated in a growth chamber (TGC-130H, Espec Mic Corp., Aichi, Japan), where the seeds were cultured at a constant temperature of 20 °C. The experiment was performed with four replicates and 25 seeds per treatment group. If microorganisms were present during the culture period, they were disinfected by being soaked for 24 h in

1000 mg·L^{-1} of benomyl (FarmHannong), followed by being washed in distilled water; the distilled water was replenished in the Petri dishes to prevent the drying of the filter paper.

The seeds were considered germinated when the radicle emergence through the seed coat was >1 mm; the germination percentages were recorded at 1-week intervals [23]. Seeds that died due to decay during the experiment were removed immediately and excluded while calculating the germination percentage.

2.4. Water Imbibition Test

To determine whether the *B. koreana* seeds were physically dormant, the moisture absorption percentages of the seeds were measured. A Petri dish containing two sheets of filter paper (Whatman No. 1) soaked in distilled water was plated with 100 seeds; three replicates were maintained. The initial weight before water absorption and the weights at 3, 6, 9, 12, 24, 36, and 48 h after settling were measured. The water absorption percentage was calculated using the following formula [24]:

$$\%\text{Ws} = [(\text{Wh} - \text{Wi})/\text{Wi}] \times 100 \quad (1)$$

where Ws is the relative increase in the weight of the seeds due to moisture absorption, Wh is the weight of the seeds per hour after the water supply, and Wi is the initial weight of the seeds in the dried state. The water imbibition test was conducted at the Seed Germination Laboratory of Baekdudaegan National Arboretum from 2–4 December 2019.

2.5. Effect of Temperature on Germination: A Move-Along Experiment

According to Baskin and Baskin [25], the "move-along experiment" provides the dormancy-breaking temperature required for germination in most species. Based on the move-along experiment, the temperature for each season was set and the changes in the germination rates were measured. For the temperature treatments, the four seasons of natural environmental conditions were set as spring (15 °C), summer (25 °C), autumn (20 °C), and winter (5 °C). The treatment groups were subjected to temperature changes from winter to spring and summer (T1: 5→15→20→25 °C) and temperature changes from summer to autumn and winter (T2: 25→20→15→5 °C). In the T1 and T2 treatment groups, the dwell times at each temperature were set to 12, 4, 4, and 12 weeks (Table 1). For all the treatment groups, 25 seeds were used in four replicates and the measurements were obtained at 1-week intervals. In addition, to observe the changes in the embryo and endosperm based on the temperature changes, the surfaces of the seeds were cut at 1-month intervals and photographed using a digital microscope. A controlled temperature shift experiment was conducted at the Seed Germination Laboratory of Baekdudaegan National Arboretum from 19 December 2019 to 9 July 2020.

Table 1. Outline of the modified temperature treatments [25].

No. of Weeks at Treatment Temperatures		**4**	**4**	**4**	**4**	**4**	**4**
Move along	T1	5 °C winter	5 °C winter	5 °C winter	15 °C early spring	20 °C late spring	25 °C summer
	T2	25 °C summer	25 °C summer	25 °C summer	20 °C early autumn	15 °C early autumn	5 °C winter

2.6. Effect of Cold Stratification Experiment on Germination

For the cold stratification treatment, the surface-sterilized seeds placed into a Petri dish were subjected to 4 °C in a growth chamber for 0, 2, 4, 8, and 12 weeks. At the end of each low-temperature treatment duration, the Petri dishes were shifted to a 20 °C growth chamber; the germination percentage was recorded while culturing the treatment groups at 20 °C and the experiment interval was set to 1 week. For all the treatment groups, 25 seeds were used in four replicates and the measurements were obtained at 1-week intervals. The experiments were conducted at the Seed Germination Laboratory of Baekdudaegan National Arboretum from 19 December 2019 to 9 July 2020.

2.7. *Experiment to Determine the Effect of GA_3 on Germination*

The seeds were soaked in solutions with 0 (distilled water, control), 10, 100, 500, or 1000 mg·L^{-1} of GA_3 for 24 h at room temperature and then incubated in a growth chamber at 20 °C. The germination rates were determined at 1-week intervals. For all the treatment groups, 25 seeds were used in four replicates and the measurements were obtained at 1-week intervals. The GA_3 treatment experiments were conducted at the Seed Germination Laboratory of Baekdudaegan National Arboretum from 1 April 2021 to 3 June 2021.

2.8. *Effect of Light Conditions on Seed Germination*

For the cold stratification, GA_3 treatment, and move-along experiment, similar light (12 h light/dark photoperiod) and dark conditions (24 h dark) were utilized. For all the treatment groups, 25 seeds were used in four replicates and the measurements were obtained at 1-week intervals.

The light conditions inside the growth chamber were maintained using fluorescent lamps with 40 ± 10 μmol·m^{-2}·s^{-1} PPFD. The dark conditions were maintained by wrapping the Petri dishes with aluminum foil.

2.9. *Phenology of Embryo Growth, Germination, and Seedling Emergence under Natural Environmental Conditions*

To observe the seasonal changes in the seeds under natural environmental conditions, the ground (almost loamy sand) was dug to a depth of 5 cm and a tray was planted in the nursery field of the Baekdudaegan National Arboretum. The tray was filled with potting soil with a mixing ratio of 64.3% cocopeat, 15% peatmoss, 2.5% nitrogen, 10% pearlite, 8% zeolite, 0.19% fertilizer, and 0.01% of a wetting agent. The phenology of the embryo growth, germination, and seedling emergence was investigated from 1 December 2019 to 1 August 2020.

2.9.1. Embryo Growth

Approximately 400 seeds were placed in fine-mesh polyester bags filled with river sand and buried in a tray filled with potted soil. The trays were placed at ground level in the nursery field. Every 2 or 4 weeks, a bag was exhumed, and 10 seeds were randomly selected for the embryo growth measurement. The seeds were cut into thin sections using a razor blade; the lengths of the seeds and embryos were measured using a digital microscope. The ratio of the embryo length to the seed length (E:S ratio) was calculated to correct for the positive correlation between the seed and embryo lengths.

2.9.2. Germination

Four replicates of 25 seeds were sown in 8 cm plastic pots filled with potting soil and placed in trays filled with the same potting soil. The trays were placed at the ground level in the experimental garden. The seeds with emerged radicles were counted and removed every week. The seeds were considered to have "germinated" when the radicle protrusion length was at least 1 mm. The intact seeds that did not germinate were buried in the field.

2.9.3. Seedling Emergence

The timing of the seedling emergence was monitored by sowing four replicates of 25 seeds at a depth of 3 cm in plastic pots filled with potting soil, which were placed in the trays described above. The emerged seedlings were counted and removed every week during the field experiments. The pots were covered with nets to prevent disturbance from wild animals.

2.10. Statistical Analyses

The statistical analyses were performed using SPSS version 21 (SPSS Inc., Chicago, IL, USA). The results of the germination experiment were subjected to an analysis of variance and Duncan's multiple range test ($p \leq 0.05$).

3. Results

3.1. Investigation of Internal and External Characteristics of Seeds

To investigate the morphological characteristics of the *B. koreana* seeds, they were photographed using scanning electron and digital microscopes (Figure 1). The color of the seed coat was reddish-brown (Figure 1c). When part of the seed coat was magnified and photographed using an electron microscope, it revealed a curvature (Figure 1a,b). The examination of the dissected seeds using a digital microscope confirmed the development of the embryo in the seed of the ripe fruit (Figure 1d).

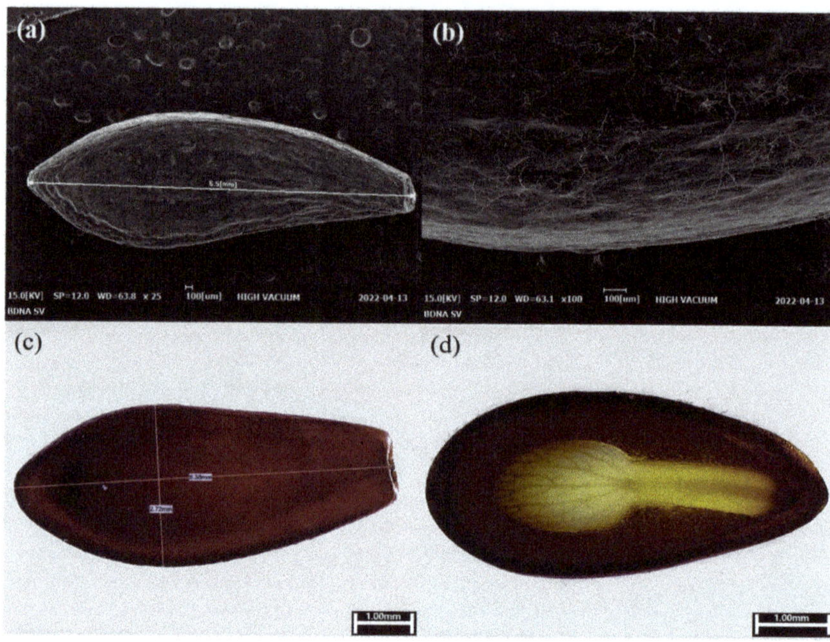

Figure 1. Seed morphology of *Berberis koreana*. (**a**) Scanning electron micrograph of the seed. (**b**) Scanning electron micrograph of an enlarged part of the seed coat. (**c**) Digital image of the seed. (**d**) The cutting plane of the seed showing the developed embryo. Scale bars are 100 μm (**a**,**b**) and 1 mm (**c**,**d**).

The mean length of the embryo was 4.00 ± 0.04 mm (mean ± standard error); the mean seed length was 6.48 ± 0.03 mm; and the E/S ratio (embryo: seed ratio) was 0.62 ± 0.05. In addition, the mean weight of 1000 seeds was 11.514 ± 0.392 g.

3.2. Water Imbibition Test

The water imbibition test, performed to evaluate the permeability of the *B. koreana* seeds, showed that the seed weight increased by 34.10 ± 0.54% in 24 h and 57.13 ± 0.58% in 48 h (Figure 2).

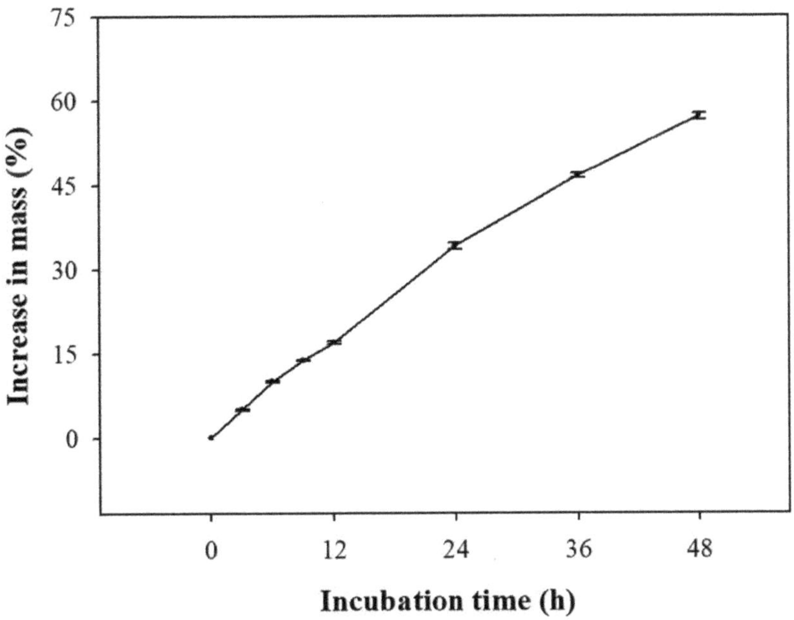

Figure 2. Water uptake by intact seeds of *Berberis koreana* represented by an increase in weight. Seeds were incubated at ambient conditions (22–25 °C) on filter paper moistened with distilled water for 48 h. Vertical error bars represent standard error (n = 3).

3.3. Effect of GA_3 Treatment on Seed Germination

The germination percentages of the *B. koreana* seeds subjected to the GA_3 treatment for 30 weeks at 20 °C under light/dark conditions were as follows: at GA_3 concentrations of 0, 10, 100, 500, and 1000 mg/L, the final germination percentages under the light/dark cycle conditions were 7.00 ± 3.00, 5.00 ± 1.91, 4.00 ± 1.63, 7.00 ± 1.91, and 13.00 ± 2.52%, respectively, and the final germination percentages under the dark conditions were 4.00 ± 1.63, 1.00 ± 1.00, 0, 11.00 ± 3.42, and 6.00 ± 3.46%, respectively. Both conditions showed a significant difference when the GA_3 concentration was 500 mg/L or higher (Figure 3).

3.4. Effect of Cold Stratification Experiment on Seed Germination

The seeds treated with cold stratification for 0, 2, 4, 8, and 12 weeks were cultured in a growth chamber (light/dark conditions) at 20 °C for 30 weeks. The final germination percentage of the seeds treated with cold stratification for 12 weeks was the highest at 71.00 ± 1.91%, regardless of the light conditions (Figure 4a).

Under the light conditions, the final germination percentage of the seeds was the highest at 71 ± 1.91% when treated with cold stratification for 12 weeks. The germination rates at 4, 8, and 12 weeks of cold stratification were statistically identical. Under the dark conditions, the final germination percentage of the seeds was the highest at 70.00 ± 3.83% when treated with cold stratification for 12 weeks.

In addition, the cold stratification for 12 weeks under the light conditions and the cold stratifications for 8 and 12 weeks under the dark conditions were observed to initiate germination during the cold stratification treatment period.

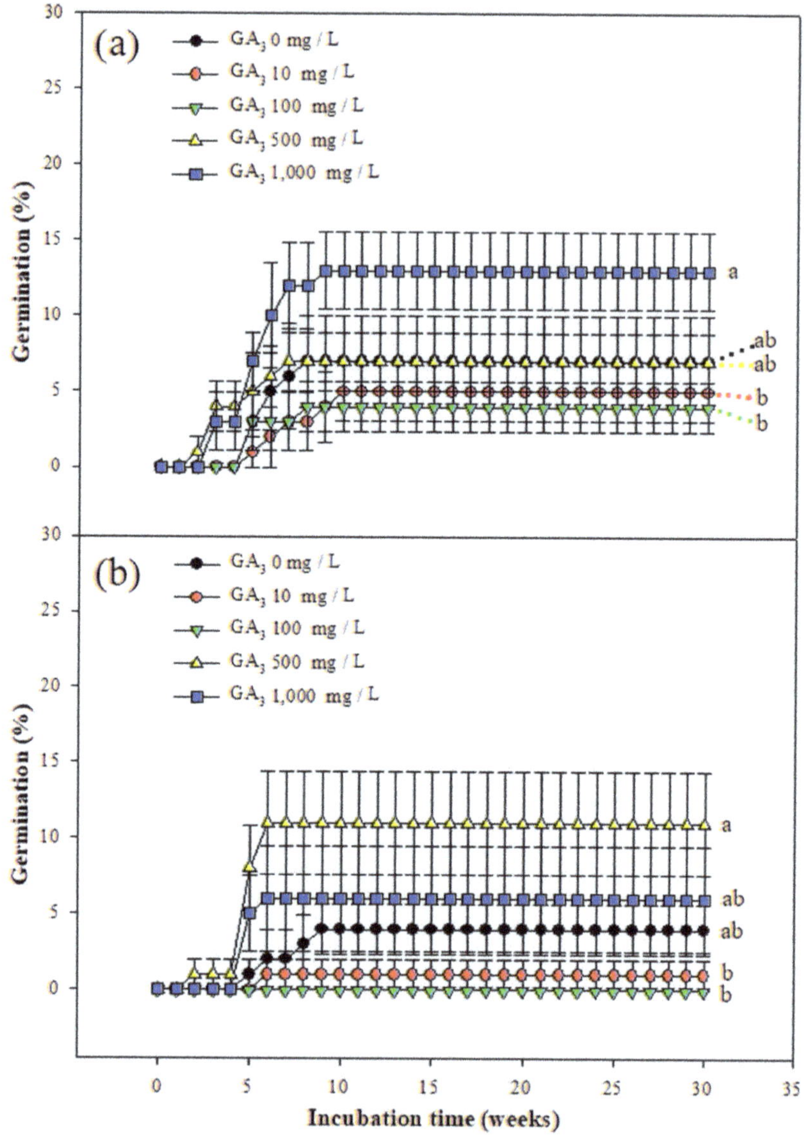

Figure 3. Germination of *Berberis koreana* seeds subjected to GA_3 treatment (0, 10, 100, 500, or 1000 $mg \cdot L^{-1}$). Seeds were soaked in a GA_3 solution for 24 h and then incubated for 30 weeks at 20 °C. (**a**) Seeds incubated under a light/dark cycle of 12 h/12 h. (**b**) Seeds incubated under dark conditions. Vertical error bars represent standard error (n = 4). Final germination percentages designated by different small letters are significantly different at $p \leq 0.05$ (Duncan's multiple range test).

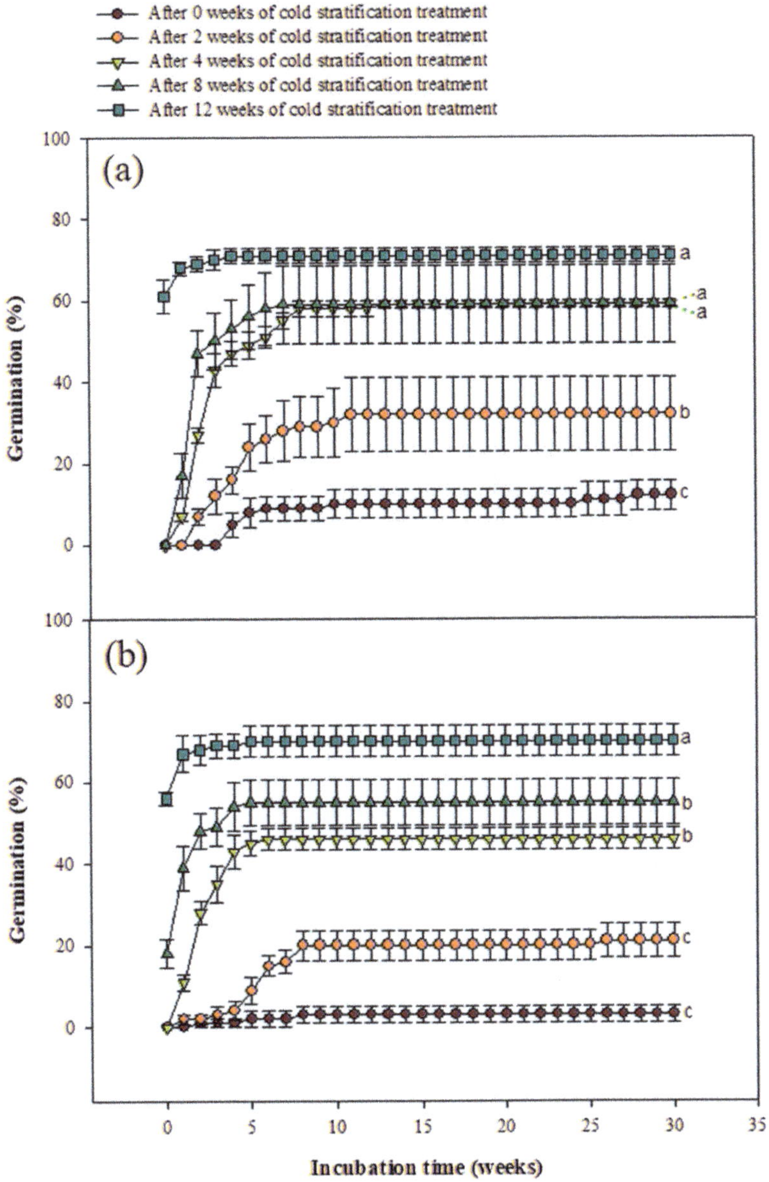

Figure 4. Germination of *Berberis koreana* seeds subjected to cold stratification (0, 2, 4, 8, or 12 weeks) treatment. (**a**) Seeds incubated under a light/dark (12 h/12 h) cycle. (**b**) Seeds incubated under dark conditions. Vertical error bars represent standard error (n = 4). Final germination percentages designated by different small letters are significantly different at $p \leq 0.05$ (Duncan's multiple range test).

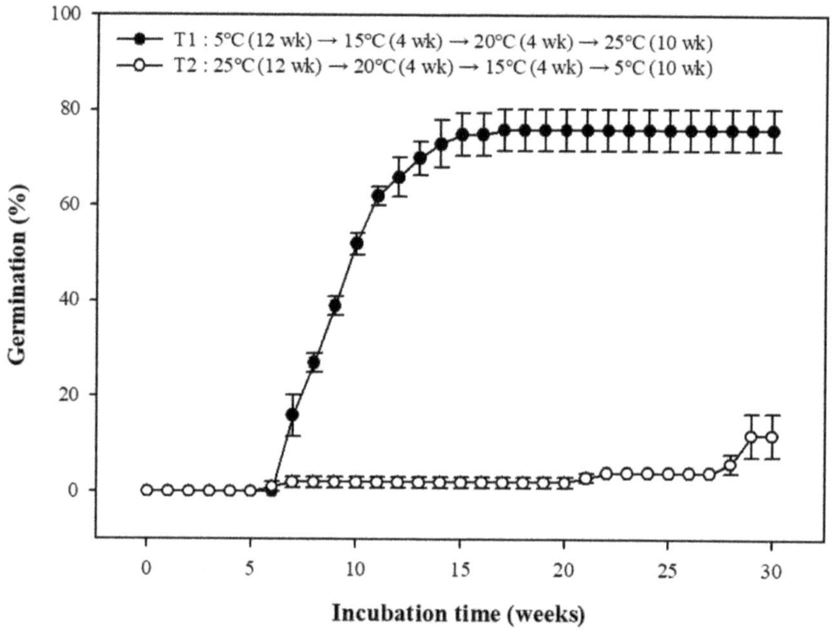

Figure 5. Seed germination in *Berberis koreana* incubated under a temperature sequence beginning at 5 °C (T1) or 25 °C (T2). Vertical error bars represent standard error (n = 4).

3.5. Seed Germination Based on Temperature Conditions: A Move-along Experiment

A move-along experiment indicated final germination percentages of 76.00 ± 4.32% at T1 (5→15→20→25 °C) and 12.00 ± 4.62% at T2 (25→20→15→5 °C), indicating that the temperature change from winter to spring and summer showed better results (Figure 5). The *Berberis koreana* seeds germinated under the T1 temperature conditions as follows: (1) the seeds started to germinate in 6 weeks under the winter temperature conditions (5 °C); the germination percentage was 66.0 ± 4.16% until 12 weeks; (2) the germination percentage increased to 73.00 ± 4.43% under the early spring temperature conditions (15 °C, 13–16 weeks); (3) after changing the temperature conditions in late spring (20 °C), the final germination percentage was 76.00 ± 4.32% at 17 weeks; and (4) further germination of the seeds was not observed.

3.6. Phenology of Embryo Growth, Germination, and Seedling Emergence

The cutting planes of the seeds observed at 2–4-week intervals showed that there was no increase in the E:S ratio (Figure 6). The E:S ratio was measured for the ungerminated seeds. The germination rate measurements based on seasonal changes showed that germination began in the 13th week (19 March 2020) and continued until the 18th week (23 April 2020); the final germination percentage was 80.00 ± 1.65%. The seedling emergence rate measurements were observed from the 17th week (16 April 2020) and were completed at 47.00 ± 3.42% in the 25th week (11 June 2020; Figure 7).

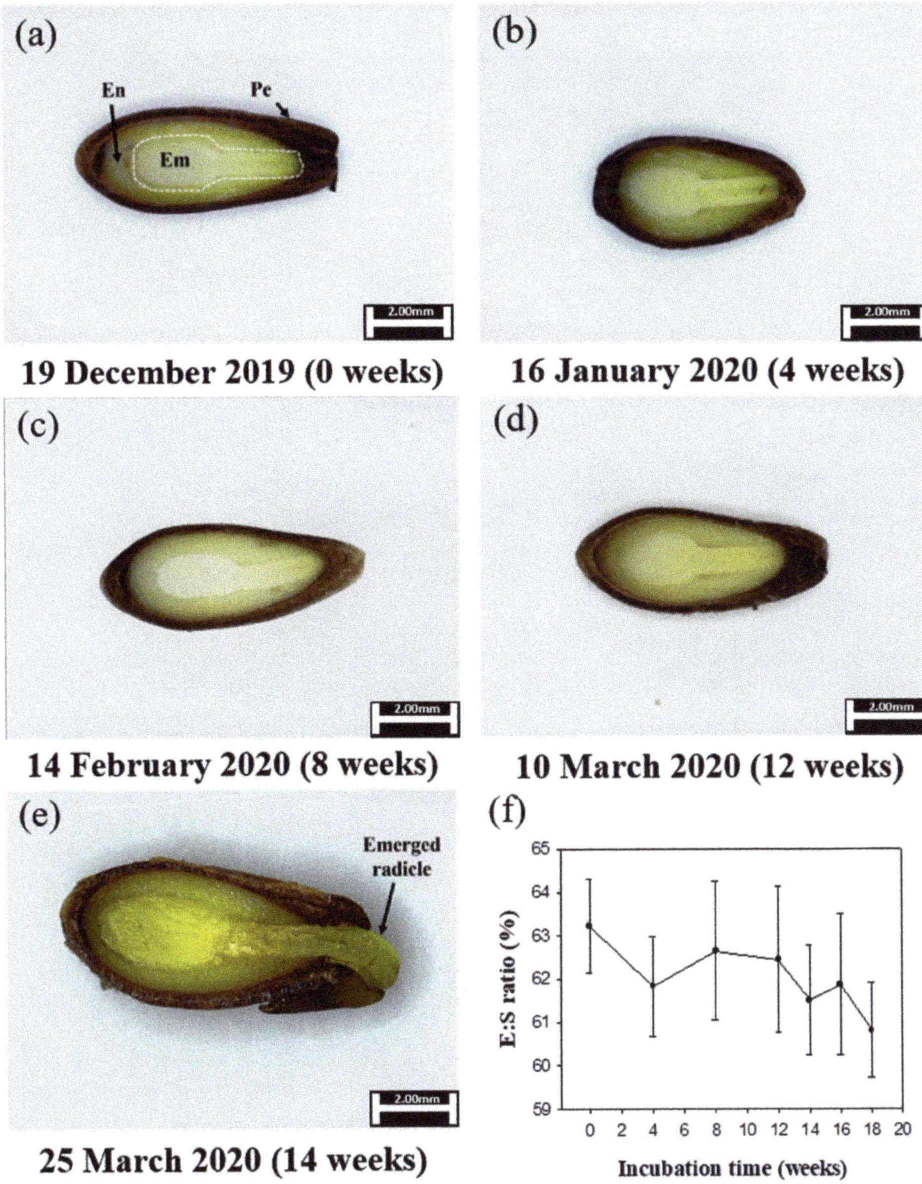

Figure 6. Embryo growth (E:S ratio) and radicle emergence of *Berberis koreana* seeds kept outdoors in BongHwa, Republic of Korea, in 2020. Scale bars are 2.0 mm. (**a**–**e**). Em, embryo; En, endosperm; and Pe, pericarp. Vertical bars represent standard error (n = 10) (**f**).

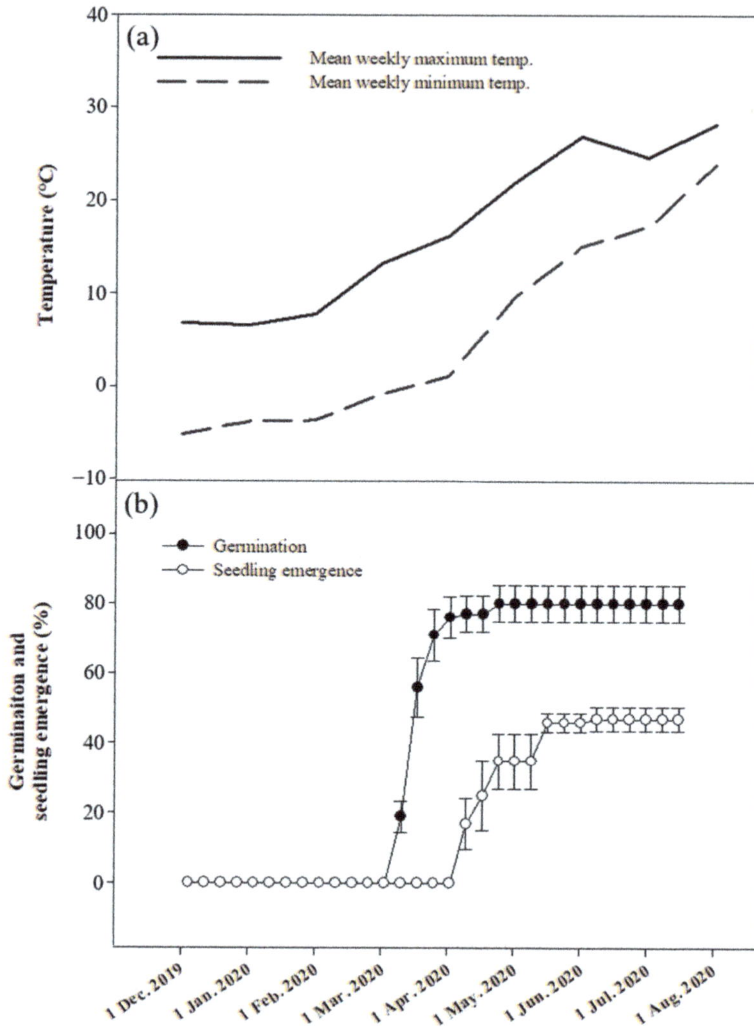

Figure 7. Temperature variations and phenology of *Berberis koreana* seeds buried at a depth of 3 cm in 2019. (**a**) Mean weekly maximum and minimum temperatures; and (**b**) germination and seedling emergence. Vertical bars represent standard error (n = 10).

4. Discussion

The measurements of the cutting planes of the *B. koreana* seeds with time showed that there was no increase in the E/S ratio, and no embryo growth was observed (Figure 6). According to Baskin and Baskin [25], if a seed has an immature embryo, it will grow and germinate within 30 days of incubation under appropriate conditions, referred to as MD. The results of our study indicate that *B. koreana* seeds do not show MD.

The *B. koreana* seeds showed water permeability as the seed weight increased by 134% within 24 h of water imbibition (Figure 2). According to Baskin and Baskin [25], if the water absorption rate of the seed is ≤20% of its dry weight within 24 h, it is considered to be impermeable; this phenomenon is termed PY. Our results indicate that *B. koreana* seeds do not have PY.

According to Baskin and Baskin [26], the seeds of most *Berberis* plants have PD; seeds with PD can be divided into three types: deep PD, intermediate PD, and non-deep PD. In the case of non-deep PD, it has been reported that dormancy can be broken using a GA_3 treatment. The GA_3 treatment of the *B. koreana* seeds showed that there was no correlation between the GA_3 concentration and germination percentage, regardless of the light/dark cycle or dark conditions (Figure 3). In addition, the maximum germination percentage was 13.0%, which showed that GA_3 had a limited influence on the seed germination. Therefore, it was concluded that GA_3 did not affect the dormancy breakage; *B. koreana* seeds do not have non-deep PD.

The germination percentage of the *B. koreana* seeds increased with an increase in the period of the cold stratification treatment (Figure 4), indicating that it ended their dormancy. According to Zheng et al. [27], among the dormant types, dormancy can be broken with 2–3 months of a cold stratification treatment in the case of intermediate PD and 3–4 months in the case of deep PD. Our results indicate that *B. koreana* seeds are of the "intermediate PD" type. Similar results have been obtained by previous studies conducted using seeds of the same genus [15,17].

However, in all the experiments, there was no significant difference in the germination percentages between the seeds subjected to the light/dark cycles and dark conditions. Therefore, it appears that *B. koreana* seeds germinate regardless of the light or dark conditions. In addition, some seeds started germinating during the cold stratification treatment. From these results, it can be hypothesized that some *B. koreana* seeds have a life cycle that begins within the ground under natural conditions during winter and then germinate as the temperature rises.

Seeds with PD break their dormancy with a cold stratification treatment, but this dormancy may also be broken using a warm stratification treatment [26]. A move-along experiment indicated an increase in the germination percentage under the T1 treatment, whereas the seeds under the T2 treatment showed a poor germination percentage (Figure 5). These results indicate that the dormancy release of *B. koreana* seeds is affected by cold and not warm stratification treatment.

The phenology experiment showed that most seeds germinated during the period of 6–12 weeks at 5 °C, which is characteristic of winter (Figure 7). These results suggest that natural *B. koreana* seeds germinate in the ground during winter and seedling emergence occurs when the temperature rises in spring. These observations were similar to those observed in the cold stratification experiment and consistent with those of previous reports [9,12,28–30].

5. Conclusions

We hypothesized that (b) *B. koreana* seeds will absorb water, (c) a low-temperature wet treatment will break the dormancy of these *B. koreana* seeds, and (d) a hormone treatment (GA_3) will break the dormancy of these *B. koreana* seeds. Our results show that our (b) and (c) hypotheses were consistent and that a GA_3 treatment had little effect on the seed dormancy (hypothesis d). The most effective method for breaking the dormancy was a cold stratification treatment at 5 °C for 12 weeks. Collectively, the physiological characteristics indicated that, in the natural environment, the seeds of *B. koreana* begin to germinate at an average temperature of 20 °C after experiencing low-temperature conditions for over 8 weeks in the soil. Thus, the seeds of *B. koreana* Palibin exhibit intermediate PD among the dormant seed types (hypothesis a).

Author Contributions: Conceptualization, D.-H.K., C.-S.N. and D.-H.L.; methodology, D.-H.K., C.-S.N. and D.-H.L.; validation, D.-H.K., C.-S.N. and D.-H.L.; formal analysis, D.-H.K.; investigation, D.-H.K. and S.-G.K.; resources, S.-G.K.; data curation, D.-H.K.; writing—original draft preparation, D.-H.K.; writing—review and editing, D.-H.K., C.-S.N. and D.-H.L.; visualization, D.-H.K.; supervision, C.-S.N. and D.-H.L.; project administration, H.L.; funding acquisition, H.L. All authors have read and agreed to the published version of the manuscript.

Funding: This work was supported by a National Research Foundation of Korea grant funded by the Korean Government (NRF-2019R1I1A2A01062559).

Data Availability Statement: Not applicable.

Conflicts of Interest: The authors declare no conflict of interest.

References

1. Finch-Savage, W.E.; Leubner-Metzger, G. Seed dormancy and the control of germination. *New Phytol.* **2006**, *171*, 501–523. [CrossRef] [PubMed]
2. Kim, K.S.; Lee, S.Y.; Rhie, Y.H.; Kim, Y.J. Morphological and morphophysiological dormancy in seeds of several spring ephemerals native to Korea. *Flower Res. J.* **2012**, *20*, 193–199. [CrossRef]
3. Baskin, C.C.; Baskin, J.M. *Seeds: Ecology, Biogeography and Evolution of Dormancy and Germination*, 2nd ed.; Elsevier Academic Press: San Diego, CA, USA, 2014.
4. Lang, G.A. Dormancy: A new universal terminology. *HortScience* **1987**, *22*, 817–820. [CrossRef]
5. Yoo, S.J.; Lee, K.B.; Kwak, J.H. Studies on the seasonal variation of berberine contents in *Berberis koreana*. *Korean J. Pharmacogn.* **1986**, *17*, 123–128.
6. Hyun, M.S.; Woo, W.H.; Hur, J.M.; Kim, D.; Mun, Y.J. The role of ROS and p38 MAP kinase in berberine-induced apoptosis on human hepatoma Hep G2 cells. *J. Korean Soc. Appl. Biol. Chem.* **2008**, *51*, 129–135.
7. Ling, J.; Ha, J.; Choi, Y.; Seo, Y.; Kim, J.; Kim, Y.; Cha, S.; Kim, J.; Lee, H. Enhancement of cosmeceutical activities of *Berberis koreana* bark by high pressure and ultrasonification extraction processes. *Korean J. Med. Crop Sci.* **2011**, *19*, 54–65. [CrossRef]
8. Verma, N.; Singh, J.P.; Mishra, D.; Nath, A.; Kashyap, P.; Dutta, D. Indian barberry (*Berberis aristata* D. C). In *Underutilized Fruit Crops: Importance and Cultivation*; Jaya Publishing House: New Delhi, India, 2017; pp. 143–154.
9. Belwal, T.; Bisht, A.; Bhatt, I.D.; Rawal, R.S. Influence of seed priming and storage time on germination and enzymatic activity of selected Berberis species. *Plant Growth Regul.* **2015**, *77*, 189–199. [CrossRef]
10. Thakur, A.; Thakur, P.S.; Mehta, R. Studies on germination, viability and vigour in Indian barberry (*Berberis aristata* DC.)—An endangered medicinal plant species of western Himalayas. *Indian For.* **2006**, *132*, 485–492.
11. Wang, J.H.; Du, G.Z.; Cui, X.L.; Zheng, X.F.; Qi, W. Germination characteristics of 61 common woody species from the eastern Qinghai-Tibet Plateau of China and their life history correlates. *J. Plant Ecol.* **2009**, *33*, 171–179.
12. Wang, J.H.; Baskin, C.C.; Chen, W.; Du, G.Z. Variation in seed germination between populations of five sub-alpine woody species from eastern Qinghai-Tibet Plateau following dry storage at low temperatures. *Ecol. Res.* **2010**, *25*, 195–203. [CrossRef]
13. Deb, C.R.; Sangtam, T.L.; Jamir, N.S. Seed Biology of *Berberis manipurana* Ahrendt: A Threatened Natural Dye Yielding Plant. *Am. J. Plant Sci.* **2017**, *8*, 1285–1295. [CrossRef]
14. Bahuguna, V.K.; Rawart, M.M.S.; Joshi, S.R. Preliminary studies on seed germination behaviour of *Berberis lycium* Royle—An important shrub for reclamation of wastelands in the Himalaya. *Indian For.* **1988**, *114*, 181–183.
15. Sathyakumar, S.; Viswanath, S. Observations on food habits of Asiatic black bear in Kedarnath Wildlife Sanctuary, India: Preliminary evidence on their role in seed germination and dispersal. *Ursus* **2003**, *14*, 99–103.
16. Rudolf, P.O.; Buckeye, A.L. Horse chestnut. Seeds of Woody Plants in the United States. *Agric. Handb.* **1974**, *450*, 195–199.
17. Taylor, J. Propagation successes, failures and lessons learned. In *Proceedings of the Conference: Native Plant Propagation and Restoration Strategies*; Nursery Technology Cooperative and Western Forestry and Conservation Association: Eugene, Oregon, 2002; pp. 45–54.
18. Jannatizadeh, A.; Khadivi-Khub, A. Morphological variability of *Berberis integerrima* from Iran. *Erwerbs-Obstbau* **2016**, *58*, 247–252. [CrossRef]
19. Karlović, K.; Kremer, D.; Liber, Z.; Šatović, Z.; Vršek, I. Intra- and interpopulation variability and taxonomic status of *Berberis croatica* Horvat. *Plant Biosyst.* **2009**, *143*, 40–46. [CrossRef]
20. Akbulut, M.; Çalişir, S.; Marakoğlu, T.; Çoklar, H. Some physicomechanical and nutritional properties of barberry (*Berberis vulgaris* L.) fruits. *J. Food Process Eng.* **2009**, *32*, 497–511. [CrossRef]
21. Allen, R.B.; Wilson, J.B. Fruit and seed production in *Berberis darwinii* Hook., a shrub recently naturalised in New Zealand. *N. Z. J. Bot.* **1992**, *30*, 45–55. [CrossRef]
22. Kremer, D.; Jurišić Grubješić, R.; Popović, Z.; Karlović, K. Fruit and seed traits of *Berberis croatica* Horvat and *Berberis vulgaris* L. *Acta Bot. Croat.* **2012**, *71*, 115–123. [CrossRef]
23. ISTA. International Seed Testing Association. International rules for seed testing. *Seed Sci. Technol.* **1999**, *27*, 333.
24. Lee, S.Y.; Rhie, Y.H.; Kim, K.S. Non-deep simple morphophysiological dormancy in seeds of *Thalictrum rochebrunianum*, an endemic perennial herb in the Korean Peninsula. *Hortic. Environ. Biotechnol.* **2015**, *56*, 366–375. [CrossRef]
25. Baskin, C.C.; Baskin, J.M. When breaking seed dormancy is a problem: Try a move-along experiment. *Nat. Plants J.* **2003**, *4*, 17–21. [CrossRef]
26. Baskin, J.M.; Baskin, C.C. A classification system for seed dormancy. *Seed Sci. Res.* **2004**, *14*, 1–16. [CrossRef]
27. Zheng, R.; Ma, Z.; Jiang, L.; Zhao, Z.; Shi, X.; Wang, L. Germination characteristics of plump and shriveled seeds of *Tamarix ramosissima* matured in different seasons. *Seed Sci. Technol.* **2022**, *50*, 21–25. [CrossRef]

28. Bravo, C.; Chamorro, D.; Hiraldo, F.; Speziale, K.; Lambertucci, S.A.; Tella, J.L.; Blanco, G. Physiological dormancy broken by endozoochory: Austral parakeets (*Enicognathus ferrugineus*) as legitimate dispersers of calafate (*Berberis microphylla*) in the Patagonian Andes. *J. Plant Ecol.* **2020**, *13*, 538–544. [CrossRef]
29. Khudonogova, E.; Zatsepina, O.; Polovinkina, S.; Rachenko, M.; Tyapaeva, M. Seed germination of woody and shrubby introduced species. *IOP Conf. Ser. Earth Environ. Sci.* **2019**, *316*, 12–21. [CrossRef]
30. Larsen, S.U.; Eriksen, E.N. Delayed release of primary dormancy and induction of secondary dormancy in seeds of woody taxa caused by temperature alternations. *Acta Hortic.* **2004**, *630*, 91–100. [CrossRef]

horticulturae

MDPI

Article

Chitosan Soaking Improves Seed Germination of *Platycodon Grandiflorus* and Enhances Its Growth, Photosynthesis, Resistance, Yield, and Quality

Hai Liu [1], Zhihong Zheng [1], Xue Han [1], Cheng Zhang [2], Haitao Li [3,*] and Mingkai Wu [1,*]

1 Institute of Modern Chinese Herbal Medicines, Institute of Crop Germplasm Resources, Guizhou Academy of Agricultural Sciences, Guiyang 550025, China
2 School of Public Health, Guizhou Medical University, Guiyang 550025, China
3 Department of Food and Medicine, Guizhou Vocational College of Agriculture, Qingzhen 551400, China
* Correspondence: lisea02@126.com (H.L.); bywmk1999@163.com (M.W.)

Abstract: *Platycodon grandiflorus* is a medical, ornamental, and edible traditional Chinese medicine whose seed germination and plant growth are frequently restricted by dormancy and stresses. In this study, we investigated how chitosan soaking affected seed germination, growth, photosynthesis, resistance, yield, and quality of *P. grandiflorus*. The results indicated that chitosan soaking had a preferable enhancing effect on seed germination of *P. grandiflorus*, which significantly ($p < 0.05$) promoted its germination rate, energy, and index, as well as cotyl and radicle length. Furthermore, 0.15–0.20% chitosan soaking effectively enhanced the leaf growth, height, stem diameter, and overground part dry weight of *P. grandiflorus* and reliably improved their leaves' chlorophyll, photosynthetic rate, transpiration rate, and water use efficiency. Moreover, 0.15–0.20% chitosan soaking effectively enhanced the stress resistance and adaptability of *P. grandiflorus* via increasing its resistance substances and triggering its defense enzyme activity. Meanwhile, 0.15–0.20% chitosan soaking effectively improved the underground part growth and medical quality of *P. grandiflorus*. This study highlights that chitosan can be used as a favorable, efficient, and economical candidate or promoter for enhancing seed germination of *P. grandiflorus* and improving its growth, photosynthesis, resistance, yield, and quality; it also highlights that 0.15–0.20% chitosan is a suitable concentration.

Keywords: chitosan; *Platycodon grandiflorus*; seed germination; growth; resistance; yield; quality

Citation: Liu, H.; Zheng, Z.; Han, X.; Zhang, C.; Li, H.; Wu, M. Chitosan Soaking Improves Seed Germination of *Platycodon Grandiflorus* and Enhances Its Growth, Photosynthesis, Resistance, Yield, and Quality. *Horticulturae* **2022**, *8*, 943. https://doi.org/10.3390/horticulturae8100943

Academic Editors: Konstantinos Bertsouklis, Epaminondas Kartsonas and Angela Carra

Received: 22 September 2022
Accepted: 11 October 2022
Published: 14 October 2022

1. Introduction

Platycodon grandiflorus (Jacq) A.DC., a medical, ornamental, and edible perennial plant, has been widely planted in China, North Korea, South Korea, Japan, and Russia [1,2]. It is rich in saponins, polysaccharides, vitamins, amino acids, flavonoids, phenolics, anthocyanins, and various minerals, etc. [1–3]. As an oriental traditional herb, it is widely used for treating sore throats, excessive phlegm, coughing, amnesia, dementia, and inflammatory diseases due to its notably anti-asthmatic, anti-tumor, anti-oxidant, anti-inflammatory, anti-obesity, hepatoprotective and immunoregulative, and other characteristics [1,2,4–6]. Meanwhile, its flower buds are found in different colors, including blue, purple, red, violet, white, or pink, which have an extremely high ornamental value [2,7]. Moreover, as a functional food ingredient or wild vegetable, its tender seedlings and roots are usually processed as soup, salads, sauces, pickles, noodles, preserved fruits, and health drinks, etc. [8,9]. Recently, the *P. grandiflorus* industry in China has made a major contribution to alleviating poverty and revitalizing rural aspects, and its annual output reaches 1 million kg, of which exports account for half [7]. Due to the major development values and prospects of *P. grandiflorus*, artificial cultivation technology to enhance its yield and medicinal quality has been paid growing attention.

Unfortunately, the seed size of *P. grandiflorus* is small, its germination and emergence rates are also weak, and it easily dies from drought under the traditional artificial cultivation conditions. Furthermore, *P. grandiflorus* seeds frequently show dormancy as they contain endogenous germination inhibitory substances with high activity [10,11]. Additionally, its germination is also significantly influenced by several factors, such as various abiotic and biotic stresses [12,13]. Currently, stratification, chemical, physical, and hormone treatments are often used to relieve dormancy and promote germination of *P. grandiflorus* seeds [14]. For example, Wang et al. [15] reported that 0.005 g mL^{-1} KNO_3 and 0.003 g mL^{-1} $KMnO_4$ had the best enhanced influence on seed germination and seedling growth of *P. grandiflorus*. Zhang et al. [16] reported that 0.10–0.40 g L^{-1} of 15% alginate water-soluble fertilizer could effectively promote seed germination and plant growth of *P. grandiflorus*. Subsequently, the authors also found that 25,000–20,000 folds liquid of 0.136% gibbenic acid·indoleacetic acid·brassicin wettable powder could also significantly promote its seed germination and plant growth [17]. Considering the extremely limiting factors for seed germination of *P. grandiflorus* and the potentially low efficiency or restricting use of the reported techniques or chemicals, more novel candidate strategies or promoters need to be developed or found to enable meeting the sustainable development of the *P. grandiflorus* industry.

Chitosan, a linear copolymer of β-(1-4)-2-amido-2-deoxy-D-glucan (glucosamine) and β-(1-4)-2-acetamido-2-deoxy-D-glucan (acetylglucosamine), is not toxic to humans or other organisms [18,19]. As an ideal natural polymer, it has been widely utilized in agriculture, medical care, food, cosmetics, and other fields due to its favorable bioactivity, film-forming ability, biodegradability, non-toxicity, and renewability [20,21]. In agriculture, it has emerged as a promising resource used as a plant growth enhancer, resistance inductor, and bio-fungicide [21–23]. It can effectively stimulate and enhance plant growth by affecting plant physiological processes, such as seed germination, nutrient uptake, cell division and elongation, stress resistance, and so on, eventually leading to increased yield and improved quality [21,24–28]. Pan et al. [29] reported that 50 mg L^{-1} chitosan effectively enhanced the seed germination, radicle and germ length, plant height, fresh and dry weight, and root–shoot ratio of *Trifolium repens* under salt stress. Additionally, Li et al. [30] found that 0.15% chitosan significantly improved seed germination of *Sctellaria baicalensis* and its seedlings' drought resistance. To date, there has been no documentation of whether chitosan enhances seed germination and growth of *P. grandiflorus*. In this case, whether chitosan can be used as an enhancer or candidate for enhancing seed germination and plant growth of *P. grandiflorus* is very worthy of intensive study.

In this study, the influences of chitosan soaking on seed germination of *P. grandiflorus* were primarily investigated. Moreover, the influences of chitosan soaking on growth, photosynthesis, resistance, yield, and quality of *P. grandiflorus* were also evaluated. This work provides a safe, effective, and economical candidate or enhancer for promoting seed germination of *P. grandiflorus* and improving its growth, underground part weight, and quality.

2. Materials and Methods

2.1. Experimental Materials

Chitosan with deacetylation of 90.00% and molecular weight of 50 KDa was purchased by Mingrui Bioengineering Co. Ltd. (Zhengzhou, China). $KMNO_4$ was produced from Kemiou Chemical Reagent Co. Ltd. (Tianjin, China). *P. grandiflorus* seeds were collected from the medicinal botanical garden at the Guizhou Academy of Agricultural Sciences and identified as seeds of *P. grandiflorus* (Jacq.) A.DC. by Prof. Mingkai Wu, who is a researcher of Guizhou Academy of Agricultural Sciences. The potted soils were also collected from the above-mentioned medicinal garden, and the soil fertility information is shown in Table 1.

Table 1. The fertility information of the potted soils.

Indices	Content	Indices	Content
Organic matter	15.36 g kg^{-1}	Available zinc	0.83 mg kg^{-1}
Total nitrogen	1.45 g kg^{-1}	Available iron	7.04 mg kg^{-1}
Total phosphorus	1.73 g kg^{-1}	Available boron	0.16 mg kg^{-1}
Total potassium	1.18 g kg^{-1}	Available manganese	16.31 mg kg^{-1}
Available nitrogen	57.33 mg kg^{-1}	Exchangeable magnesium	315.78 mg kg^{-1}
Available phosphorus	4.92 mg kg^{-1}	Exchangeable calcium	16.85 cmol kg^{-1}
Available potassium	28.27 mg kg^{-1}	pH value	6.42

2.2. Seed Germination Experiment of *P. grandiflorus*

Six soaking concentrations of chitosan were designed: 0.00%, 0.05%, 0.10%, 0.15%, 0.20%, and 0.25% chitosan solution. Distilled water was used to dissolve and dilute chitosan. The full seeds of *P. grandiflorus* were first sterilized with 0.5%$KMNO_4$ for 30 min and then washed with distilled water and subsequently soaked in a 50 mL chitosan solution for 12 h. The soaked seeds were washed with distilled water and then arranged evenly on the germination bed, which was a 9 cm diameter Petri dish with a covering of two wet filter papers. Each germination bed had one hundred seeds and each treatment was repeated three times. The germinating bed was incubated in a dark incubator at 25 °C; an appropriate amount of distilled water was added every day to ensure that the filter paper was moist. Germination rate, energy, and index of *P. grandiflorus* seeds were investigated according to Equations (1)–(3), respectively. Meanwhile, cotyl length and radicle length of *P. grandiflorus* were determined after 18 d of incubation.

$$\text{Germination rate (\%)} = 100 \times (\text{Number of normally germinated seeds within 22 days/Total seed number}) \quad (1)$$

$$\text{Germination energy (\%)} = 100 \times (\text{Number of normally germinated seeds in the first 10 days/Total seed number}) \quad (2)$$

$$\text{Germination index} = (\text{Number of germinated seeds at a given incubation time/Incubation time}) \quad (3)$$

2.3. Growth Experiment of *P. grandiflorus*

The operation of soaking the *P. grandiflorus* seeds in different concentrations of the chitosan solution was the same as above. The soaked seeds were washed with distilled water and then evenly scattered in a flowerpot containing 4 kg of soil for growth. The outer diameter, inner diameter, and height of each flowerpot were 26 cm, 22 cm, and 17 cm, respectively. Each pot was sown with 50 soaked seeds, and each treatment consisted of three replicates. After emergence, 10 strong seedlings were kept in each pot for further growth. Irrigation was timely during growth of the *P. grandiflorus* to avoid drought. The growth, photosynthesis, and resistance parameters of the *P. grandiflorus* leaves were investigated 110 days after sowing. Moreover, the overground part growth, underground part growth, and quality parameters of *P. grandiflorus* were determined 300 days after sowing.

2.4. Determination Methods

The length and width of the fourth leaves of the *P. grandiflorus* plants were measured by a ruler, and the leaf area was approximately equal to the product of its length and width. The chlorophyll, photosynthetic rate (Pn), transpiration rate (Tr), and water use efficiency (WUE) in the third and fourth leaves of *P. grandiflorus* were determined following Zhang et al. [31]. The chlorophyll was determined using an SPDA-502 Plus chlorophyll analyzer (Konica-Minolta, Tokyo, Japan), and the Pn, Tr, and WUE were monitored by portable LI-6400XT photosynthesis determination equipment (LI-COR Inc., Lincoln, NE, USA) between 8:00 and 10:00 a.m. The resistance parameters of the *P. grandiflorus* leaves, including the soluble sugar and protein, proline (Pro), malonaldehyde (MDA), superoxide dismutase activity (SOD), and peroxidase (POD) activity, were analyzed following Wang et al. [32] and Zhang et al. [33]. The plants' dry height, stem diameter at the second segment, and root diameter were measured using a ruler or digital caliper, and their overground part weight and the fresh and dry weight of the root were determined by an analytical balance. Meanwhile, the platycodin D, total platycodin, extractum, and polysaccharides of the *P. grandiflorus* roots were analyzed following Liu et al. [34] and Zhang et al. [35].

2.5. Statistical Analyses

The data were displayed as the mean value ± standard deviation (SD) of three replicates, and their significant differences were determined by a one-way analysis of variance (ANOVA) with Duncan's test in SPSS 18.0 software (SPSS Inc., Chicago, IL, USA). The figures were edited using Origin 10.0 software (OriginLab, Northampton, MA, USA).

3. Results

3.1. Effects of Chitosan Soaking on Seed Germination of P. grandiflorus

The influences of chitosan soaking on seed germination, cotyl length, and radicle length of *P. grandiflorus* are shown in Table 2. Indeed, 0.05–0.25% chitosan soaking significantly ($p < 0.05$) enhanced the germination rate, germination energy, germination index, cotyl length, and radicle length of *P. grandiflorus*, which effectively increased by 1.18–1.46, 1.41–2.10, 1.18–1.46, 1.04–1.23, and 1.09–1.54 times compared to the control (0.00% chitosan), respectively. The germination rate, energy, and index, as well as the cotyl length of *P. grandiflorus* seeds treated by 0.15% chitosan soaking, were, respectively, 92.33%, 61.67%, 4.20, and 7.48 mm, which were significantly ($p < 0.05$) superior to those of other chitosan concentrations. Meanwhile, the radicle length of *P. grandiflorus* seeds treated by 0.15% chitosan soaking was also significantly ($p < 0.05$) superior to that of 0.05%, 0.10%, and 0.25% chitosan treatments. Moreover, the seed germination, cotyl length, and radicle length were not significantly different between 0.20% chitosan and 0.25% chitosan treatments, but they were significantly ($p < 0.05$) higher than those of 0.05% chitosan and 0.10% chitosan treatments. The present results show that 0.05–0.25% chitosan had a preferable enhancing effect on seed germination of *P. grandiflorus*.

Table 2. Influences of chitosan soaking on seed germination, cotyl length, and radicle length of *P. grandiflorus*.

Chitosan (%)	Germination Rate (%)	Germination Energy (%)	Germination Index	Cotyl Length (mm)	Radicle Length (mm)
0.00	63.33 ± 2.52^{d}	29.33 ± 1.53^{e}	2.88 ± 0.11^{d}	6.08 ± 0.16^{e}	7.18 ± 0.14^{e}
0.05	74.67 ± 3.51^{c}	41.33 ± 1.53^{d}	3.39 ± 0.16^{c}	6.32 ± 0.12^{d}	7.84 ± 0.26^{d}
0.10	82.67 ± 3.21^{b}	49.00 ± 2.00^{c}	3.76 ± 0.15^{b}	6.75 ± 0.13^{c}	9.46 ± 0.27^{c}
0.15	92.33 ± 3.51^{a}	61.67 ± 3.51^{a}	4.20 ± 0.16^{a}	7.48 ± 0.14^{a}	11.04 ± 0.24^{a}
0.20	85.67 ± 2.52^{b}	55.00 ± 3.00^{b}	3.89 ± 0.11^{b}	7.05 ± 0.06^{b}	10.65 ± 0.28^{ab}
0.25	84.00 ± 2.65^{b}	53.67 ± 2.08^{b}	3.82 ± 0.12^{b}	6.83 ± 0.16^{bc}	10.37 ± 0.27^{b}

The mean ± SD of three replicates represents value. A one-way analysis of variance (ANOVA) followed by Duncan's test was used, and the significant differences at a 5% level ($p < 0.05$) in the six treatments are represented by different letters.

3.2. Effects of Chitosan on Overground Part Growth of P. grandiflorus Plants

The influences of chitosan soaking on leaf growth of *P. grandiflorus* plants are depicted in Table 3. Indeed, 0.15–0.25% chitosan soaking significantly ($p < 0.05$) enhanced the leaf length, leaf width, and leaf area of *P. grandiflorus* plants compared to those of 0.05–0.10% chitosan soaking or control (0.00% chitosan). However, the leaf length, leaf width, and leaf area of *P. grandiflorus* plants were not significant different among 0.15%, 0.20%, and 0.25% chitosan treatments, and there were also not significant differences among 0.00% (control), 0.05%, and 0.10% chitosan treatments. Further, 0.15% chitosan soaking exhibited a relatively good promoting effect on leaf growth of *P. grandiflorus* plants, which effectively increased the leaf length by 11.21%, the leaf width by 11.17%, and the leaf area by 23.64% compared to non-soaked chitosan, respectively. These results showed that 0.15–0.25% chitosan soaking could effectively enhance leaf growth of *P. grandiflorus* plants.

Table 3. Influences of chitosan soaking on leaf growth of *P. grandiflorus* plants.

Chitosan (%)	Leaf Length (mm)	Leaf Width (mm)	Leaf Area (mm^2)
0.00	36.38 ± 0.46 ^{b}	23.90 ± 0.33 ^{b}	869.46 ± 22.92 ^{b}
0.05	36.34 ± 0.69 ^{b}	23.86 ± 0.44 ^{b}	867.27 ± 32.48 ^{b}
0.10	36.95 ± 1.06 ^{b}	24.26 ± 0.68 ^{b}	897.01 ± 50.45 ^{b}
0.15	40.46 ± 0.58 ^{a}	26.57 ± 0.36 ^{a}	1075.02 ± 29.90 ^{a}
0.20	40.15 ± 0.56 ^{a}	26.34 ± 0.37 ^{a}	1057.82 ± 29.72 ^{a}
0.25	40.13 ± 0.83 ^{a}	26.33 ± 0.54 ^{a}	1057.06 ± 43.30 ^{a}

The mean ± SD of three replicates represents value. A one-way analysis of variance (ANOVA) followed by Duncan's test was used, and the significant differences at a 5% level ($p < 0.05$) in the six treatments are represented by different letters.

The influences of chitosan soaking on height, stem diameter, and overground part dry weight of *P. grandiflorus* plants are depicted in Table 4. Compared with non-soaked chitosan, 0.05–0.25% chitosan soaking significantly ($p < 0.05$) enhanced the plant height of *P. grandiflorus*, and 0.15–0.25% chitosan soaking significantly ($p < 0.05$) promoted its stem diameter, as well as 0.10–0.25% chitosan soaking significantly ($p < 0.05$) increased its overground part dry weight. The height and overground part weight of *P. grandiflorus* plants treated by 0.15% chitosan soaking were respectively 35.13 cm and 2.86 g plant^{-1}, which were significantly ($p < 0.05$) superior to those of other chitosan concentrations. The plants' dry height of *P. grandiflorus* was not significant differences among 0.05%, 0.10%, 0.20%, and 0.25% chitosan treatments, its stem diameter was not significant differences among 0.15%, 0.20%, and 0.25% chitosan treatments. These results further indicate that 0.15–0.20% chitosan soaking could effectively enhance the growth and biomass formation of *P. grandiflorus* plants.

Table 4. Influences of chitosan soaking on height, stem diameter, and overground part dry weight of *P. grandiflorus* plants.

Chitosan (%)	Plant Height (cm)	Stem Diameter (mm)	Overground Part Dry Weight (g Plant^{-1})
0.00	27.13 ± 1.32 ^{c}	1.80 ± 0.15 ^{c}	1.48 ± 0.14 ^{e}
0.05	30.32 ± 0.07 ^{b}	1.79 ± 0.13 ^{c}	1.53 ± 0.08 ^{e}
0.10	29.68 ± 1.09 ^{b}	1.86 ± 0.18 ^{bc}	1.69 ± 0.07 ^{d}
0.15	35.13 ± 1.01 ^{a}	2.16 ± 0.12 ^{a}	2.86 ± 0.08 ^{a}
0.20	31.68 ± 1.30 ^{b}	2.10 ± 0.14 ^{ab}	2.24 ± 0.10 ^{b}
0.25	31.32 ± 1.19 ^{b}	2.05 ± 0.03 ^{ab}	2.07 ± 0.04 ^{c}

The mean ± SD of three replicates represents value. A one-way analysis of variance (ANOVA) followed by Duncan's test was used, and the significant differences at a 5% level ($p < 0.05$) in the six treatments are represented by different letters.

3.3. *Effects of Chitosan Soaking on the Chlorophyll and Photosynthesis of P. grandiflorus Leaves*

The influences of chitosan soaking on the chlorophyll, Pn, Tr and WUE of *P. grandiflorus* leaves are displayed in Figure 1. Compared with non-soaked chitosan, 0.05–0.25% chitosan soaking notably ($p < 0.05$) enhanced the chlorophyll content of *P. grandiflorus* leaves, and 0.10–0.25% chitosan soaking could significantly ($p < 0.05$) improve its Pn and Tr, as well as 0.15–0.25% chitosan soaking could significantly ($p < 0.05$) promote its WUE. The Pn, and Tr of *P. grandiflorus* leaves treated by 0.15% chitosan soaking were 8.54 μmol CO_2 m^{-2} s^{-1} and 2.31 mmol H_2O m^{-2} s^{-1}, which were significantly ($p < 0.05$) superior to those of other chitosan concentrations. Moreover, the chlorophyll of *P. grandiflorus* leaves treated by 0.15% chitosan soaking was slightly higher than that of 0.10%, 0.20%, and 0.25% chitosan treatments, whereas there were no significant differences. Simultaneously, the WUE of *P. grandiflorus* was not significant differences among all chitosan soaking treatments. These results demonstrate that 0.15–0.20% chitosan soaking could effectively improve the chlorophyll, Pn, Tr and WUE of *P. grandiflorus* leaves, thereby enhancing its favorable growth and development.

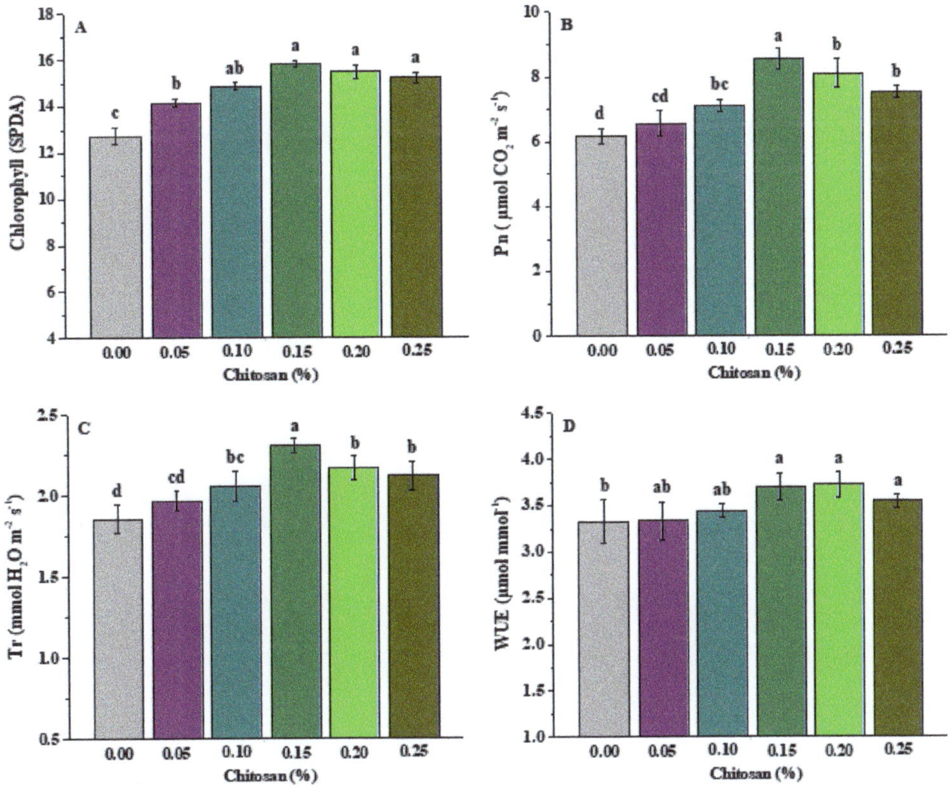

Figure 1. Influences of chitosan soaking on chlorophyll (**A**), Pn (**B**), Tr (**C**), and WUE (**D**) of *P. grandiflorus* leaves. The SD of three replicates was indicated by error bar. A one-way analysis of variance (ANOVA) followed by Duncan's test was used, and the significant differences at a 5% level ($p < 0.05$) in the six treatments are represented by different letters.

3.4. Effects of Chitosan Soaking on Resistance of *P. grandiflorus* Plants

The influences of chitosan soaking on sugar, protein, Pro, and MDA of *P. grandiflorus* leaves are shown in Figure 2. Soluble sugar, soluble protein, and Pro are indispensable regulating compounds to maintain permeability equilibrium in various organs, while MDA reflects the intensity of membrane lipid peroxidation. Indeed, 0.05–0.25% chitosan soaking notably ($p < 0.05$) increased soluble sugar, soluble protein, and Pro of *P. grandiflorus* leaves compared with non-soaked chitosan and decreased MDA content. The soluble sugar value of *P. grandiflorus* leaves treated by 0.15% chitosan soaking was slightly superior to that of 0.10%, 0.20%, and 0.25% chitosan treatments, but there were no significant differences. Moreover, the soluble protein and Pro of *P. grandiflorus* leaves treated by 0.15% chitosan soaking were significantly ($p < 0.05$) superior to those of other chitosan concentrations. Further, the Pro of *P. grandiflorus* leaves treated by 0.20% chitosan soaking was significantly ($p < 0.05$) superior to that of 0.05%, 0.10%, and 0.25% chitosan treatments. Meanwhile, the soluble protein of *P. grandiflorus* leaves treated by 0.15% chitosan soaking was slightly inferior to that of 0.20% chitosan soaking, and significantly ($p < 0.05$) lower than that of 0.05%, 0.10%, and 0.25% chitosan treatments. The soluble sugar of *P. grandiflorus* was not significantly different among 0.05%, 0.10%, 0.20%, and 0.25% chitosan treatments, and its Pro was not significantly different among 0.00%, 0.05%, 0.10%, and 0.25% chitosan treatments, as well as its MDA was not significantly different between 0.20% and 0.25% chitosan treatments. These findings emphasize that 0.15–0.20% chitosan soaking could effectively enhance soluble sugar, soluble protein, and Pro of *P. grandiflorus* leaves, as well as reduce their MDA, thereby effectively promoting their stress resistance, environmental adaptability, and healthy growth.

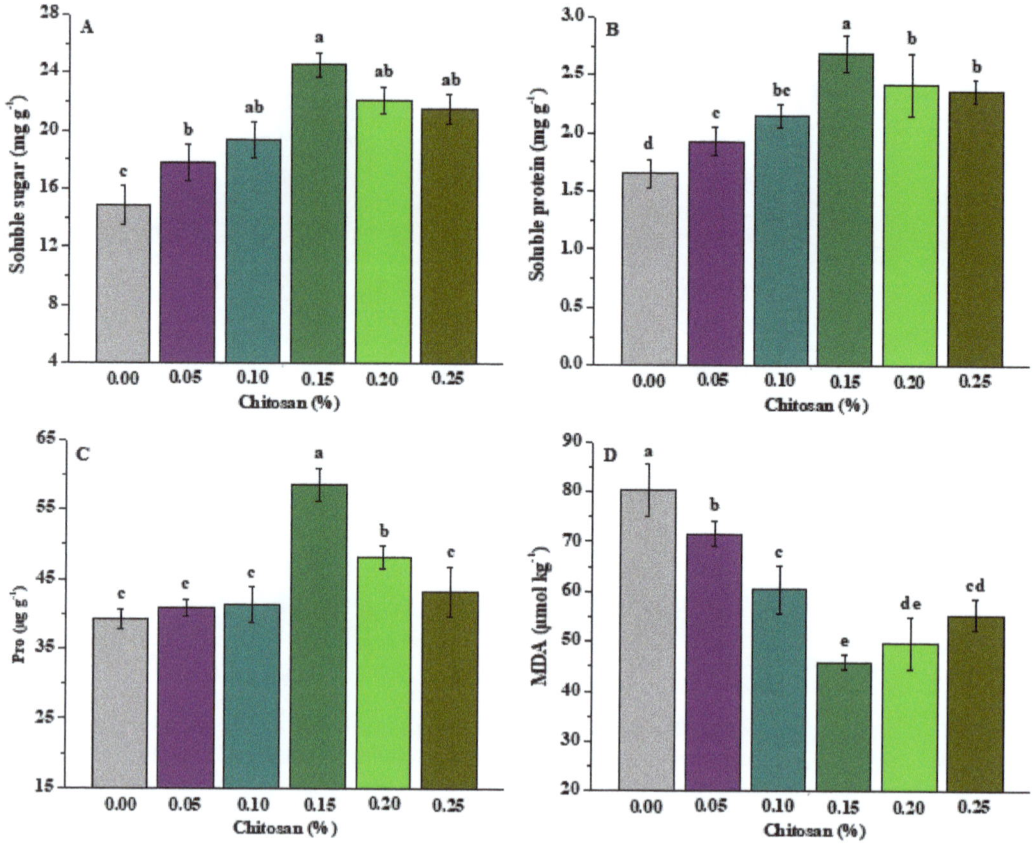

Figure 2. Influences of chitosan soaking on soluble sugar (**A**), soluble protein (**B**), Pro (**C**), and MDA (**D**) of *P. grandiflorus* leaves. The SD of three replicates was indicated by error bar. A one-way analysis of variance (ANOVA) followed by Duncan's test was used, and the significant differences at a 5% level ($p < 0.05$) in the six treatments are represented by different letters.

The influences of chitosan soaking on SOD and POD activities of *P. grandiflorus* leaves are shown in Figure 3. SOD and POD are important defense enzymes participating in plants' stress resistance. Compared with non-soaked chitosan, 0.15–0.20% chitosan soaking significantly ($p < 0.05$) enhanced the SOD activity of *P. grandiflorus* leaves, and 0.10–0.25% chitosan soaking could significantly ($p < 0.05$) promote its POD activity. The SOD activity of *P. grandiflorus* leaves treated by 0.05%, 0.10%, and 0.25% chitosan soaking was slightly lower than that of non-soaked chitosan, but there were no significant differences. Meanwhile, 0.15% chitosan soaking displayed an optimal promotion effect on the SOD and POD activities of *P. grandiflorus* leaves. The present results further demonstrate that 0.15–0.20% chitosan soaking could effectively enhance defense enzyme activity, stress resistance, and adaptability of *P. grandiflorus*.

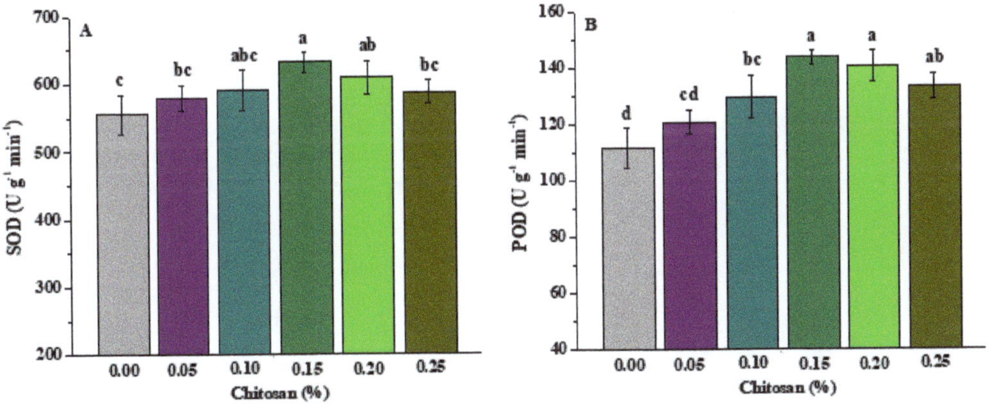

Figure 3. Influences of chitosan soaking on SOD (**A**) and POD (**B**) activities of *P. grandiflorus* leaves. The SD of three replicates was indicated by error bar. A one-way analysis of variance (ANOVA) followed by Duncan's test was used, and the significant differences at a 5% level ($p < 0.05$) in the six treatments are represented by different letters.

3.5. Effects of Chitosan Soaking on Growth and Yield of P. grandiflorus Roots

The influences of chitosan on diameter, fresh weight, and dry weight of *P. grandiflorus* roots are shown in Table 5. Compared with non-soaked chitosan, 0.05–0.25% chitosan soaking significantly ($p < 0.05$) enhanced the plant height of *P. grandiflorus*, and 0.15–0.25% chitosan soaking could obviously ($p < 0.05$) improve the fresh and dry weight of roots. The diameter, fresh weight, and dry weight of *P. grandiflorus* roots treated by 0.15% chitosan soaking were, respectively, 7.13 mm, 1.43 g plant^{-1}, and 0.38 g plant^{-1}, which were obviously ($p < 0.05$) superior to those of other chitosan concentrations and effectively increased by 1.35-, 1.96-, and 1.90-fold compared to non-spray chitosan, respectively. Moreover, those treated by 0.20% chitosan soaking were also obviously ($p < 0.05$) superior to those of 0.05%, 0.10%, and 0.25% chitosan treatments. The diameter, fresh weight, and dry weight of *P. grandiflorus* roots were not significantly different between 0.10% and 0.25% chitosan treatments, and its fresh weight and dry weight were not significantly different between 0.05% and 0.10% chitosan treatments. These findings indicate that 0.15–0.20% chitosan soaking could effectively enhance growth and yield formation of *P. grandiflorus* roots.

Table 5. Influences of chitosan on diameter, fresh weight, and dry weight of *P. grandiflorus* roots.

Chitosan (%)	Root Diameter (mm)	Fresh Weight of Root (g Plant^{-1})	Dry Weight of Root (g Plant^{-1})
0.00	5.28 ± 0.20^{d}	0.73 ± 0.03^{d}	0.20 ± 0.01^{d}
0.05	5.35 ± 0.10^{d}	0.79 ± 0.10^{d}	0.22 ± 0.01^{cd}
0.10	5.88 ± 0.23^{c}	0.87 ± 0.03^{cd}	0.23 ± 0.03^{cd}
0.15	7.13 ± 0.12^{a}	1.43 ± 0.07^{a}	0.38 ± 0.03^{a}
0.20	6.71 ± 0.17^{b}	1.25 ± 0.09^{b}	0.32 ± 0.03^{b}
0.25	5.84 ± 0.08^{c}	0.98 ± 0.13^{c}	0.26 ± 0.02^{c}

The mean ± SD of three replicates represents value. A one-way analysis of variance (ANOVA) followed by Duncan's test was used, and the significant differences at a 5% level ($p < 0.05$) in the six treatments are represented by different letters.

3.6. Effects of Chitosan Soaking on Medical Quality of P. grandiflorus

The influences of chitosan soaking on the platycodin D, total platycodin, extractum, and polysaccharides of *P. grandiflorus* are depicted in Figure 4. Compared with non-soaked chitosan, 0.05–0.25% chitosan soaking significantly ($p < 0.05$) increased the platycodin D, extractum, and polysaccharides of *P. grandiflorus*, and 0.10–0.25% chitosan soaking could obviously ($p < 0.05$) improve its total platycodin. The platycodin D and polysaccharides of *P.*

grandiflorus treated by 0.15% chitosan soaking were obviously ($p < 0.05$) superior to those of 0.05%, 0.10%, and 0.25% chitosan treatments; meanwhile, its total platycodin and extractum were obviously ($p < 0.05$) superior to those of other chitosan concentrations. The platycodin D, total platycodin, extractum, and polysaccharides of *P. grandiflorus* treated by 0.1–0.20% chitosan soaking effectively increased by 1.20–1.18-fold, 1.19–1.31-fold, 1.19–1.28-fold, and 1.29–1.31-fold compared to non-soaked chitosan, respectively. The results presented here reveal that soaking in a suitable concentration of chitosan could effectively improve the medical quality of *P. grandiflorus*.

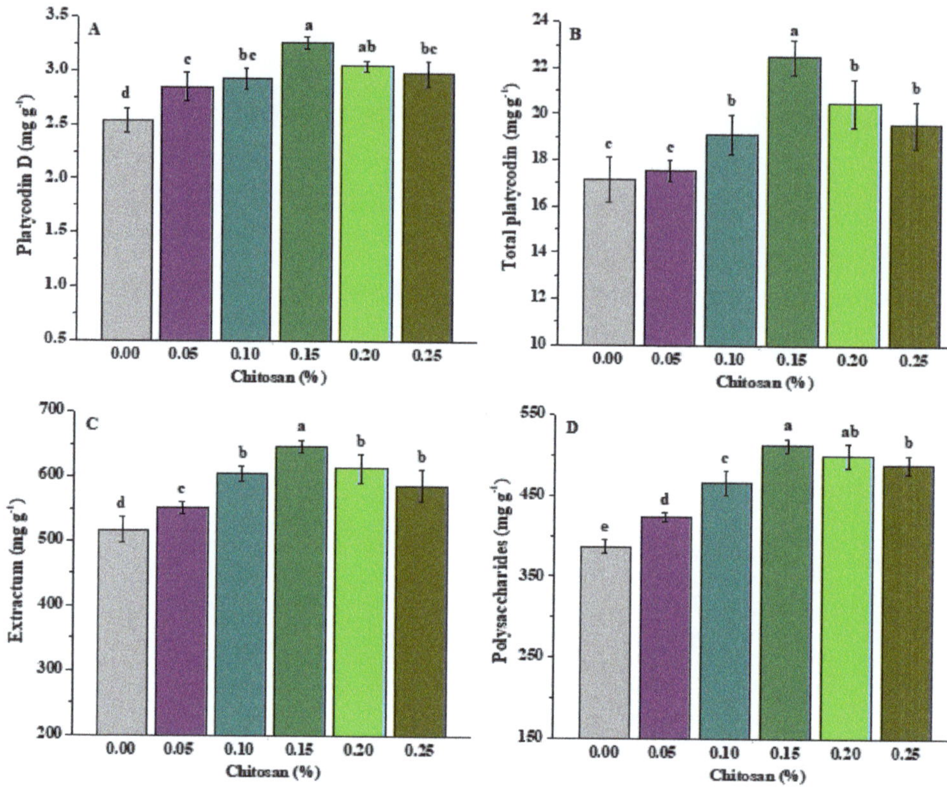

Figure 4. Influences of chitosan soaking on platycodin D (**A**), total platycodin (**B**), extractum (**C**), and polysaccharides (**D**) of *P. grandiflorus*. The SD of three replicates was indicated by error bar. A one-way analysis of variance (ANOVA) followed by Duncan's test was used, and the significant differences at a 5% level ($p < 0.05$) in the six treatments are represented by different letters.

4. Discussion

Germination rate, energy, and index represent the vigor index of seeds, which are important metrics to evaluate seed quality. Pan et al. [29] found that 50 mg L^{-1} chitosan effectively enhanced the germination rate, energy, and index, as well as the vigor index, cotyl length, and radicle length of *Trifolium repens* under salt stress. Li et al. [30] found that 0.15% chitosan significantly improved the germination rate, energy, and index of *Sctellaria baicalensis* under drought stress. Moreover, 1000–3000 mg L^{-1} chitosan could also effectively increase the germination energy of *Cucumis sativus* [36]. The results here exhibited that 0.05–0.25% chitosan soaking obviously ($p < 0.05$) enhanced the germination rate, energy, and index, as well as the cotyl and radicle length of *P. grandiflorus*, and 0.15% chitosan soaking displayed an optimal enhancing effect. These results were similar to previous studies and extended application of chitosan in traditional Chinese medicine cultivation. We also found that, when the chitosan concentration was higher than 0.15%, the germination index of *P. grandiflorus* seeds decreased. This indicated that chitosan could significantly improve the respiration rate of seed germination, which increased with an increase in chitosan concentration, accelerating transformation of the substances in seeds and then promoting seed germination. However, with the increasing concentration of chitosan, the respiration rate of seeds was too fast, which might cause loss of a large amount of energy in the form of heat, decreasing the promotion effects of chitosan.

Chlorophyll is a very important pigment in photosynthesis, which is the physiological basis of plant growth. Chitosan can substantially enhance plants' growth and yield via promoting the photosynthetic rate via improving chlorophyll [21]. Meanwhile, chitosan also acts as a growth promoter to enhance division and elongation of cells via activating signal transduction and gene expression of auxin and cytokinin, which promotes nutrient intake and plant growth [21,24–28]. Pan et al. [29] found that chitosan effectively increased the plant height, fresh and dry weight, and root–shoot ratio of *Trifolium repens* under salt stress, and Li et al. [30] also found that chitosan significantly improved the chlorophyll content of *Sctellaria baicalensis* under drought stress. The present results indicate that 0.15–0.20% chitosan soaking could effectively enhance the leaf growth, plant height, stem diameter, and overground part dry weight of *P. grandiflorus* and increase its leaf's chlorophyll, Pn, Tr, and WUE, which was consistent with the previous reports on other plants. These findings imply that 0.15–0.20% chitosan soaking could reliably enhance growth, chlorophyll, Pn, Tr, and WUE of *P. grandiflorus* leaves, thereby improving growth and biomass formation of *P. grandiflorus* plants.

Soluble sugar and Pro are important osmoregulation substances in plant cells, and soluble protein is the metabolism basis of material and energy, which are closely related to plants' stress resistance [37,38]. Pro can effectively maintain the osmotic balance between cytoplasmic matrix and environment, prevent water loss, and protect the structural integrity of protective film [38]. Soluble sugar can maintain plant growth under adversity by regulating osmotic potential of plant tissues [38]. Moreover, MDA reflects the degrees of membrane damage and stress resistance [38]. SOD plays a role in scavenging free radicals in plants, and POD is an important protective enzyme for catalyzing H_2O_2 decomposition in lignin biosynthesis, which is also connected to a plant's slow death and stress resistance [37,38]. Many studies have also shown that chitosan can enhance the sugar, protein, and Pro contents in plants and reduce their MDA content, as well as boost their defense enzyme activity [39–41]. In this study, 0.15–0.20% chitosan soaking significantly ($p < 0.05$) enhanced soluble sugar, soluble protein, Pro, SOD activity, and POD activity of *P. grandiflorus* leaves as well as reduced their MDA, demonstrating that chitosan soaking could effectively improve the stress resistance, environmental adaptability, and healthy growth of *P. grandiflorus* plants.

Generally, *P. grandiflorus* is used as medication, especially its root and rhizome, and Platycodin D, total Platycodin, and polysaccharides are considered to be its main medical bioactive components [5–9]. Meanwhile, good growth determines the underground part yield and medical quality of *P. grandiflorus*. In this study, 0.15–0.20% chitosan soaking could effectively enhance the diameter, fresh weight, dry weight, platycodin D, total platycodin, extractum, and polysaccharides of *P. grandiflorus* roots. This favorable effect probably derived from the effects of chitosan on seed germination and plant growth of *P. grandiflorus*. Chitosan is a natural biopolymer with nontoxic, antibacterial, antioxidant, renewable, and biodegradable superiorities, and the safe interval time of more than 10 months for *P. grandiflorus* was very long; therefore, the quality and safety risks of the medicinal materials caused by chitosan were almost nonexistent [21,22,42,43]. This work emphasizes that chitosan can be applied as a favorable candidate or promoter for enhancing seed germination of *P. grandiflorus* and improving its growth, photosynthesis, resistance, yield, and quality, and 0.15–0.20% chitosan is a safe, cost-efficient, and suitable soaking concentration.

5. Conclusions

In conclusion, the present work indicates that chitosan soaking had a preferable enhancing effect on seed germination of *P. grandiflorus*, which could significantly ($p < 0.05$) enhance its germination rate, energy, and index, as well as cotyl and radicle length. Indeed, 0.15–0.20% chitosan soaking could effectively enhance the leaf growth, height, stem diameter, and overground part dry weight of *P. grandiflorus* plants, as well as reliably improve their leaves' chlorophyll, Pn, Tr, and WUE. Moreover, 0.15–0.20% chitosan soaking notably enhanced the soluble sugar, soluble protein, and Pro contents, as well as the SOD and POD activities in leaves of *P. grandiflorus*, and reduced their MDA, thereby effectively promoting its stress resistance and adaptability. Meanwhile, 0.15–0.20% chitosan soaking effectively improved the underground part growth and medical quality of *P. grandiflorus*. This work highlights that chitosan can be used as a favorable promoter to enhance seed germination of *P. grandiflorus* and improve its growth, photosynthesis, resistance, yield, and quality.

Author Contributions: M.W. and H.L. (Haitao Li) constructed the project; M.W., H.L. (Haitao Li) and H.L. (Hai Liu) designed the experiments; H.L. (Hai Liu), Z.Z. and X.H. performed the experiments; C.Z. and H.L. (Hai Liu) analyzed the data; H.L. (Hai Liu) and C.Z. wrote the paper. All authors have read and agreed to the published version of the manuscript.

Funding: This work was supported by the National Key Research and Development Program of China (No. 2021YFD1601002) and the Modern Industrial Technology System of Chinese Medicinal Materials in Guizhou Province (No. GZCYTX2021-0202).

Data Availability Statement: The datasets used or analyzed during the current study available from the corresponding author upon reasonable request.

Conflicts of Interest: We declare that we do not have any commercial or associative interest that represents a conflict of interest in connection with the work submitted.

References

1. Zhang, L.; Wang, Y.; Yang, D.; Zhang, C.; Zhang, N.; Li, M.; Liu, Y. Platycodon Grandiflorus—An Ethnopharmacological, Phytochemical and Pharmacological Review. *J. Ethnopharmacol.* **2015**, *164*, 147–161. [CrossRef] [PubMed]
2. Jaiswal, V.; Cho, Y.I.; Lee, H.J. Preliminary Study to Explore the Immune-Enhancement Mechanism of *Platycodon grandiflorus* Extract through Comparative Transcriptome *Analysis. Appl. Sci.* **2021**, *11*, 226. [CrossRef]
3. Kim, G.; Rim, Y.; Cho, H.; Hyun, T.K. Identification and Functional Characterization of FLOWERING LOCUS T in *Platycodon grandiflorus. Plants* **2022**, *11*, 325. [CrossRef] [PubMed]
4. Bae, N.; Ahn, T.; Chung, S.; Oh, M.S.; Ko, H.; Oh, H.; Park, G.; Yang, H.O. The Neuroprotective Effect of Modified Yeoldahanso-tang via Autophagy Enhancement in Models of Parkinson's Disease. *J. Ethnopharmacol.* **2011**, *134*, 313–322. [CrossRef] [PubMed]
5. Nyakudya, E.; Jeong, J.H.; Lee, N.K.; Jeong, Y.S. Platycosides from the Roots of *Platycodon Grandiflorum* and Their Health Benefits. *Prev. Nutr. Food Sci.* **2014**, *19*, 59–68. [CrossRef]
6. Tamura, K.; Teranishi, Y.; Ueda, S.; Suzuki, H.; Kawano, N.; Yoshimatsu, K.; Saito, K.; Kawahara, N.; Muranaka, T.; Seki, H. Cytochrome P450 Monooxygenase CYP716A141 is a Unique β -amyrin C-16 β Oxidase Involved in Triterpenoid Saponin Biosynthesis in *Platycodon grandiflorus. Plant Cell Physiol.* **2017**, *58*, 874–884. [CrossRef] [PubMed]
7. Kim, J.; Jeon, S.G.; Kim, K.A.; Kim, J.J.; Song, E.J.; Jeon, Y.; Kim, E.; Lee, K.B.; Kwak, J.H.; Moon, M. *Platycodon grandiflorus* Root Extract Improves Learning and Memory by Enhancing Synaptogenesis in Mice Hippocampus. *Nutrients* **2017**, *9*, 794. [CrossRef] [PubMed]
8. Ji, M.; Bo, A.; Yang, M.; Xu, J.; Jiang, L.; Zhou, B.; Li, M. The Pharmacological Effects and Health Benefits of *Platycodon grandiflorus*—A Medicine Food Homology Species. *Foods* **2020**, *9*, 142. [CrossRef]
9. Kim, M.H. Contemplation on the Emergency Foods in Korea under the Japanese Occupation. *J. East Asian Soc. Diet. Life* **2015**, *25*, 721–738. [CrossRef]
10. Zhao, M.; Xu, Z. Preliminary Study on the Characteristics of Endogenous Inhibitory Substances in *Platycodon grandiflorum* Seeds. *J. Northeast Forestry Univ.* **2000**, *28*, 51–54. [CrossRef]
11. Liu, Z.; Zhang, Y.; Wang, X.; Yu, H. Research Rrogress on Seed Vigor of *Platycodon grandiflorus. Seed* **2006**, *25*, 48–51. [CrossRef]
12. Han, C.; Yang, P. Studies on the Molecular Mechanisms of Seed Germination. *Proteomics* **2015**, *15*, 1671–1679. [CrossRef]
13. Vishal, B.; Kumar, P.P. Regulation of Seed Germination and Abiotic Stresses by Gibberellins and Abscisic Acid. *Front. Plant Sci.* **2018**, *9*, 838. [CrossRef] [PubMed]
14. Wang, Y.; Zhang, X.; Zhu, X.; Zhang, Z.; Li, Z.; Geng, X.; Cai, Q.; Liu, Z. Research Progress on Dormancy and Germination Mechanism of Forest Seeds. *Sci. Silvae Sin.* **2021**, *57*, 128–144. [CrossRef]

15. Chen, J.; Lei, J.; Wang, L.; Zhao, W.; Zhang, L.; Mu, X.; Cheng, Q. Effects of Four Chemical Reagents on Seed Germination and Seedling Growth of *Platycodon grandiflorus*. *J. Northwest Agric.* **2010**, *19*, 100–105. [CrossRef]
16. Zhang, C.; Long, Y.; He, G.; Wang, Q. Effects of Growth Regulator Soaking on the Seed Germination and Growth of *Platycodon grandiflorus*. *J. Chin. Med. Mater.* **2015**, *38*, 21–24. [CrossRef]
17. Zhang, C.; Wang, Q.; Long, Y.; Li, M.; Wu, X.; He, G. Effects of Bihu Soaking on on the Seed Germination and Growth of *Platycodon grandiflorus*. *Seed* **2016**, *35*, 42–44. [CrossRef]
18. Anitha, A.; Sowmya, S.; Kumar, P.T.S.; Deepthi, S.; Chennazhi, K.P.; Ehrlich, H.; Tsurkan, M.; Jayakumar, R. Chitin and chitosan in selected biomedical applications. *Prog. Polym. Sci.* **2014**, *39*, 1644–1667. [CrossRef]
19. Malerba, M.; Cerana, R. Chitosan Effects on Plant Systems-A Review. *Int. J. Mol. Sci.* **2016**, *17*, 996. [CrossRef] [PubMed]
20. Dzung, N.A.; Khanh, V.T.P.; Dzung, T.T. Research on Impact of Chitosan Oligomers on Biophysical Characteristics, Growth, Development and Drought Resistance of Coffee. *Carbohydr. Polym.* **2011**, *84*, 751–755. [CrossRef]
21. Chakraborty, M.; Hasanuzzaman, M.; Rahman, M.; Khan, M.; Bhowmik, P.; Mahmud, N.U.; Tanveer, M.; Islam, T. Mechanism of Plant Growth Promotion and Disease Suppression by Chitosan Biopolymer. *Agriculture* **2020**, *10*, 624. [CrossRef]
22. Lopez-Moya, F.; Suarez-Fernandez, M.; Lopez-Llorca, L.V. Molecular Mechanisms of Chitosan Interactions with Fungi and Plants. *Int. J. Mol. Sci.* **2019**, *20*, 332. [CrossRef]
23. Maluin, F.N.; Hussein, M.Z. Chitosan-Based Agronanochemicals as a Sustainable Alternative in Crop Protection. *Molecules* **2020**, *25*, 1611. [CrossRef] [PubMed]
24. Salachna, P.; Zawadzinska, A. Effect of Chitosan on Plant Growth, Flowering and Corms Yield of Potted Freesia. *J. Ecol. Eng.* **2014**, *15*, 93–102. [CrossRef]
25. Pirbalouti, A.G.; Malekpoor, F.; Salimi, A.; Golparvar, A. Exogenous Application of Chitosan on Biochemical and Physiological Characteristics, Phenolic Content and Antioxidant Activity of Two Species of Basil (*Ocimum ciliatum* and *Ocimum basilicum*) under reduced irrigation. *Sci. Hortic.* **2017**, *217*, 114–122. [CrossRef]
26. Choudhary, R.C.; Kumaraswamy, R.V.; Kumari, S.; Sharma, S.S.; Pal, A.; Raliya, R.; Biswas, P.; Saharan, V. Cu-chitosan Nanoparticle Boost Defense Responses and Plant Growth in Maize (*Zea mays* L.). *Sci. Rep.* **2017**, *7*, 9754–9765. [CrossRef] [PubMed]
27. Li, J.; Li, R.; Zhang, C.; Guo, Z.; Wu, X.; An, H. Co-Application of Allicin and Chitosan Increases Resistance of Rosa roxburghii against Powdery Mildew and Enhances Its Yield and Quality. *Antibiotics* **2021**, *10*, 1449. [CrossRef]
28. Wang, Q.; Li, H.; Lei, Y.; Su, Y.; Long, Y. Chitosan as an Adjuvant to Improve Isopyrazam Azoxystrobin against Leaf Spot Disease of Kiwifruit and Enhance Its Photosynthesis, Quality, and Amino Acids. *Agriculture* **2022**, *12*, 373. [CrossRef]
29. Pan, L.; Wei, H.; Zhang, H.; Wang, Y. Effects of Chitosan on Seed Germination and Seedling Growth of *Trifolium repens* under Salt Stress. *Mol. Plant Breed.* **2018**, *16*, 3740–3744. [CrossRef]
30. Li, X.; Liu, R.; Hua, Z. Effects of Chitosan on Seed Germination and Seedling Drought Resistance of *Sctellaria baicalensis*. *Acta Agric. Jiangxi* **2020**, *32*, 75–81. [CrossRef]
31. Zhang, C.; Li, H.; Wu, X.; Su, Y.; Long, Y. Co-Application of Tetramycin and Chitosan in Controlling Leaf Spot Disease of Kiwifruit and Enhancing Its Resistance, Photosynthesis, Quality and Amino Acids. *Biomolecules* **2022**, *12*, 500. [CrossRef]
32. Wang, Q.; Zhang, C.; Long, Y.; Wu, X.; Su, Y.; Lei, Y.; Ai, Q. Bioactivity and Control Efficacy of the Novel Antibiotic Tetramycin against Various Kiwifruit Diseases. *Antibiotics* **2021**, *10*, 289. [CrossRef]
33. Zhang, C.; Long, Y.; Wang, Q.; Li, J.; Wu, X.; Li, M. The Effect of Preharvest 28.6% Chitosan Composite Film Sprays for Controlling the Soft Rot on Kiwifruit. *Hortic. Sci.* **2019**, *46*, 180–194. [CrossRef]
34. Liu, Q.; Cui, T.; Cheng, F. Effects of Soil Microorganism on Output and Quality of *Platycodon grandiflorum*. *Guangdong Agric. Sci.* **2009**, *2*, 36–38. [CrossRef]
35. Zhang, X.; Deng, H.; Hua, Z. Effects of Soil Compaction Stress on Growth, Quantity and Quality of *Platycodon grandiflorum*. *J. Northwest AF Univ.* **2013**, *41*, 176–181. [CrossRef]
36. Li, Y.; Liu, X.; Kong, L.; Cao, H.; Lv, Y. Effects of Four Kinds of Active Organic Substances on Cucumber Seed Germination. *Humic Acid.* **2022**, *4*, 13–19. [CrossRef]
37. Caffall, K.H.; Mohnen, D. The Structure, Function, and Biosynthesis of Plant Cell Wall Pectic Polysaccharides. *Carbohydr. Res.* **2009**, *344*, 1879–1900. [CrossRef]
38. Vlot, A.C.; Sales, J.H.; Lenk, M.; Bauer, K.; Brambilla, A.; Sommer, A.; Nayem, S. Systemic Propagation of Immunity in Plants. *New Phytol.* **2020**, *229*, 1234–1250. [CrossRef]
39. Wang, Q.; Long, Y.; Ai, Q.; Su, Y.; Lei, Y. Oligosaccharins Used Together with Tebuconazole Enhances Resistance of Kiwifruit against Soft Rot Disease and Improves Its Yield and Quality. *Horticulturae* **2022**, *8*, 624. [CrossRef]
40. Xoca-Orozco, L.Á.; Cuellar-Torres, E.A.; Gonz á lez-Morales, S.; Gutiérrez-Martínez, P.; López-García, U.; Herrera-Estrella, L.; Vega-Arreguín, J.; Chacón-López, A. Transcriptomic Analysis of Avocado Hass (*Persea americana* Mill) in the Interaction System Fruit-Chitosan-*Colletotrichum*. *Front. Plant Sci.* **2017**, *8*, 956. [CrossRef]
41. Rahman, M.; Mukta, J.A.; Sabir, A.A.; Gupta, D.R.; Mohi-ud-din, M.; Hasanuzzaman, M.; Miah, M.G.; Rahman, M.; Islam, M.T. Chitosan Biopolymer Promotes Yield and Stimulates Accumulation of Antioxidants in Strawberry Fruit. *PLoS ONE* **2018**, *13*, e0203769. [CrossRef]

42. Emami Bistgani, Z.; Siadat, S.A.; Bakhshandeh, A.; Ghasemi Pirbalouti, A.; Hashemi, M. Interactive Effects of Drought Stress and Chitosan Application on Physiological Characteristics and Essential Oil Yield of *Thymus daenensis* Celak. *Crop J.* **2017**, *5*, 407–415. [CrossRef]
43. Zhang, C.; Li, Q.; Li, J.; Su, Y.; Wu, X. Chitosan as an Adjuvant to Enhance the Control Efficacy of Low-Dosage Pyraclostrobin against Powdery Mildew of *Rosa roxburghii* and Improve Its Photosynthesis, Yield, and Quality. *Biomolecules* **2022**, *12*, 1304. [CrossRef]

horticulturae

MDPI

Communication

Effect of Parental Components and Pollination Frequency on the Setting and Germination of Chrysanthemum Seeds

Natalia Miler and Dariusz Kulus *

Laboratory of Ornamental Plants and Vegetable Crops, Faculty of Agriculture and Biotechnology, Bydgoszcz University of Science and Technology, Bernardyńska 6, 85-029 Bydgoszcz, Poland
* Correspondence: dkulus@gmail.com

Abstract: The chrysanthemum is one of the most economically important ornamental plants in the *Asteraceae* family. Unfortunately, the efficacy of breeding through the traditional crossing in this species is highly limited due to inefficient seed setting. Therefore, this study aimed to investigate the effect of parental components and crossing frequency on the set and germination of chrysanthemum seeds. For this purpose, seven chrysanthemum cultivars were used as parental components in 10 crossing combinations. The crossing was performed either once or twice a week, for three successive weeks, starting from November. Next, the obtained chrysanthemum seeds were collected, sown in pots in a greenhouse, and seedling growth was observed. The efficacy of the seed set, germination rate, and plant development was evaluated. The plants of the F_1 generation were brought to flowering and evaluated phenotypically in the three successive vegetative propagation cycles. Both the arrangement of parental components and pollination frequency affected the production of seeds. More seeds were obtained if pollination was performed twice a week and if the 'Wda' cultivar was used as the maternal component. Approximately 50–100% of the seeds were able to germinate in the greenhouse, depending on the parental components, which also affected the developmental pace of the seedlings. Nearly all of the seedlings (80–100%) developed into properly growing plants. Out of 10 parental combinations tested, 7 produced the F_1 offspring. The obtained plants varied in the shape, size, and color of their flowers. A total of eight new phenotypes were found, among which six new cultivars granted plant breeders' rights, so far. The present research expands knowledge on how effective crossing should be performed.

Keywords: breeding; crossing; *Chrysanthemum* × *morifolium* (Ramat.) Hemsl.; hybridization; seed germination

Citation: Miler, N.; Kulus, D. Effect of Parental Components and Pollination Frequency on the Setting and Germination of Chrysanthemum Seeds. *Horticulturae* **2022**, *8*, 827. https://doi.org/10.3390/horticulturae8090827

Academic Editors: Konstantinos Bertsouklis, Epaminondas Kartsonas and Angela Carra

Received: 21 August 2022
Accepted: 5 September 2022
Published: 8 September 2022

1. Introduction

The chrysanthemum (*Chrysanthemum* × *morifolium*/Ramat./Hemsl., syn. *Dendranthema grandiflorum* (Ramat.) Kitamura) was already used as an ornamental garden plant in ancient China. Currently, it is one of the four major cut flowers in the world, with a high economic value [1]. Depending on the cultivar, it is suitable for bedding, rock gardens, or as a pot and cut flower. Inflorescences of chrysanthemum are compound, i.e., consist of centrally arranged, evolutionarily older, bisexual tubular/disc florets, surrounded by female ligulate/ray florets, placed on a plate-shaped extension of the stem axis, known as the settler. The inflorescences exhibit various colors and shapes, which makes them so successful in the market [2].

Breeders constantly introduce new cultivars to meet the requirements of customers and the increasing popularity of the chrysanthemum [3]. The primary goal in chrysanthemum breeding is the creation of cultivars with varied shapes and colors of inflorescences, as well as cultivars with a fast-growing capacity and increased resistance. Currently, mutation breeding using physical and chemical mutagens is often used with chrysanthemums [4,5]. However, access to physical mutagens is limited for plant breeders, while chemical ones are user- and environmentally dangerous. Genetic transformation and protoplast fusion are modern biotechnological tools used in the creation of new cultivars, but their application is technically difficult and expensive [6,7]. Classic breeding based on cross-breeding is relatively cheap and easy to perform in numerous ornamental plant species. Cross-breeding done between individuals representing desired traits of distant values can be a source of new variability, being of high value in the world of ornamentals [8]. However, in chrysanthemums, it is ineffective due to the plants' sporophytic self-incompatibility (SI) and lack of viable pollen or inability of pollen to germinate in the style [9–11]. Most of the modern cultivated chrysanthemums are self-incompatible and their trait inheritance and genetic background are extremely complex [12,13]. Therefore, it is very difficult to create pure lines [13]. Self-incompatibility, also called self-infertility, means the inability of a fully fertile hermaphroditic plant to produce zygotes when self-pollinated [14]. Moreover, vastly applied mutation breeding resulted in inbreeding depression and a negative genetic load in modern greenhouse cultivars, which reduces the overall fertility of the species [15]. The development of female and male reproductive organs is well explored, as is the role of their mutual interaction in effective fertilization [16]. Nonetheless, little research has been conducted to gain better seed production in the common greenhouse chrysanthemum.

The aim of this research was to study the effect of pollination frequency and the arrangement of parental genotypes on the setting and germination of chrysanthemum seeds. Following the experiment, the in vivo seed set efficiency was examined and several new cultivars of chrysanthemum were identified and registered. The scientific hypothesis assumed that by increasing the frequency of pollination and choosing the parental genotypes correctly, it will be possible to obtain more seeds and, consequently, more new phenotypes in chrysanthemum cross-breeding.

2. Materials and Methods

2.1. Plant Material

The plant material included paternal and mother plants of seven chrysanthemum cultivars (*Chrysanthemum* × *morifolium*/Ramat./Hemsl.). Five of them were bred in the Laboratory of Ornamental Plants and Vegetable Crops of the Bydgoszcz University of Science and Technology, i.e., 'Polka' (P), 'Bydgoszczanka' (BG), 'Wda' (W), 'Brda' (B), and 'Łuczniczka' (Ł), and were granted plant breeder's rights (PBR). The other two cultivars, namely UTP4 and JTY, were selected from the Laboratory's perennial chrysanthemums collection. All of the cultivars (except JTY) used in the experiment have medium-size semi-full inflorescences with a diameter of 5–10 cm and yellow to green disc florets. Chrysanthemums 'Polka', UTP4, and 'Brda' have white ray florets of flat and tubular ('Polka') shape, 'Bydgoszczanka' and 'Łuczniczka' have light- and dark-pink, curved ray florets, while 'Wda' has short, red, flat ray florets. Chrysanthemum JTY is characterized by yellow, flat ray florets and a full type of inflorescence (Figure 1).

2.2. Cultivation in the Greenhouse and Crossing

The plants were grown in a greenhouse of the Laboratory of Ornamental Plants and Vegetable Crops (53°07′12.0″ N 18°00′29.4″ E) in pots filled with peat substrate (Hartman, Poznań, Poland) in a natural photoperiod. The dates of pollination were adjusted to the dates of maturation of selected parental components, the first crossing was performed on 10 November (Table 1). The starting point for crossing was: for the paternal component—when most of the disc florets matured and produced an abundance of bright yellow pollen; for the maternal component—when the outer whorls of disc florets of the mother plants were mature (started to open and produce pollen). The paternal components that were the pollen source for the selected maternal plants were not used in other crosses. All of the cultivars used in the experiment were previously confirmed as self-incompatible.

Figure 1. Parental components used for crossing in the experiment. Names of cultivars and abbreviations (in brackets) are indicated in the photographs.

Pollination was performed once (1×) or twice (2×) a week. In each set of parental components, pollen was transferred to a total of 12 inflorescences of the mother plants (6 inflorescences were pollinated once a week and 6 inflorescences twice a week). Each inflorescence was considered a single repetition. Pollination was carried out for a total of three weeks for each parental combination.

Pollen from the paternal components was collected on the pollination day in the morning hours (9:00–11:00 a.m.) with a brush and applied to the stigmas of the pistils in the ligulate and tubular florets of the mother plants. During the crossing, the air temperature in the greenhouse was maintained at 14 ± 2 °C. The parental components and pollination dates are given in Table 1

Table 1. Dates of chrysanthemum pollination.

No.	Parental Components (Maternal × Paternal)	Dates of Crossing									
1.	B × Ł	10 November		18 November		25 November		2 December			
		10 November	15 November	18 November	22 November	25 November	29 November	2 December			
2.	Ł × B	10 November		18 November		25 November		2 December			
		10 November	15 November	18 November	22 November	25 November	29 November	2 December			
3.	JTY × UTP4	10 November		18 November		25 November		2 December			
		10 November	15 November	18 November	22 November	25 November	29 November	2 December			
4.	W × Ł			18 November		25 November		2 December			
				18 November	22 November	25 November	29 November	2 December	6 December		
5.	JTY × P			18 November		25 November		2 December			
				18 November	22 November	25 November	29 November	2 December	6 December		
6.	W × BG			18 November		25 November		2 December			
				18 November	22 November	25 November	29 November	2 December	6 December		
7.	W × P			18 November		25 November		2 December			
				18 November	22 November	25 November	29 November	2 December	6 December		
8.	JTY × Ł					25 November		2 December		9 December	
						25 November	29 November	2 December	6 December	9 December	13 December
9.	JTY × W					25 November		2 December		9 December	
						25 November	29 November	2 December	6 December	9 December	13 December
10.	W × UTP4					25 November		2 December		9 December	
						25 November	29 November	2 December	6 December	9 December	13 December

From 2 December to 13 December, the shoots with maternal inflorescences were cut, placed in water in a light, airy room, under natural light and photoperiod, at 16 ± 2 °C, and left for 10 weeks to set seeds. Next, the seeds were collected and counted (Figure 2A,B). Based on the number of seeds obtained and the number of ligulate and tubular flowers present in the mother inflorescences, the efficiency of seeds set in individual parental systems was calculated as:

$$\text{seed setting efficacy} = \frac{\text{number of seeds obtained}}{\text{total number of flowers in the inflorescences}} \times 100\%$$

Sowing was performed in the second week of March in a greenhouse. Each seed was sown in a separate pot (Ø 6 cm), filled with Gramoflor Culvito substrate (Vechta, Germany), pH 4.5–6.5, mixed with perlite (2:1). The pots were covered with transparent perforated film and mesh, to protect the seedlings from excessive sunlight and loss of humidity (Figure 2C). The seeds were grown for 30 days. During that period, they were sprayed with water, aired, and watered. The seedling growth phases were recorded every two days according to the key:

- phase I—the beginning of germination—seedling emergence (Figure 2D),
- phase II—fully developed cotyledons (Figure 2E),
- phase III—the first fully developed leaf (Figure 2F).

Next, the F_1 plants were grown and brought to full flowering, as described earlier, and the quality of their inflorescences was studied in search of new potential cultivars. To confirm the distinctness, uniformity, and stability (DUS) of the novel traits, chrysanthemums were propagated vegetatively through stem cuttings and cultivated in the following three years.

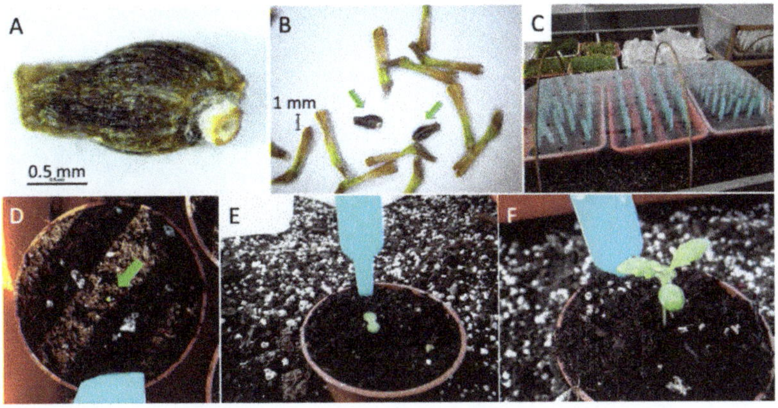

Figure 2. Fruit (achenes, considered as seed) of chrysanthemum (**A**); fruits (seeds) of chrysanthemums (pointed with an arrow) among dried tubular florets (**B**); seeds planted in pots covered with perforated film (**C**); plant growth phase I—beginning of germination (**D**); phase II—fully developed cotyledons (**E**); phase III—the first fully developed leaf (**F**).

2.3. Statistical Analysis

The experiment was set in a completely randomized design and repeated six times. Nine parental combinations and two crossing frequencies were included. Data were statistically verified using Statistica 12.0 (StatSoft Polska, Cracow, Poland) software. The analysis of variance (ANOVA) was performed and means were compared with a post-hoc test (Newman-Keuls' or Fisher's exact test) at the significance level of $p \leq 0.05$. To obtain the normal distribution of the data expressed as a percentage, the Freeman-Tukey double-arcsine transformation was used. Tables with results provide real, untransformed numerical data, with the alphabet indicating the homogeneous groups.

3. Results

A total of 15,507 ligulate and tubular florets were counted in 120 maternal inflorescences, including 93 seeds, which gives an average of 0.8 seeds per inflorescence (Table 2). The highest number of florets (178.7–182.5) was produced by the 'Brda' and 'Wda' cultivars, while the lowest (79.2–112.2) was by the JTY chrysanthemum. The highest number of seeds (13–22) was produced by crossing plants twice a week, especially in the experimental objects where the 'Wda' cultivar was used as the maternal component (W × P, W × UTP4, W × Ł). In contrast, no seeds were found in three parental combinations: JTY × UTP4, JTY × P, and JTY × W, which suggests that JTY does not work well as a mother plant (Table 2 and Figure 2).

Table 2. Total and mean numbers of flowers in inflorescences, produced seeds, and percentage efficiency of seed set.

Parental Components (Maternal × Paternal)	Frequency of Pollination (per Week)	No. of Maternal Inflorescences	No. of Florets in Inflorescences		No. of Seeds in Inflorescences		Efficacy of Seed Setting [%]
			Total	Mean	Total	Mean	
B × Ł	1×	6	994	$165.7^{a,b}$	9	$1.5^{b,c}$	$0.9^{b–d}$ *
	2×	6	1095	182.5^{a}	9	$1.5^{b,c}$	$0.8^{c,d}$
Ł × B	1×	6	722	$120.3^{c–f}$	1	0.2_{c}	0.1^{d}
	2×	6	781	$130.2^{b–f}$	1	0.2_{c}	0.1^{d}
JTY × UTP4	1×	6	475	79.2^{h}	0	0.0_{c}	0.0^{d}
	2×	6	591	$98.5^{f–h}$	0	0.0_{c}	0.0^{d}
W × Ł	1×	6	875	$145.9^{a–d}$	0	0.0_{c}	0.0^{d}
	2×	6	942	$157.0^{a–c}$	13	2.2_{b}	$1.4^{b–d}$
JTY × P	1×	6	608	$101.3^{e–h}$	0	0.0_{c}	0.0^{d}
	2×	6	725	$120.8^{c–f}$	0	0.0_{c}	0.0^{d}
W × BG	1×	6	836	$139.3^{b–e}$	3	$0.5_{b,c}$	$0.4^{c,d}$
	2×	6	810	$135.0^{b–f}$	1	0.2_{c}	0.1^{d}
W × P	1×	6	907	$151.2^{a–c}$	2	$0.3_{b,c}$	$0.2^{c,d}$
	2×	6	1072	178.7^{a}	22	3.7_{a}	$2.1^{a,b}$
JTY × Ł	1×	6	619	$103.2^{e–h}$	3	$0.5_{b,c}$	$0.5^{c,d}$
	2×	6	673	$112.2^{d–g}$	1	0.2_{c}	0.1^{d}
JTY × W	1×	6	530	$88.4^{g,h}$	0	0.0_{c}	0.0^{d}
	2×	6	504	$84.0^{g,h}$	0	0.0_{c}	0.0^{d}
W × UTP4	1×	6	824	$137.3^{b–e}$	7	$1.2^{b,c}$	$0.8^{c,d}$
	2×	6	924	$154.0^{a–c}$	21	3.5_{a}	2.3^{a}
Total			**15,507**	**129.2**	**93**	**$0.8^{b,c}$**	**0.5**

Means in columns marked with the same letter do not differ significantly according to Newman–Keuls' test at $p \leq 0.05$.

The seeds started germinating five days after sowing. Among the 93 seeds obtained in total, 67 seeds germinated (72%), however, three plants did not develop properly. The share of germinating seeds ranged from 50 to 100%, depending on the parental components (Table 3). The highest share of sprouted seeds was obtained from the following parental components: Ł × B, W × Ł, W × P, and JTY × Ł, while the lowest was in the experimental object W × BG.

Table 3. Number and share of germinating seeds and growing plants depending on the parental components.

Parental Components (Maternal × Paternal)	Seeds			Properly Growing Plants	
	Obtained	Germinating	Share [%]	Number	Share [%]
B × Ł	18	10^{b}	$55.5^{a,b}$	$8^{b,c}$	80.0^{a}
Ł × B	2	2^{c}	100.0^{a}	2^{c}	100.0^{a}
W × Ł	13	$11^{a,b}$	84.6^{a}	$11^{a–c}$	100.0^{a}
W × BG	4	2^{c}	50.0^{b}	2^{c}	100.0^{a}
W × P	24	21^{a}	87.5^{a}	20^{a}	95.2^{a}
JTY × Ł	4	$4^{b,c}$	100.0^{a}	4^{c}	100.0^{a}
W × UTP4	28	$17^{a,b}$	$60.7^{a,b}$	$17^{a,b}$	100.0^{a}
Total	**93**	**67**	**72.0**	**64**	**95.5**

Means in columns marked with the same letter do not differ significantly according to Fisher's exact test at $p \leq 0.05$.

The beginning of germination (phase I) was the shortest in seedlings from the W × BG parental object (4 days), and the longest in seedlings obtained from the crossing of Ł × B (10.5 days; Figure 2). Phase II (fully developed cotyledons) was achieved first by the seedlings which were the offspring of the W × BG components (in less than 7 days). On the other hand, in the seedlings derived from the Ł × B crossing, phase II was only achieved after an average of 3 weeks. The seedlings that were the B × Ł offspring reached their full maturity (phase III) the earliest, at 10 days after sowing, whereas for the seedlings from the Ł × B crossing, it was after 30 days. The mean results show that the experimental object Ł × B needed the most time to achieve all of the recorded stages of seedling development. As for chrysanthemum seedlings derived from B × Ł and W × BG parent components, the successive growing phases followed quickly one after another, at 2–3 days intervals (Figure 3).

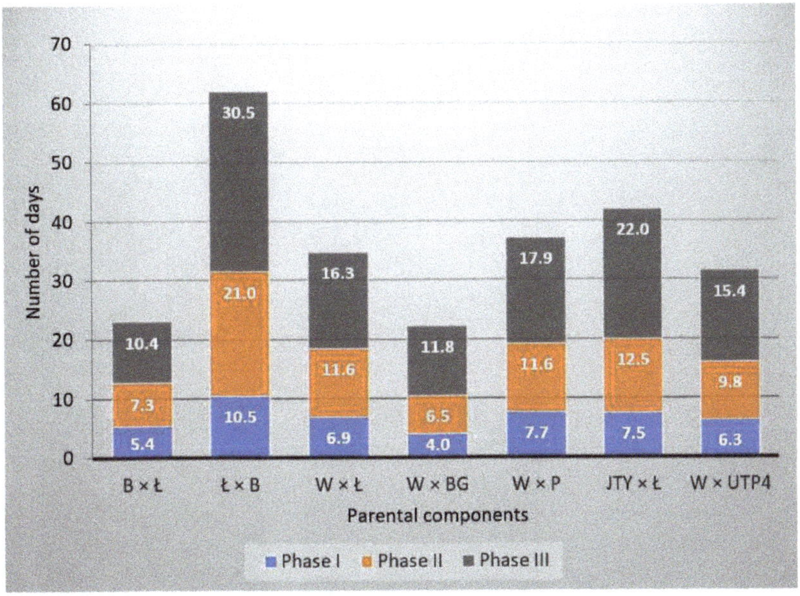

Figure 3. The number of days needed to obtain the three following stages of development in F_1 seedlings, depending on the arrangement of the parental components.

Out of 10 parental combinations tested, 7 produced the F_1 offspring (Table 4).

Table 4. Number of F1 plants obtained in the experiment from different parental combinations; number of promising phenotypes preliminarily selected for further observations after vegetative propagation; current number of phenotypes which granted PBRs and number of phenotypes being submitted to grant PBRs (DUS tests under proceedings).

Parental Components (Maternal × Paternal)	F1 Plants			
	Obtained	Preliminary Selected	Granted PBRs	PBR Submission, Under Proceedings
B × Ł	8	3	0	0
Ł × B	2	1	1	0
W × Ł	11	3	0	0
W × BG	2	0	0	0
W × P	20	5	2	1
JTY × Ł	4	2	1	0
W × UTP4	17	6	2	1
Total	**64**	**20**	**6**	**2**

The obtained plants varied in the shape, size, and color of their flowers. A total of 64 new phenotypes were found, from which 20 were selected for further observations due to their promising traits. Most of the selected phenotypes showed single or semi-full inflorescences, except one full inflorescence coming from the JTY × Ł crossing. The most abundant in new interesting phenotypes were crosses W × UTP4 and W × P. They gave six and five promising phenotypes, respectively. The crosses Ł × B and JTY × Ł produced only one and two interesting phenotypes, respectively. No new phenotype was found from W × BG. The remaining two crosses both resulted in three selected phenotypes. The distinctness, uniformity, and stability of the new phenotypes of chrysanthemums were confirmed in the successive three years of greenhouse cultivation and annual vegetative propagation of all the obtained phenotypes. Eight of them were selected based on the inflorescences' attractiveness, novelty, type of growth, and resistance, and submitted to the Community Plant Variety Office (CPVO) in Poland for official distinctness, uniformity, and stability (DUS) testing. As a result, six new, valuable cultivars were granted PBRs in the years 2019–2022, while with the two other submissions, the procedure is ongoing (Figure 4).

Figure 4. New chrysanthemum cultivars obtained as a result of the experiment and submitted for granting PBRs. Denomination and grant number are followed by the parental composition from which the genotype originates.

4. Discussion

A number of factors determine the possibility of obtaining new cultivars through the crossing. In the common greenhouse chrysanthemums, one of the bottle-necks is seed establishment, since they perform a low seed-setting capability, which limits the breeding efficiency [10]. In the present study, the influence of selected factors (pollination frequency and parental genotypes) on the setting and germination of chrysanthemum seeds was analyzed.

Our research has shown that, despite a large number of florets, inflorescences of chrysanthemums contain a relatively small amount of seeds, sometimes even none. Likewise, no seeds were obtained in the crossing between *Chrysanthemum* × *morifolium* 'Yuhuaxingchen' and *Chrysanthemum* × *nankingense* Hand.-Mazz. [16]. On the other hand, in the cross between *C. grandiflorum* and *C. indicum*, the seed set level reached 59%. In the cross between *C. grandiflorum* and *C. zawadskii* Herbich, the seed set was only 9% [16]. In chrysanthemums, the incubation of pollen at −80 °C could significantly increase the pollen germinability and seed set efficiency [17]. On the other hand, according to Deng et al. [18], pollen viability has no significant effects on seed set in chrysanthemums.

Nonetheless, very few germinated pollen grains on stigmas and the abnormal growth of most pollen tubes before fertilization, as well as embryo abortion, are the main factors which cause the failure of cross-breeding [16]. This could be explained by the fact that the chrysanthemum is a hexaploid hybrid. Moreover, modern cultivars are commonly a result of mutation breeding, which decreases their reproductive capacities [15]. Therefore, embryo rescue under in vitro conditions appears to be a promising solution to this problem and could be considered in future studies [19].

The highest number of seeds was produced in those crossings where the 'Wda' cultivar was used as the maternal component (W × P, W × UTP4, W × Ł). This could be explained by the fact that it originates from garden chrysanthemums, which establish the seeds more easily [9]. On the other hand, no seeds were found in the following three experimental objects: JTY × Ł, JTY × P, and JTY × W, which indicates that the JTY genotype is not suitable as a mother plant, possibly due to low fertility. Such a low seed setting efficiency may be related to the fact that most modern greenhouse cultivars of chrysanthemums are created through induced mutagenesis, which results in the loss of their natural reproductive abilities [20].

Various seed setting efficiency in the investigated combinations of parental plants may result from the genetic background related to sporophytic self-incompatibility, which is ubiquitous in chrysanthemums [21]. It prevents inbreeding, but S alleles are widely distributed in the genomes of the greenhouse cultivars, which lead to the decrease in successful fertilization and poor seed production [21,22]. Matching genetically distinct individuals, such as 'Wda' (possessing garden chrysanthemums' genetic load), for crosses with the typical greenhouse cultivars could increase the seed set in our experiment. More seeds were obtained by crossing plants twice a week than once a week. Thus, it can be concluded that the frequency of crossing has a significant influence on seed establishment in chrysanthemums. The more often the crosses are repeated, the greater the likelihood of obtaining more seeds, which is related to the structure of the chrysanthemum inflorescence. In Asteraceae, numerous single florets are collected within a head, which was developed in the span of the evolutionary process as the most complex, yet effective inflorescence type [9]. During pollination, the pollen is transferred to the pistil stigma in the florets of a mother plant, which mature gradually from the outer whorls to the center of the inflorescence [23]. The maturation of all the florets within the head from the edges to the center takes about two to three weeks [24]. The developed flowers stay receptive (retain the ability to absorb pollen) for several days, therefore, the inflorescence can be pollinated recurrently, every two or three days to the next developing whorls. Consequently, the chance of pollinating more receptive flowers is higher, resulting in a better seed setting.

Despite several obstacles, it is worth investing in chrysanthemum crossbreeding, as it is a good way to create a pool of novel interesting cultivars, useful in further breeding programs, for example, through mutagenesis [25]. It is worth mentioning that a complex chrysanthemum genome consisting of six sets of chromosomes, as well as high heterozygosity, are the source of great F1 progeny variation [9]. In the present study, it was possible to obtain eight new phenotypes of inflorescences that varied in shape, size, and color, and each one was different from another. Among these new phenotypes, six received PBRs, so far, and the remaining two are under DUS tests. These new cultivars received great interest, were awarded several prestigious awards during international innovation shows, and are currently being introduced to the Polish market. Chrysanthemums produced in our Laboratory are the sole Polish cultivars available in the market.

5. Conclusions

Greenhouse cultivars of chrysanthemums produce a relatively small number of seeds concerning the number of flowers in the inflorescence. Therefore, classical cross-breeding in this species is a considerable challenge. The frequency of chrysanthemum pollination has a significant effect on the outcome of the breeding process. Pollination performed twice a week increases the chance of obtaining more seeds. The number of seeds produced also depends on the genotype of the parental components. Chrysanthemum 'Wda' was the most effective mother plant, as it produced the highest number of seeds, regardless of the paternal component used. The present research confirmed the usefulness of cross-breeding in the creation of novel attractive chrysanthemum cultivars. Future studies should focus on the verification of other chemical and physical factors that could improve the effectiveness of chrysanthemum pollination.

Author Contributions: Conceptualization, N.M.; methodology, N.M.; software, N.M. and D.K.; validation, N.M. and D.K.; formal analysis, N.M. and D.K.; investigation, N.M.; resources, N.M.; data curation, N.M.; writing—original draft preparation, D.K.; writing—review and editing, N.M. and D.K.; visualization, N.M. and D.K.; supervision, N.M. and D.K.; project administration, N.M.; funding acquisition, D.K. All authors have read and agreed to the published version of the manuscript.

Funding: This research received no external funding.

Informed Consent Statement: Not applicable.

Data Availability Statement: Data available by e-mail on reasonable request.

Acknowledgments: The authors wish to acknowledge Jagoda Dorota Gołach (student of PBS) for the technical support in performing the experiments.

Conflicts of Interest: The authors declare no conflict of interest.

References

1. Zhao, K.; Li, S.; Jia, D.; Xing, X.; Wang, H.; Song, A.; Jiang, J.; Chen, S.; Chen, F.; Ding, L. Characterization of the MADS-Box Gene *CmFL3* in chrysanthemum. *Agronomy* **2022**, *12*, 1716. [CrossRef]
2. Ohmiya, A. Molecular mechanisms underlying the diverse array of petal colors in chrysanthemum flowers. *Breed. Sci.* **2018**, *68*, 17075. [CrossRef] [PubMed]
3. Zhao, Y.; Huo, B.; Lin, S.; Zhang, S.; Mao, C.; Jiang, J.; Chen, S.; Fang, W.; Guan, Z.; Liao, Y.; et al. Germplasm innovation and establishment of comprehensive evaluation system for hedgerow garden chrysanthemum. *Agronomy* **2022**, *12*, 1736. [CrossRef]
4. Latado, R.R.; Adames, A.H.; Neto, A.T. In vitro mutation of chrysanthemum (*Dendranthema grandiflora* Tzvelev) with ethylmethanesulphonate (EMS) in immature floral pedicels. *Plant Cell Tiss. Organ Cult.* **2004**, *77*, 103–106. [CrossRef]
5. Yamaguchi, H.; Shimizu, A.; Hase, Y.; Tanaka, A.; Shikazono, N.; Degi, K.; Morishita, T. Effects of ion beam irradiation on mutation induction and nuclear DNA content in chrysanthemum. *Breed. Sci.* **2010**, *60*, 398–404. [CrossRef]
6. Haider, S.; Gao, Y.; Gao, Y. Standardized genetic transformation protocol for chrysanthemum cv. 'Jinba' with TERMINAL FLOWER 1 homolog *CmTFL1a*. *Genes* **2020**, *11*, 860. [CrossRef]
7. Adedeji, O.S.; Naing, A.H.; Kim, C.K. Protoplast isolation and shoot regeneration from protoplast-derived calli of chrysanthemum cv. White ND. *Plant Cell Tiss. Organ Cult.* **2020**, *141*, 571–581. [CrossRef]
8. Teixeira da Silva, J.A.; Kulus, D. Chrysanthemum biotechnology: Discoveries from the recent literature. *Folia Hortic.* **2014**, *26*, 67–77. [CrossRef]
9. Anderson, N.O. Chrysanthemum. In *Flower Breeding and Genetics*; Anderson, N.O., Ed.; Springer: Dordrecht, The Netherlands, 2007; pp. 389–437.
10. Wang, F.; Zhang, F.J.; Chen, F.D.; Fang, W.M.; Teng, N.J. Identification of chrysanthemum (*Chrysanthemum morifolium*) self-incompatibility. *Sci. World J.* **2014**, *2014*, 625658. [CrossRef]
11. Yang, J.S.; Endo, M. In vitro germination and viability of *Dendranthema* pollen. *Asian J. Plant Sci.* **2005**, *4*, 673–677. [CrossRef]
12. Spaargaren, J.; Van Geest, G. Chrysanthemum. In *Ornamental Crops, Handbook of Plant Breeding*; Van Huylenbroeck, J., Ed.; Springer: Berlin/Heidelberg, Germany, 2018; Volume 11, pp. 319–348.

13. Wang, F.; Zhong, X.; Wang, H.; Song, A.; Chen, F.; Fang, W.; Jiang, J.; Teng, N. Investigation of differences in fertility among progenies from self-pollinated chrysanthemum. *Int. J. Mol. Sci.* **2018**, *19*, 832. [CrossRef] [PubMed]
14. Saumitou-Laprade, P.; Vernet, P.; Vekemans, X.; Billiard, S.; Gallina, S.; Essalouh, L.; Mhaïs, A.; Moukhli, A.; Bakkali, A.E.; Barcaccia, G.; et al. Elucidation of the genetic architecture of self-incompatibility in olive: Evolutionary consequences and perspectives for orchard management. *Evol. Appl.* **2017**, *10*, 867–880. [CrossRef] [PubMed]
15. Datta, S.K. Induced mutations: Technological advancement for development of new ornamental varieties. *Nucleus* **2020**, *63*, 119–129. [CrossRef]
16. Sun, C.Q.; Chen, F.D.; Teng, N.J.; Liu, Z.L.; Fang, W.M.; Hou, X.L. Factors affecting seed set in the crosses between *Dendranthema grandiflorum* (Ramat.) Kitamura and its wild species. *Euphytica* **2010**, *171*, 181–192. [CrossRef]
17. Miler, N.; Wozny, A. Effect of pollen genotype, temperature and period of storage on in vitro germinability and in vivo seed set in chrysanthemum—Preliminary study. *Agronomy* **2021**, *11*, 2395. [CrossRef]
18. Deng, Y.; Teng, N.; Chen, S.; Chen, F.; Guan, Z.; Song, A.; Chang, Q. Reproductive barriers in the intergeneric hybridization between *Chrysanthemum grandiflorum* (Ramat.) Kitam. and *Ajania przewalskii* Poljak. (Asteraceae). *Euphytica* **2010**, *174*, 41–50. [CrossRef]
19. Puglisi, D.; Las Casas, G.; Ferlito, F.; Nicolosi, E.; Di Guardo, M.; Scollo, F.; Saitta, G.; La Malfa, S.; Gentile, A.; Distefano, G. Parents' selection affects embryo rescue, seed regeneration and the heredity of seedless trait in table grape breeding programs. *Agriculture* **2022**, *12*, 1096. [CrossRef]
20. Zalewska, M.; Tymoszuk, A.; Miler, N. New chrysanthemum cultivars as a result of in vitro mutagenesis with the application of different explant types. *Acta Sci. Pol. Hort. Cult.* **2011**, *10*, 109–123.
21. Anderson, N.O.; Ascher, P.D. Inheritance of seed set, germination, and day neutrality/heat delay insensitivity of garden chrysanthemums (*Dendranthema* × *grandiflora*) under glasshouse and field conditions. *J. Am. Soc. Hort. Sci.* **2004**, *129*, 509–516. [CrossRef]
22. Pu, Y.; Huo, R.; Lin, Q.; Wang, F.; Chun, X.; Huang, H.; Dai, S. Investigation and screening of chrysanthemum resources to identify self-compatible mutants. *Sci. Hortic.* **2021**, *281*, 2021. [CrossRef]
23. Mekapogu, M.; Vasamsetti, B.M.K.; Kwon, O.-K.; Ahn, M.-S.; Lim, S.-H.; Jung, J.-A. Anthocyanins in floral colors: Biosynthesis and regulation in chrysanthemum flowers. *Int. J. Mol. Sci.* **2020**, *21*, 6537. [CrossRef] [PubMed]
24. Xu, L.; Liu, C.L.; Wang, H.D.; Chen, K.L. Study on the pollen viability and stigma receptivity of *Chrysanthemum morifolium* 'Fubaiju'. *Zhong Yao Cai* **2012**, *35*, 1546–1550. [PubMed]
25. Din, A.; Qadri, Z.A.; Wani, M.A.; Rather, Z.A.; Iqbal, S.; Malik, S.A.; Hussain, P.R.; Rafiq, S.; Nazki, I.T. Congenial in vitro γ-ray-induced mutagenesis underlying the diverse array of petal colours in chrysanthemum (*Dendranthema grandiflorum* Kitam.) cv. 'Candid'. *Biol. Life Sci. Forum* **2021**, *4*, 21. [CrossRef]

horticulturae

MDPI

Article

Asexual Propagation of Greek *Salvia officinalis* L. Populations Selected for Ornamental Use

Christos Nanos [1,†], Parthena Tsoulpha [2,†], Stefanos Kostas [3,*,†], Stefanos Hatzilazarou [3], Ioanna Michail [3], Vasiliki Anastasiadi [3], Elias Pipinis [4], Evangelos Gklavakis [5], Angelos K. Kanellis [6] and Irini Nianiou-Obeidat [1,*]

1 Laboratory of Genetics and Plant Breeding, School of Agriculture, Aristotle University, 54124 Thessaloniki, Greece; nanos95@hotmail.com
2 Laboratory of Forest Genetics and Plant Breeding, School of Forestry and Natural Environment, Aristotle University, 54124 Thessaloniki, Greece; thena@for.auth.gr
3 Laboratory of Floriculture, School of Agriculture, Aristotle University, 54124 Thessaloniki, Greece; hatzilaz@agro.auth.gr (S.H.); ioannamichail@hmu.gr (I.M.); vasakianas19@hotmail.com (V.A.)
4 Laboratory of Silviculture, School of Forestry and Natural Environment, Aristotle University, 54124 Thessaloniki, Greece; epipinis@for.auth.gr
5 Evangelos Gklavakis Nurseries, 58400 Aridea, Greece; evan.glavakis@gmail.com
6 Group of Biotechnology of Pharmaceutical Plants, Laboratory of Pharmacognosy, Department of Pharmaceutical Sciences, Aristotle University, 54124 Thessaloniki, Greece; kanellis@pharm.auth.gr
* Correspondence: skostas@agro.auth.gr (S.K.); nianiou@agro.auth.gr (I.N.-O.)
† These authors contributed equally to this work.

Abstract: *Salvia officinalis*, commonly known as sage, is highly valued for its medicinal and ornamental properties. In the present work, 12 native sage populations of north-west Greece were evaluated for eight ornamental traits. Populations from the locations of Aristi, Kefalovryso and Igoumenitsa were selected as the best performing and for their preservation and availability in the market, their asexual propagation was investigated by (a) shoot cutting and (b) in vitro techniques. Propagation by cuttings was investigated during the four seasons. Aristi exhibited the highest rooting (65%) in spring with a well-developed root system (4.7 root number and 5.0 cm length) by applying 0.5 $g{\cdot}L^{-1}$ Indole-3-butyric acid, potassium salt (K-IBA), established on perlite under a fog system. However, the rooting performance of Aristi spring cuttings was not affected by different substrates of peat:perlite (0:1, 1:1, 1:2 v/v) or rooting systems (mist, fog) tested. Furthermore, the in vitro propagation of the selected sage populations was investigated using shoot tips as explants. After successful disinfection, the effect of Murashige and Skoog (MS) medium in ten different combinations of Indole-3-acetic acid (IAA), 6-Benzylaminopurine (BAP) and Thidiazuron (TDZ) were tested on shoot multiplication. Aristi presented the highest number of newly formed shoots on MS9 (0.1 $mg{\cdot}L^{-1}$ IAA and 0.8 $mg{\cdot}L^{-1}$ TDZ) and MS5 (0.1 $mg{\cdot}L^{-1}$ IAA and 0.8 $mg{\cdot}L^{-1}$ BAP) (3.35 and 3.21 new shoots/explant, respectively) with the highest shoot length (2.23 cm and 3.2 cm) and unexpected spontaneous root formation (64%) at MS5. The rooting ability of Aristi microshoots was further investigated in order to enhance their response. Of the three rooting variants tested, optimal rooting formation (100%) was observed on 0.9 $mg{\cdot}L^{-1}$ IAA (R3) combined with successful acclimatization (100%). Aristi exceeded the other populations in both the tested propagation systems, thus holding a strong potential for its introduction in the market as a competitive ornamental variety.

Keywords: aesthetic use; cuttings; IBA; micropropagation; sage; season

Citation: Nanos, C.; Tsoulpha, P.; Kostas, S.; Hatzilazarou, S.; Michail, I.; Anastasiadi, V.; Pipinis, E.; Gklavakis, E.; Kanellis, A.K.; Nianiou-Obeidat, I. Asexual Propagation of Greek *Salvia officinalis* L. Populations Selected for Ornamental Use. *Horticulturae* **2023**, *9*, 847. https://doi.org/10.3390/horticulturae9070847

Academic Editors: Konstantinos Bertsouklis, Epaminondas Kartsonas and Angela Carra

Received: 19 June 2023
Revised: 20 July 2023
Accepted: 21 July 2023
Published: 24 July 2023

1. Introduction

Salvia officinalis L., an evergreen, perennial species, is one of the most important of the genus *Salvia* and has been widely recognized for its medicinal, aromatic and culinary uses since ancient times [1–3]. The species is well known for its high content of essential oils [4–8]. Moreover, due to the beautiful natural leaf, inflorescence variations and the species cold hardiness, *S. officinalis* is also an important ornamental plant [9].

The first step toward the improvement of a plant species by its desired traits is based on the analysis of the species morphology, mainly by assessing leaf, flower, fruit and plant shape variability [10], which is a useful tool for the selection of populations suitable as ornamentals [11,12]. However, the reliability of the evaluation by phenotypic characteristics is affected by environmental factors. The elimination of this effect is possible using a randomized experimental design and cultivating the plants in the same environment, field [12].

The need for readily available vegetative material directs research toward investigating and applying rapid and efficient propagating techniques, such as asexual propagation in vivo and in vitro.

For the successful rooting of sage cuttings, among the main factors investigated were the influence of season and the application of the most efficient indole-3-butyric acid (IBA) concentration [13–15]. Loconsole et al. [16] studied the effect of IBA dosage on the improvement of adventitious root quality on wild sage and some of its cultivars.

Concerning tissue culture, it is a useful, modern method of producing certified propagation material, for commercial use exploiting elite populations for their desired characteristics, having the advantage of retaining genetic stability and conserving plant genetic resources [17–19]. Especially for *S. officinalis*, several studies on its micropropagation using different explants have been conducted, such as from shoot tips [20,21], nodal segments [3,22,23], and axillary and apical buds [24]. Bolta et al. [25] investigated the cell cultures of the species. However, most of the previous works used in vitro propagation for the production of high value endogenous substances or for their bioactive molecules [17,21,25,26].

S. officinalis is a first-class alternative crop, with very good yields and a prominent position in the Greek market, holding the potential for further expansion. Although the majority of studies on sage focuses on the production of essential oil and other useful medicinal endogenous metabolites, there are no previous investigations of wild sage population morphological traits, selection or propagation of the best plant material for ornamental purposes.

The aim of this study was to evaluate the morphological traits of twelve sage populations from different regions of north-western Greece and select the best ones. For the populations of *S. officinalis* that stood out for their aesthetic morphology, asexual propagation by cuttings and in vitro techniques were studied in order to establish effective and functional propagation protocols.

2. Materials and Methods

2.1. Plant Material

Twelve wild-grown populations of *S. officinalis,* from different habitats of north-west Greece, were selected for evaluation. The code names and geographical coordinates of the central point of each population are given in Table 1. The sampling of plant material was conducted as follows: shoot cuttings were collected and transferred for propagation to the Evangelos Gklavakis nurseries (Piperia, Pella, Greece, latitude 40.964263 N, longitude 22.017363 E). The cuttings were treated with 0.5 $g \cdot L^{-1}$ Indole-3-butyric acid, potassium salt (K-IBA) (Sigma-Aldrich, St. Louis, MO, USA) and maintained under the fog system to root successfully. The young rooted plants were planted and grown for two years in an outdoor experimental collection of the same nursery (Table 1).

Shoots with leaves and flowers from the twelve populations were collected in June and transferred and kept in the herbarium (code numbers SO201-SO392) of the Floriculture Laboratory.

Table 1. Code names, coordinates of the central point of 12 *S. officinalis* populations and number of plants per population.

	Population	Latitude (North)	Longitude (East)	Number of Plants per Population
1	ARISTI	39.933689	20.679384	21
2	ARNISSA	40.798127	21.828355	9
3	ELAFOTOPOS	39.901731	20.692177	26
4	IGOUMENITSA	39.485731	20.264105	10
5	KEFALOVRYSO	40.003702	20.558476	11
6	KALPAKI	39.902897	20.641573	18
7	KALYBIA	39.902781	20.641872	15
8	KATO PEDINA	39.877418	20.670230	31
9	KERKYRA	39.770129	19.697890	10
10	MAYROBOUNI	39.954294	20.619267	18
11	MESOBOUNI	39.942593	20.646485	26
12	MIKROBALTOS	40.078276	21.872652	22

2.2. Analysis of Morphological Traits

The evaluated ornamental characteristics of the above plant material are analytically shown in Table 2. Traits No. 2–4 were measured in the laboratory with a ruler, and all the rest were measured directly at the outdoor experimental collection site (in June) [12].

Table 2. Morphological traits for the 12 selected populations of *S. officinalis* related to their ornamental value.

Morphological Traits	Description
1. Leaf number	Number of leaves per branch, 20 terminal branches (15 cm from the shoot tip) per population
2. Leaf length	In cm, measured from the base to the tip of adult /mature leaf, 50 leaves per population
3. Leaf width	In cm, measured at the widest part of adult leaf/mature, 50 leaves per population
4. Inflorescence length	In cm, measured from the base to the tip of inflorescence, 20 inflorescences per population
5. Node number per inflorescence	Number of nodes per inflorescence, 20 inflorescences per population
6. Flower number	Number of flowers per inflorescence, 20 inflorescences per population
7. Branch number	Number of terminal branches per plant, was measured in all plants
8. Branch length	Length of branches per plant, in cm, 20 terminal branches per population

2.3. Asexual Propagation of Selected *S. officinalis* Populations

Terminal shoots of the three best populations (Aristi, Igoumenitsa, Kefalovryso), selected for their ornamental traits, were harvested from the mother plants of the field collection in Piperia Aridea and used as starting material for both propagation techniques.

2.3.1. Propagation by Shoot Cuttings

Effect of K-IBA and Season on the Rooting of Cuttings

Terminal cuttings collected during the four seasons of 2021 were tested for their rooting ability. This type of cutting was reported as the most suitable for the propagation of *Salvia* [27]. The basal portion of each shoot cutting (8–10 cm, 5–6 leaves on the apical part) was dipped into aqueous solutions of 0, 0.5, or 1 g·L^{-1} of K-IBA for 10 s and planted in 10 L plastic trays (40 cm × 25 cm × 10 cm) filled with perlite (Isocom, Athens, Greece). The plastic trays were then established for rooting in a fog system, with the relative humidity (RH) adjusted to 95 ± 1%. Forty shoot cuttings were used for each treatment and population. After four weeks, the rooting ratio (%), as well as the number and length (cm) of roots, were recorded.

Effect of Substrate and the Mist or Fog System on Rooting of Cuttings

Terminal shoots of the best performing Aristi population were harvested in spring of the following year and used as cuttings. The effect of substrate was evaluated on three different mixtures of peat TS2 Klasmann® (Klasmann-Deilmann, Geeste, Germany) and perlite: 0:1, 1:1, or 1:2 v/v, after treating the cuttings as previously described with 0.5 g·L^{-1} K-IBA. Finally, the cuttings were placed for rooting either under the fog or intermittent mist system. The RH in the fog system was adjusted to 95 ± 1%, while in the intermittent mist system, water was sprayed for 30 s every 30 min, from 06:00 to 22:00. In both rooting systems, the temperature at the bottom of the benches was set at 20 ± 1 °C using electrical cables. In each treatment, 40 shoot cuttings were used. Four weeks after planting, the rooting ratio (%) and the number and length (cm) of roots were measured.

2.3.2. In Vitro Propagation of *S. officinalis*

Explant Preparation and Disinfection

For the tissue culture experiments, shoot tips (2–3 cm) of the selected populations of *S. officinalis* were used as explants. First, they were pretreated with a mild dish soap and washed under running tap water for 20 min. Seven different disinfection treatments were tested for all populations (twenty explants each) (Table 3), followed by three successive washings with double distilled water (ddH_2O). The explants were established in vitro on MS [28] (Murashige and Skoog, 1962) medium, free of growth regulators, supplemented with sucrose (3%) and agar (0.8% Plant Agar, Duchefa Biochemie, The Netherlands). The pH was adjusted to 5.8. Cultures were maintained in plant growth chamber conditions: 23 ± 2 °C, 16 h photoperiod and light intensity at 50 µmol·m^{-2}·s^{-1} provided by cool-white fluorescent lamps. The same environmental conditions were applied to all subsequent experiments.

Table 3. Disinfection treatments applied to *S. officinalis* explants.

Treatment No.	(%) EtOH v/v *	Ascorbic and Citric Acid **	NaOCl (%)	Time Duration (min)
D1	70	−	0.06	10
D2	70	+	0.06	11
D3	70	−	0.06	12
D4	60	+	0.06	11
D5	50	+	0.06	12
D6	70	−	0.08	7
D7	70	−	0.04	17

* Stirring with ethanol solution for 30 s; ** A solution of ascorbic acid (300 mg·L^{-1}) and citric acid (200 mg·L^{-1}) for 15 min.

Effect of Growth Regulators on the Multiplication of *S. officinalis*

Clean explants were transferred on ten variants of MS nutrient media supplemented with different growth regulators: 6-Benzylaminopurine (BAP), Thidiazuron (TDZ) (Sigma-Aldrich, St. Louis, MO, USA) at concentrations of 0, 0.2, 0.5, 0.8 and 1.1 mg·L^{-1} combined with Indoleacetic acid (IAA) (0 or 0.1 mg·L^{-1}) (Sigma-Aldrich, St. Louis, MO, USA) (Table 4). A total of 600 shoots, 20 per population and substrate were used. The multiplication frequency, average number of shoots induced per explant and mean height of shoots were recorded after five weeks in culture.

In Vitro Rooting and Plantlet Acclimatization

Newly formed shoots (~3.5 cm) from the two best performing populations (Aristi, Kefalovryso originating from MS5 and MS9) were tested for rooting on MS supplemented with three variants of IAA: R1 (0.3 mg·L^{-1}), R2 (0.6 mg·L^{-1}), and R3 (0.9 mg·L^{-1} IAA). Fifteen microshoots per population and variant were used. For rooted plantlets with at least one root (>0.5 cm), the following were recorded: percentage (%) of rooted shoots and root length after five weeks. Rooted plantlets were transferred to acclimatization after thoroughly washing off the agar residue and established in propagating trays of 24 cells on TS-2 Klasmann® peat in a Styrofoam structure covered with plexiglass airtightly. The initial environmental conditions were as follows: 25 ± 2 °C, relative humidity 95 ± 1%,

12 h photoperiod (60 μmol·m^{-2}·s^{-1}) provided by artificial light. Gradually, in order to harden off the plants in a four-week period, the relative humidity was reduced to 65% ± 1 and the light intensity was adjusted to ambient (180 μmol·m^{-2}·s^{-1}). Plant survival was assessed after six weeks.

Table 4. Nutrient substrate composition for shoot multiplication.

Medium	IAA (mg·L^{-1})	BAP (mg·L^{-1})	TDZ (mg·L^{-1})
MS 1	0	0	0
MS 2	0.1	-	-
MS 3	0.1	0.2	-
MS 4	0.1	0.5	-
MS 5	0.1	0.8	-
MS 6	0.1	1.1	-
MS 7	0.1	-	0.2
MS 8	0.1	-	0.5
MS 9	0.1	-	0.8
MS 10	0.1	-	1.1

2.4. Statistical Analysis

For the morphological traits, the analysis of variance (ANOVA) was performed, while the separation of means was conducted by Duncan's multiple range test at $p \leq 0.05$. In all of the asexual propagation procedures, a complete randomized design was used. Four replicates of ten cuttings were used for the first asexual propagation technique, whereas three replications were used for each of the in vitro experiments. For in vitro and cuttings measurements, the mean ± standard deviation (SD) was calculated, while for morphological traits, the mean ± standard error (SE) was used. Mean comparisons were conducted using ANOVA and Duncan's test ($p \leq 0.05$). All percentages were subjected to arcsine transformation. Analyses were conducted using the SPSS V. 27 (IBM, Armonk, NY, USA) statistical package.

3. Results and Discussion

3.1. Morphological Analysis of Ornamental Traits

The results of the phenotypic evaluation indicated that the populations with the best ornamental properties were Aristi, Igoumenitsa, Kalpaki and Kefalovryso. According to the assessment, the trait "Leaf number" of the Kefalovryso population had the highest value, up to 30 leaves per branch and differed significantly from the others (Figure 1). Respectively, the lowest values were recorded for Kato Pedina and Kerkyra with less than seven leaves, although there was no statistical difference from all the other populations, apart from Kefalovryso, Arnissa, Igoumenitsa and Kalpaki. The populations Aristi and Kalpaki had the largest leaves concerning their "Leaf length", apart from Kato Pedina, which were statistically similar. In relation to "Leaf width", Igoumenitsa, Kerkyra and Arnissa had the widest leaves without being statistically different from most of the populations, whereas Kalybia was the narrowest. For "Inflorescence length", the lowest value was measured for Mikrobaltos (approximately 7.6 cm), followed by Kerkyra. In addition, the populations with the lowest values for the trait "Node number per inflorescence" were Kerkyra and Mikrobaltos. For the trait "Flower number", the lowest measurements were recorded for Arnissa, Mikrobaltos and Kerkyra. The highest values for the last two traits, "Branch Number and Length" per plant were observed for Igoumenitsa and Aristi, without statistical differences compared to the rest of the populations (Figure 1).

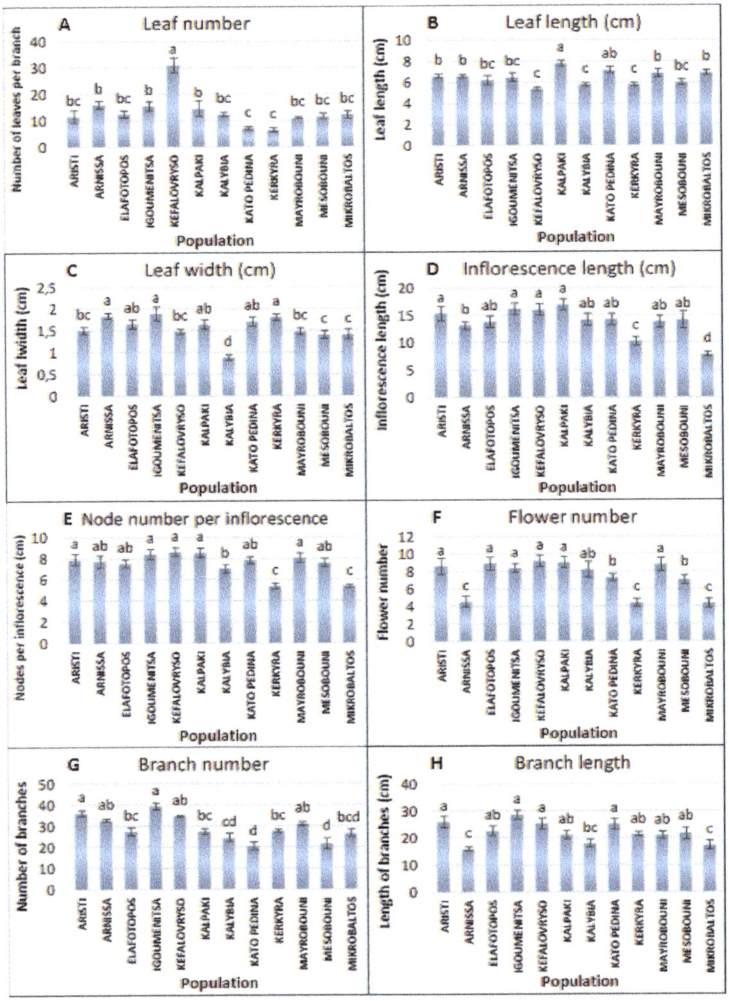

Figure 1. Mean values ± SE for eight morphological traits of ornamental interest [(**A**) Leaf number, (**B**) Leaf length (cm), (**C**) Leaf width (cm), (**D**) Inflorescence length (cm), (**E**) Node number per inflorescence, (**F**) Flower number, (**G**) Branch number and (**H**) Branch length] recorded from plants of the twelve populations of *S. officinalis* growing in an experimental field in Piperia (Pella, Greece). Different letters indicate statistically significant differences according to Duncan's multiple range test at $p \leq 0.05$, error bars indicate standard errors.

In general, the evaluation of morphological data revealed significant phenotypic variability among the twelve populations of *S. officinalis*. More specifically, Igoumenitsa was the population with the better morphological traits, while Kefalovryso presented better values not only in the number of nodes and flowers per inflorescence but also in the number of leaves per branch. However, better values in both leaf and inflorescence length were recorded for the Aristi population. According to the above, the Igoumenitsa, Kefalovryso and Aristi populations were selected as the most suitable for decorative use and their asexual propagation was further investigated by tissue culture and shoot cutting techniques.

Thus far, there has been limited research on the morphological characterization of *S. officinalis* for ornamental purposes. However, several studies have been conducted to estimate the phenotypic diversity of different species of *Salvia* using both qualitative and quantitative morphological traits [29–31]. Leontaritou et al. [32] evaluated the morphological diversity (leaf and floral traits) of 49 individuals from *Salvia pomifera* subsp. *calycina* (Sm.) Hayek (Apple sage), originating from five natural populations of the Peloponnese (Greece). In that study, leaf length ranged from 3.88 to 5.23 cm, while in the present, it ranged from 5.3 to 7.8 cm. Concerning leaf width and inflorescence length, the range of values was higher than in our recorded measurements. These could be attributed to variations between *Salvia* species [32].

In another study, the morphological traits of *Salvia fruticosa* (Greek sage) were also evaluated. Ten populations of *S. fruticosa*, from different locations of the Peloponnese, were evaluated and found to differ significantly for both leaf and floral traits. The highest value of leaf length was 4.32 cm instead of 7.8 cm in our study, while the leaf width presented the same size. The researchers concluded that this morphologic variability could be attributed to environmental parameters, such as altitude, latitude and climatic type [33].

Similar studies on the morphological traits of other plants of the Lamiaceae family, such as *R. officinalis,* indicated a significant phenotypic variability among the seven rosemary populations tested [12]. Using 15 qualitative traits, Zigene et al. [34] recorded the phenotypic diversity of 45 Ethiopian rosemary accessions from different growing regions. Morphological traits were also used in *Mentha longifolia* to describe a significant positive correlation between morphological and phytochemical characteristics [35]. Furthermore, the phenotypic and genetic diversity among 19 different populations of *Mentha longifolia* from various altitudes were also examined [36].

In general, a significant amount of phenotypic diversity exists in morphological traits among populations of the Lamiaceae plant species, which could be used to distinguish accessions of different growth regions for future selection and characterization work for various uses [34].

3.2. Asexual Propagation of Selected S. officinalis Populations

3.2.1. Propagation by Shoot Cuttings

Effect of K-IBA and Season on Cutting Rootability

The season of cutting collection, as well as the application of K-IBA, influenced the rooting of the three selected populations of *S. officinalis* (Figure 2). Spring proved to be the best season for rooting for all studied populations, with figures up to 65% (Aristi treated with 0.5 $g{\cdot}L^{-1}$ K-IBA), while no rooting was noticed in summer, except for Aristi cuttings treated with 1 $g{\cdot}L^{-1}$ K-IBA (10% rooting). In autumn and winter, rooting percentages reached up to 30 and 15%, respectively, for Aristi. Season influences the rooting ability of shoot cuttings due mainly to the different physiological status and the different lignification levels of the tissues among the four seasons. As Nikola et al. [13] reported, the best period for rooting of *S. officinalis* shoot cuttings was from spring to the end of autumn. These results are partly in agreement with our findings that the best season for rooting sage shoot cuttings was spring.

The application of K-IBA, regardless of concentration, significantly increased the rooting rate, even up to five-fold, as compared with the control. The concentration of 0.5 $g{\cdot}L^{-1}$ K-IBA was more effective on rooting than 1 $g{\cdot}L^{-1}$ in the spring, autumn and winter collection of cuttings for Aristi, whereas it was less effective in summer and winter for Kefalovryso and in spring for Igoumenitsa. In all other cases, both concentrations of K-IBA were similarly effective in rooting (Figures 2 and 3).

Thus, the application of K-IBA (0.5 $g{\cdot}L^{-1}$) in spring increased the rooting of Aristi from 12.5 to 65%, while in the Kefalovryso population from 10% (control) to 47%. For the Igoumenitsa population, rooting reached 22.5% only in the presence of 1 $g{\cdot}L^{-1}$ K-IBA.

The number and length of the roots of *S. officinalis* cuttings were influenced by both studied factors. In particular, the number of roots increased significantly in the presence of K-IBA compared to the control, especially its higher concentration (1 g·L^{-1}) (Table 5). The highest number of roots was recorded in Aristi during spring and autumn, with 1 g·L^{-1} K-IBA (5.3 and 5.4 roots per cutting, respectively) (Table 5). The populations of Igoumenitsa and Kefalovryso also formed many roots on the same auxin level during winter, spring and autumn (4.4, 4.7, 4.8 and 4.2, 4.4, 4.2 roots per cutting, respectively) (Table 5). Concerning the length of roots, the control showed the best response for Aristi in spring and autumn (5.9 and 5.5 cm, respectively) but for Kefalovryso, only in spring (5.8 cm) (Table 5). For Igoumenitsa, the best result was observed during spring on 0.5 g·L^{-1} K-IBA (5.3 cm), followed by autumn on 1 g·L^{-1} K-IBA (4.8 cm) (Table 5).

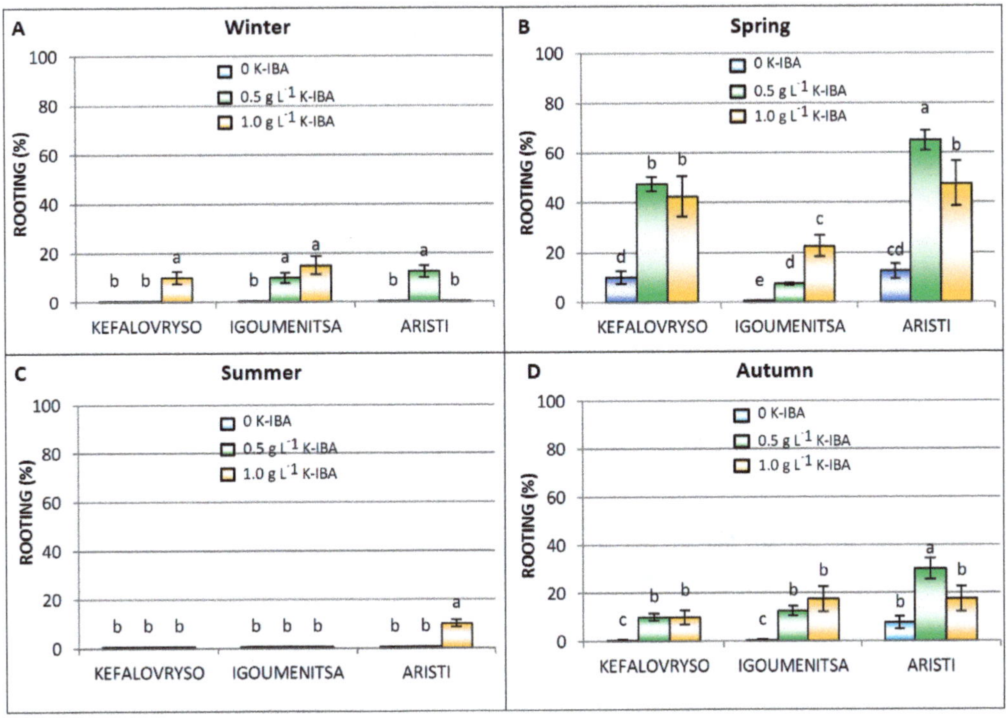

Figure 2. Effect of season, (**A**) winter, (**B**) spring, (**C**) summer, (**D**) autumn, and K-IBA on rooting (%) of *S. officinalis* Kefalovryso, Igoumenitsa and Aristi shoot cuttings (±standard deviation). Different letters indicate statistical differences, according to Duncan's multiple range test ($p \leq 0.05$).

All previous studies on sage cuttings reported the enhancement of rooting ratio and quality of the rooting system in the presence of auxin. For the successful rooting of sage cuttings, IBA was mainly used. Ayanoglu and Ozkan [37] reported that the application of 100 ppm IBA of sage cuttings led to quick rooting formation (78.75%) on the 15th day from establishment. Kara et al. [14] reported 81% rooting by applying 4000 ppm IBA and a well-developed rooting system (10.6 roots per cutting, 5.1 cm in length). Paradikovic et al. [15] achieved optimal rooting (100%) by applying Rhizopon powder (0.5% *w*/*w* IBA) on green cuttings of *S. officinalis*. Loconsole et al. [16] found that rooting system quality of the wild sage cultivar 'Little Lucky' was improved by 5000 mg·L^{-1} IBA, but it was less effective for 'Yellow'.

The improvement in the quality of the rooting system of sage cutting with the application of auxin agrees with the results of the present study. Finally, the results of our study, which confirm that the sensitivity to IBA dosage varies among species and their cultivars [38], could be relevant to the production of high-quality cuttings in the commercial nursery industry.

Figure 3. (**A**) Shoot cuttings of *S. officinalis*, (**B**) Planted cuttings of *S. officinalis* in perlite, in a fog system for rooting, (**C**) Rooted shoot cuttings of *S. officinalis* Kefalovryso, Igoumenitsa and Aristi (from left to right) treated with 0.5 g·L^{-1} K-IBA during spring and (**D**) *S. officinalis* plants from the population Aristi growing in greenhouse, six months after rooting of shoot cuttings.

Table 5. Effect of season (winter, spring, summer and autumn) and K-IBA on number and length (cm) of roots of *S. officinalis* Kefalovryso, Igoumenitsa and Aristi shoot cuttings.

	K-IBA $g \cdot L^{-1}$	Number of Roots				Length of Roots (cm)			
		Winter	Spring	Summer	Autumn	Winter	Spring	Summer	Autumn
Kefalovryso	0	-	$2.5 \pm 0.3^{d,*}$	-	-	-	5.8 ± 0.5^{a}	-	-
	0.5	-	$3.8 \pm 0.1^{b,c,**}$	-	3.4 ± 0.3^{c}	-	5.1 ± 0.2^{b}	-	$4.8 \pm 0.2^{b,c}$
	1.0	$4.2 \pm 0.2^{a,b}$	4.4 ± 0.3^{a}	-	$4.2 \pm 0.3^{a,b}$	4.2 ± 0.4^{d}	$4.4 \pm 0.4^{c,d}$	-	$4.6 \pm 0.3^{b,c}$
Igoumenitsa	0	-	-	-	-	-	-	-	-
	0.5	3.8 ± 0.3^{b}	3.9 ± 0.3^{b}	-	$4.3 \pm 0.2^{a,b}$	$4.6 \pm 0.2^{b,c}$	5.3 ± 0.2^{a}	-	$4.7 \pm 0.4^{a,b}$
	1.0	4.4 ± 0.1^{a}	4.7 ± 0.2^{a}	-	4.8 ± 0.4^{a}	4.1 ± 0.4^{c}	$4.8 \pm 0.3^{a,b}$	-	$4.8 \pm 0.3^{a,b}$
Aristi	0	-	3.1 ± 0.1^{c}	-	2.9 ± 0.1^{c}	-	5.9 ± 0.4^{a}	-	$5.5 \pm 0.4^{a,b}$
	0.5	4.7 ± 0.3^{b}	4.7 ± 0.2^{b}	-	4.6 ± 0.4^{b}	$4.7 \pm 0.3^{b,c}$	5.0 ± 0.2^{b}	-	$4.7 \pm 0.5^{b,c}$
	1.0	-	5.3 ± 0.2^{a}	2.2 ± 0.1^{d}	5.4 ± 0.2^{a}	-	4.9 ± 0.3^{b}	3.4 ± 0.2^{d}	4.2 ± 0.3^{c}

* Standard Deviation; ** Means in each population, for number and length of roots, with different letters indicate statistically significant differences, according to Duncan's multiple range test ($p \leq 0.05$).

Effect of Substrate and Mist or Fog Systems on the Rooting of Cuttings

The rooting of shoot cuttings collected from Aristi plants was affected by both the type of substrate and the kind of rooting system (Figure 4). The highest rooting of cuttings was observed on substrates of 1:0 and 2:1 perlite and peat in the fog system (60% and 37.5%, respectively) (Figure 4). No rooting was observed when a mixture of perlite and peat (1:1 v/v) was used as a substrate.

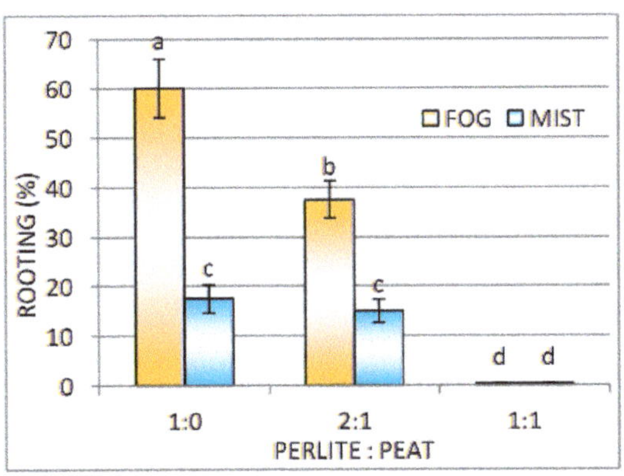

Figure 4. Effect of rooting system (fog or mist) and substrate composition (perlite:peat 1:0, 2:1 and 1:1) on rooting (%) of *S. officinalis* shoot cuttings (±Standard deviation) of population Aristi, in spring. Different letters indicate statistically significant differences, according to Duncan's multiple range test ($p \leq 0.05$).

The fog system increased the rooting ratio of the cuttings more than three-fold in the perlite: peat (1:0 v/v) substrate and more than two-fold in the perlite: peat (2:1 v/v) substrate compared with the mist system (Figure 4). The number and length of the new roots did not show any significant differences among the two tested substrates for both fog and intermittent mist systems (Table 6). Thus, 3–4 new roots 3.5–4.4 cm in length were formed (Table 6). In general, shoot cuttings of *S. officinalis* were rooted adequately on the substrate of perlite and peat (1:0, v/v) in the fog rooting system (Figure 4).

Other researchers studying the rooting of sage stem cuttings have also found differences in rooting in relation to the substrate tested [39,40]. In agreement with our results, Vârban et al. [41] concluded that perlite was the most appropriate substrate for the rooting of *S. officinalis* cuttings.

The selected sage populations were also propagated using the in vitro technique in order to optimize rooting results.

Table 6. Effect of rooting system (fog or mist) and substrate composition (perlite:peat 1:0, 2:1 and 1:1) on number and length (cm) of roots of *S. officinalis* Aristi shoot cuttings in spring.

Perlite:Peat	Fog		Mist	
	Number of Roots	Length of Roots (cm)	Number of Roots	Length of Roots (cm)
1:0	4.2 ± 0.3 a,*,**	3.9 ± 0.4 a	3.7 ± 0.7 a	4.0 ± 0.5 a
2:1	4.8 ± 0.6 a	4.7 ± 0.5 a	4.3 ± 0.5 a	3.8 ± 0.4 a
1:1	-	-	-	-

* Standard deviation; ** Different letters in the same column indicate statistically significant differences according to Duncan's multiple range test ($p \leq 0.05$).

3.2.2. In Vitro Propagation of *S. officinalis*

Effect of Disinfection Treatments on *S. officinalis*

Of the seven disinfection treatments tested, four of them (D1-D4), were equally successful for all three populations (Table 7). However, due to phenol presence in the medium, explant necrosis was observed except for the second treatment (D2), where an antioxidant solution (300 $mg \cdot L^{-1}$ ascorbic acid and 200 $mg \cdot L^{-1}$ citric acid) was applied, which provided healthy and vibrant explants (Figure 5A). Thus, it was chosen as the most appropriate for the surface sterilization of the studied sage populations: 95% for Aristi, 85% for Kefalovryso and 75% for Igoumenitsa, without necrosis.

Table 7. Survival percentage of explants of three *S. officinalis* populations.

Treatment	Percentage of Survival		
	Aristi	Kefalovryso	Igoumenitsa
D1	80.0 ± 9 a,b,*,**	75.0 ± 9 a	65.0 ± 10 a
D2	95.0 ± 5 a	85.0 ± 8 a	75.0 ± 9 a
D3	80.0 ± 8 a,b	65.0 ± 10 a,b	60.0 ± 11 a
D4	65.0 ± 10 b	60.0 ± 11 b	65.0 ± 4 a
D5	45.0 ± 11 c	30.0 ± 10 c	35.0 ± 10 b
D6	30.0 ± 10 c	20.0 ± 11 c	25.0 ± 9 b
D7	40.0 ± 11 c	40.0 ± 11 c	35.0 ± 10 b

* Standard deviation; ** Different letters in the same column indicate statistically significant differences according to Duncan's multiple range tests at $p \leq 0.05$.

Other researchers succeeded in obtaining clean explants of *S. officinalis* using a disinfection treatment similar to that of the present work. Bolta et al. [25] succeeded in disinfecting young shoots in ethanol solution (70% v/v) and NaOCl (0.5% w/v). In other works, the same procedure was followed, but either 1% NaOCl was used with a few drops of Tween-02 [42] or the explants were immersed in a 0.1% $HgCl_2$ (Mercury II chloride) solution [43]. According to the literature, a variety of disinfectants were used in other members of Lamiaceae, with the most common being commercial sodium hypochlorite (NaOCl), EtOH and mercury chloride ($HgCl_2$) solutions [26,44].

In the present work, the problem of explant browning and necrosis noticed at this stage was eliminated by immersing plant material in an antioxidant solution (ascorbic acid and citric acid). The same practice was successfully followed for another member of the Lamiaceae family, *Rosmarinus officinalis* [11] and one of the Rosaceae, *Pyrus spinosa* [45]. It is known that ascorbic acid prevents the browning and hyperhydricity of explants and improves in vitro rooting and ex vitro survival of the plants [17,46,47].

Figure 5. In vitro propagation of *Salvia officinalis* Aristi (**A–C**) multiplication on MS5 medium, (**D**,**E**) rooting ability on medium R3 and (**F**) acclimatization of plants in greenhouse. The yellow bars represent the size of 1 cm.

Effect of Growth Regulators on the Propagation of *S. officinalis*

Ten MS media, enriched with different combinations and concentrations of IAA with BAP or TDZ, were studied for Aristi, Kefalovryso and Igoumenitsa. Nutrient media MS5 and MS9 proved the best for the multiplication of explants (Tables 8 and S1, Figure S1).

In particular, Aristi exhibited the largest number of shoots on MS9 and MS5 media (3.35 and 3.21 new shoots/explant, respectively), significantly different from all other treatments (Figure 5A–C). Additionally, shoot length was the highest on the same media, i.e., 2.23 cm and 3.2 cm, respectively. For the same treatments, Kefalovryso was the second best for shoot production (2.4 and 2.3 new shoots/ explant, respectively) and shoot length (1.94 and 2.77 cm, respectively). Igoumenitsa exhibited the lowest number of new shoots on MS1 and MS7, which statistically differed from all other studied media. Concerning shoot length, it reached 2.17 cm on MS5 (Figure S2, Table S2).

Table 8. Effect of plant growth regulators on multiplication of *S. officinalis* populations.

Nutrient Medium	Number of Shoots			Length of Shoots (cm)		
	Aristi	Kefalovryso	Igoumenitsa	Aristi	Kefalovryso	Igoumenitsa
MS1	1.28 ± 0.43 b,*,**	0.57 ± 0.25 c	0.20 ± 0.16 b	0.28 ± 0.09 c	0.27 ± 0.15 d	0.42 ± 0.22 b
MS2	1.21 ± 0.36 b	0.78 ± 0.31 c	0.50 ± 0.25 a,b	0.38 ± 0.21 c	0.72 ± 0.33 c,d	0.44 ± 0.22 b
MS3	1.42 ± 0.44 b	1.07 ± 0.32 b,c	0.42 ± 0.25 a,b	0.53 ± 0.21 c	0.88 ± 0.35 c,d	0.90 ± 0.47 a,b
MS4	1.64 ± 0.42 b	1.14 ± 0.37 b,c	0.57 ± 0.30 a,b	1.25 ± 0.38 b,c	1.17 ± 0.46 b,c	1.10 ± 0.59 a,b
MS5	3.21 ± 0.40 a	2.30 ± 0.43 a	1.00 ± 0.37 a	3.20 ± 0.50 a	2.77 ± 0.54 a	2.17 ± 0.80 a
MS6	1.21 ± 0.48 b	2.00 ± 0.49 a,b	0.57 ± 0.27 a,b	0.95 ± 0.39 b,c	0.87 ± 0.34 c,d	1.17 ± 0.52 a,b
MS7	0.78 ± 0.28 b	0.70 ± 0.30 c	0.28 ± 0.19 b	1.50 ± 0.46 b,c	0.97 ± 0.38 c,d	0.42 ± 0.28 b
MS8	1.78 ± 0.49 b	1.07 ± 0.33 b,c	0.57 ± 0.30 a,b	1.39 ± 0.49 b,c	1.25 ± 0.46 b,c	0.77 ± 0.41 b
MS9	3.35 ± 0.55 a	2.40 ± 0.40 a	0.92 ± 0.40 a	2.23 ± 0.50 a,b	1.94 ± 0.47 a	1.50 ± 0.65 a,b
MS10	1.28 ± 0.39 b	1.07 ± 0.33 b,c	0.50 ± 0.27 a,b	1.49 ± 0.50 b,c	1.60 ± 0.44 a,b	1.07 ± 0.56 a,b
Average	1.71 ± 0.87	1.31 ± 0.67	0.55 ± 0.25	1.32 ± 0.89	1.24 ± 0.71	0.99 ± 0.55

* Standard deviation; ** Different letters in the same column indicate statistically significant differences according to Duncan's multiple range test at $p \leq 0.05$.

According to the bibliography, the search for an ideal combination of plant growth regulator concentrations has been extensively studied to obtain the best possible result in explant propagation. Cytokinins play the most important role in shoot development and auxins in rooting [48]. Several reports comment on the best combination of growth regulators for the multiplication of *S. officinalis*. Mohamed et al. [49] studied the effect of plant growth regulators on organogenesis of *S. officinalis* on nodal explants using MS medium with 0.1 mg·L^{-1} IAA and 1.5 mg·L^{-1} TDZ and reported the highest shoot production (7.2 shoots/explant) and shoot length (3 cm). Gostin [3] found that MS with 2.22 μM BAP was the best treatment for multiplication rate (100%) with length ranging from 4.03 to 4.59 cm in all media that contained BAP. The most relevant to our work is that of Grzegorczyk et al. [20,21], who reported that after three weeks from seed establishment in vitro, an average of 3 shoots per explant emerged on MS with 0.1 mg·L^{-1} IAA and 0.45 mg·L^{-1} BA. The results of the above-mentioned works agree with the findings of the present study.

In related research of other *Salvia* species, BA and BAP were the most frequently used cytokinins: Kintzios et al. [4] for the somatic embryogenesis of *S. officinalis*, for the multiplication *S. blancoana and S. valentine* [50], for *S. fruticosa* [51], while for *S. elegans*, *S. sinaloensis*, *S. cinnabarina* and *S. jamensis* Mascarello et al. [52] used low levels of BA. The same cytokinins have also been effectively used for the micropropagation of various plant species from the Lamiaceae family with economic interest [53]. For *Rosmarinus officinalis* adventitious shoot formation, the optimum level was 5 mg·L^{-1} 6-BAP [44]. For the same species, the highest shoot frequency was achieved on MS without growth regulators or in combinations of BAP (0.25 or 0.5 mg·L^{-1}) and IAA (0.1 mg·L^{-1}) [12]. Mehalaine and Chenchouni [26] showed that the combinations of IAA and Kin exhibited significant effects on callus and shoot proliferation in *T. algeriensis*, *R. officinalis* and *M. vulgare* in in vitro micropropagation. The culture medium is one of the most critical factors contributing to successful micropropagation [44].

Spontaneous Rooting of *S. officinalis* Explants

After the fifth week of establishment on the multiplication stage, an unexpected result was observed, i.e., the spontaneous root formation on five media of Aristi and Kefalovryso populations and on four of Igoumenitsa (Tables 9 and S3, Figure S3). The nutrient medium MS5 (0.1 mg·L^{-1} IAA and 0.8 mg·L^{-1} BAP) resulted in the highest rooting rates for Aristi (64%), Kefalovryso (57%) and Igoumenitsa (28%), the last of which, however, does not differ from MS9 (0.1 mg·L^{-1} IAA and 0.8 mg·L^{-1} TDZ) (14%) (Table 9). On the same medium (MS5), explants of two populations also exhibited the largest length of roots: for Aristi, 4.01 cm with root hairs, and for Kefalovryso, 3.03 cm. Igoumenitsa showed the lowest root length compared to the other two populations, with the best response at 1.92 cm on

MS9. Comparing the three populations, Aristi showed the best response concerning root percentage and root length. The microshoots that rooted spontaneously were immediately transferred to acclimatization.

Table 9. Effect of plant growth regulators on spontaneous rooting of *S. officinalis* populations.

Nutrient Medium	Rooting Formation (%)			Length of Roots (cm)		
	Aristi	Kefalovryso	Igoumenitsa	Aristi	Kefalovryso	Igoumenitsa
MS1	0 [b]	0 [b]	0 [b]	-	-	-
MS2	0 [b]	0 [b]	0 [b]	-	-	-
MS3	0 [b]	0 [b]	0 [b]	-	-	-
MS4	7 ± 7 [b,*,**]	7 ± 7 [b]	0 [b]	0.94 ± 0.21 [c]	0.91 ± 0.19 [c]	-
MS5	64 ± 13 [a]	57 ± 13 [a]	28 ± 12 [a]	4.01 ± 0.51 [a]	3.03 ± 0.40 [a]	1.00 ± 0.18 [b]
MS6	7 ± 7 [b]	7 ± 7 [b]	7 ± 7 [b]	1.21 ± 0.19 [c]	1.00 ± 0.17 [c]	0.57 ± 0.13 [c]
MS7	0 [b]	0 [b]	0 [b]	-	-	-
MS8	0 [b]	0 [b]	0 [b]	-	-	-
MS9	21 ± 11 [b]	14 ± 9 [b]	14 ± 9 [a,b]	3.05 ± 0.41 [b]	2.02 ± 0.35 [b]	1.92 ± 0.30 [a]
MS10	7 ± 7 [b]	7 ± 7 [b]	7 ± 7 [b]	1.01 ± 0.23 [c]	0.90 ± 0.20 [c]	0.85 ± 0.18 [b,c]
Average	10.6 ± 19.8	9.2 ± 17.4	5.6 ± 9.2	2.32 ± 1.45	1.73 ± 0.99	1.08 ± 0.58

* Standard deviation, ** Different letters in the same column indicate statistically significant differences according to Duncan's multiple range test at $p \leq 0.05$.

These findings are in complete accordance with those of Petrova et al. [17]. The aim of the study was to develop an efficient method for micropropagation of *S. officinalis*, as well as to evaluate flavonoid content and antioxidant capacity in leaves of the obtained shoots. At the multiplication stage, using nodal segments from in vitro seedlings established on MS with 0.5 mg·L^{-1} BAP and 0.1 mg·L^{-1} IAA, a rooting percentage of 40% was recorded, while on MS with 0.5 mg·L^{-1} Zeatin and 0.1 mg·L^{-1} IAA, even higher rooting was observed (75%). In a similar combination of growth regulators (0.8 mg·L^{-1} BAP and 0.1 mg·L^{-1} IAA), at the same stage, in the present work, we also obtained spontaneous rooting of all populations, with Aristi performing the best (64%). Another species of Lamiaceae, *Lavandula pedunculata*, also exhibited the best propagation rates and spontaneous rooting in MS with 0.10 mg·L^{-1} BA [54]. In a study on *S. officinalis*, Gostin [3] observed no rooting induction on a substrate with IBA. For the same species, Ioja-Boldura et al. [43] tested rooting on MS with the presence or absence of 4.92 μM IBA. In the case of MS-free medium, rooting reached 97% within two weeks, while in the presence of IBA, microshoots rooted 48% after one month. The above observations verify our results of spontaneous rooting, which is probably due to high levels of endogenous auxins in *S. officinalis* tissues, which enables rooting without applying auxins exogenously.

In Vitro Rooting and Plantlet Acclimatization

The rooting ability of microshoots of the two best performing populations in shoot multiplication was further investigated in order to elevate percentages. More specifically, microshoots from Aristi (Figure 5A–C) and Kefalovryso populations originating from MS5 and MS9 were transferred on three different MS rooting variants with IAA: R1, R2 and R3 (Tables 10 and S4, Figure S4). Optimal rooting (100%) was recorded on R3 medium (0.9 mg·L^{-1} IAA) after seven weeks for Aristi grown on MS5, while from MS9 reached 66.7% (Figure 5D,E). For Kefalovryso, the respective rooting percentages were 66.7% and 41.6%.

Both populations from all treatments showed optimal results in acclimatization, either from spontaneous or rooting experiments. In particular, both Aristi and Kefalovryso exhibited equally excellent survival rates (~88 to 100%) (Table 10). Acclimatized plants were transferred to greenhouse conditions where vibrant growth and healthy sage plants continued to grow after one month (Figure 5F).

Table 10. Rooting and acclimatization of *S. officinalis* per population and propagation medium.

Population	Propagation Medium	Rooting Medium	Rooting (%)	Acclimatization (%)
Aristi	MS5	R1	16.6 ± 11 [d,*,**]	100 [a]
	MS5	R2	33.3 ± 14 [b,c,d]	100 [a]
	MS5	R3	100 [a]	92.0 ± 5 [a]
	MS5	- ***	64.0 ± 13 [b]	88.8 ± 7 [a]
	MS9	R1	41.6 ± 14 [b,c]	100 [a]
	MS9	R2	41.6 ± 14 [b,c]	100 [a]
	MS9	R3	66.7 ± 14 [b]	100 [a]
	MS9	-	21.0 ± 11 [c,d]	100 [a]
Kefalovryso	MS5	R1	33.3 ± 14 [b,c,d]	100 [a]
	MS5	R2	33.3 ± 14 [b,c,d]	100 [a]
	MS5	R3	66.7 ± 14 [b]	87.5 ± 6 [a]
	MS5	-	57.0 ± 13 [b]	100 [a]
	MS9	R1	25 ± 13 [c,d]	100 [a]
	MS9	R2	41.6 ± 14 [b,c]	100 [a]
	MS9	R3	41.6 ± 14 [b,c]	100 [a]
	MS9	-	14.0 ± 9 [d]	100 [a]

* Standard deviation, ** Different letters in the same column indicate statistically significant differences according to Duncan's multiple range test at $p \leq 0.05$, *** i.e., spontaneous rooting.

Ghanbar et al. [55], aiming at the in vitro bud induction and shoot regeneration of *Salvia sclarea,* observed that MS in combination with IAA (0.5 mg·L^{-1}) reached the highest levels of rooting (87–100%) and acclimatization 90%, results similar to those of the present study. For *Salvia officinalis,* Jafari et al. [56] observed a 72% rooting rate and 3.9 root number on MS with IBA (1 mg·L^{-1}) after 45 days. Arikat et al. [51], for the micropropagation and accumulation of essential oils in *Salvia fruticosa,* reported a high rate of rooting (90%) by adding 2.7 μM IBA. In addition, Gostin [3] observed that the addition of kinetin (4.65 μM) promoted rooting in contrast to the effect of the same cytokinin with NAA (2.68 μM). The above results indicate the important role of the initial propagation medium in the optimization of rooting response and support the observations of spontaneous rooting and excised conclusions on rich endogenous auxin background of *S. officinalis* populations.

The propagating medium might play a key role by using it alternatively in two ways: (a) as a medium for multiplication and spontaneous rooting and (b) providing the appropriate vegetative material for the subsequent rooting stage. Acclimatization was successful for both the populations for material originating either from spontaneous or in vitro rooted plant material.

4. Conclusions

The present study will contribute to the sustainable exploitation and promotion of selected native sage populations for decorative uses with simultaneous conservation of this valuable genetic material and the provision of the market with the required vegetative plant material by either cuttings or in vitro techniques. Of the three selected wild sage populations, the results proved that both techniques adequately justified the main goals, i.e., selection and asexual propagation of the present research. Even though rooting of cuttings reached satisfactory results, in vitro propagation enhanced to the optimal the rooting and acclimatization of the selected populations. Overall, the population of Aristi has the dynamics of a new ornamental sage variety, and as such, it can be introduced in the market of plants with aesthetic value.

Supplementary Materials: The following supporting information can be downloaded at: https://www.mdpi.com/article/10.3390/horticulturae9070847/s1, Figure S1: Effect of medium and population and their interactions on in vitro shoot regeneration, Figure S2: Effect of growth regulators on the length of newly formed shoots per population, Figure S3: Spontaneous rooting (%) on multiplication media per population, Figure S4: Comparison of the two populations Aristi and Kefalovryso in rooting ability in the case of combine effect of multiplication and rooting media. 1 (MS5 + R1), 2 (MS5 + R2), 3 (MS5 + R3), 4 (MS9 + R1), 5 (MS9 + R2), 6 (MS9 + R3), Table S1: Substrate (medium) —Population interaction in terms of the number of newly formed shoots, Table S2. Substrate (medium) — population interaction in terms of shoot length, Table S3: Substrate — Variety Interaction in terms of rooting ability, Table S4: Interaction of Growth regulators and Population (Aristi and Kefalovryso) in the case of combine effect of multiplication and rooting medium.

Author Contributions: Conceptualization, S.K., S.H., I.N.-O. and A.K.K.; investigation, C.N., P.T., I.M., V.A., E.P., E.G. and I.N.-O.; plant material resources, S.K., E.G. and A.K.K.; methodology, C.N., P.T., S.K., S.H., I.M., E.P. and I.N.-O.; visualization, C.N., I.M., V.A. and E.P.; data curation, C.N., I.M., V.A. and E.P.; writing—original draft preparation, S.K., P.T., S.H. and I.N.-O.; writing—review and editing, S.K., P.T., I.N.-O., S.H. and A.K.K.; funding acquisition, S.K., A.K.K. and I.N.-O.; project administration, S.K. and A.K.K. All authors have read and agreed to the published version of the manuscript.

Funding: This research has been co-financed by the European Regional Development Fund of the European Union and Greek national funds through the Operational Programme Competitiveness, Entrepreneurship and Innovation, under the call "RESEARCH-CREATE-INNOVATE" (Project code: T1EDK-03919).

Institutional Review Board Statement: Not applicable.

Informed Consent Statement: Not applicable.

Data Availability Statement: The datasets in this paper are available from the corresponding authors on reasonable request.

Conflicts of Interest: The authors declare no conflict of interest.

References

1. Barnes, J.; Anderson, L.A.; Phillipson, J.D. *Herbal Medicines: A Guide for Healthcare Professionals*, 2nd ed.; Pharmaceutical Press: London, UK, 2002; 432p.
2. Akhondzadeh, S.; Noroozian, M.; Mohammadi, M.; Ohadinia, S.; Jamshidi, A.H.; Khani, M. *Salvia officinalis* extract in the treatment of patients with mild to moderate Alzheimer's disease: A double blind, randomized and placebo-controlled trial. *J. Clin. Pharm. Ther.* **2003**, *28*, 53–59. [CrossRef]
3. Gostin, I. Effects of different plant hormones on *Salvia officinalis* cultivated in vitro. *Int. J. Botany* **2008**, *4*, 430–436. [CrossRef]
4. Kintzios, S.; Nikolaou, A.; Skoula, M. Somatic embryogenesis and in vitro rosmarinic acid accumulation in *Salvia officinalis* and *S. fruticosa* leaf callus cultures. *Plant Cell Rep.* **1999**, *18*, 462–466. [CrossRef]
5. Azimova, S.S.; Glushenkova, A.I. *Lipids, Lipophilic Components and Essential Oils from Plant Sources*; Springer: London, UK, 2012; 1004p. [CrossRef]
6. Kontogianni, V.G.; Tomic, G.; Nikolic, I.; Nerantzaki, A.A.; Sayyad, N.; Stosic-Grujicic, S.; Tzakos, A.G. Phytochemical profile of *Rosmarinus officinalis* and *Salvia officinalis* extracts and correlation to their antioxidant and anti-proliferative activity. *Food Chem.* **2013**, *136*, 120–129. [CrossRef]
7. Modarres, M.; Esmaeilzadeh Bahabadi, S.; Taghavizadeh Yazdi, M.E. Enhanced production of phenolic acids in cell suspension culture of *Salvia leriifolia* Benth. using growth regulators and sucrose. *Cytotechnology* **2018**, *70*, 741–750. [CrossRef] [PubMed]
8. Brezoiu, A.M.; Prundeanu, M.; Berger, D.; Deaconu, M.; Matei, C.; Oprea, O.; Vasile, E.; Negreanu-Pîrjol, T.; Muntean, D.; Danciu, C. Properties of *Salvia offcinalis* L. and *Thymus serpyllum* L. extracts free and embedded into mesopores of silica and titania Nanomaterials. *Nanomaterials* **2020**, *10*, 820. [CrossRef]
9. Schmiderer, C.; Novak, J. Salvia officinalis L. and *Salvia fruticosa* Mill.: Dalmatian and three-lobed sage. In *Medicinal, Aromatic and Stimulant Plants*; Novak, J., Blüthner, W.D., Eds.; Handbook of Plant Breeding; Springer: Cham, Switzerland, 2020; Volume 12, pp. 523–527. [CrossRef]
10. Laurentin, H. Data analysis for molecular characterization of plant genetic resources. *Genet. Resour. Crop Evol.* **2009**, *56*, 277–292. [CrossRef]
11. Kostas, S.; Hatzilazarou, S.; Pipinis, E.; Vasileiadis, A.; Magklaras, P.; Smyrnioudis, I.; Vasilakis, T.; Chazakis, M.; Anastasiadi, V.; Ziogou, F.-T.; et al. Propagation of *Pistacia lentiscus* var. Chia genotypes and determination of their ornamental traits combined with a genetic analysis using ISSR markers. *Agronomy* **2021**, *11*, 205. [CrossRef]

12. Kostas, S.; Kaplani, A.; Koulaouzidou, E.; Kotoula, A.A.; Gklavakis, E.; Tsoulpha, P.; Hatzilazarou, S.; Nianiou-Obeidat, I.; Kanellis, A.K.; Economou, A. Sustainable exploitation of greek *Rosmarinus officinalis* L. populations for ornamental use through propagation by shoot cuttings and in vitro cultures. *Sustainability* **2022**, *14*, 4059. [CrossRef]
13. Nicola, S.; Fontana, E.; Hoeberechts, J.; Saglietti, D. Rooting products and cutting timing on sage (*Salvia officinalis* L.) propagation. *Acta Hortic.* **2003**, *676*, 135–141. [CrossRef]
14. Kara, N.; Baydar, H.; Erbaş, S. Effects of different cuttings periods and IBA concentrations on rooting ability of some medicinal plant. *Derim* **2011**, *28*, 71–81.
15. Paradikovic, N.; Zelkovic, S.; Tkalec, M.; Vinkovic, T.; Devic, I.; Maric, M. Influence of rooting powder on propagation of sage (*Salvia officinalis* L.) and rosemary (*Rosmarinus officinialis* L.) with green cuttings. *Poljoprivreda* **2013**, *19*, 10–15.
16. Loconsole, D.; Cristiano, G.; De Lucia, B. Image analysis of adventitious root quality in wild sage and glossy abelia cuttings after application of different indole-3-butyric acid concentrations. *Plants* **2022**, *11*, 290. [CrossRef]
17. Petrova, M.; Nikolova, M.; Dimitrova, L.; Zayova, E. Micropropagation and evaluation of flavonoid content and antioxidant activity of *Salvia officinalis* L. *Genet. Plant Physiol.* **2015**, *5*, 48–60. Available online: http://www.ifrg-bg.com (accessed on 19 June 2023).
18. Kriebel, R.; Drew, B.; González-Gallegos, J.; Celep, F.; Heeg, L.; Malik, M.; Sytsma, K. Pollinator shifts, contingent evolution, and evolutionary constraint drive floral disparity in *Salvia* (Lamiaceae): Evidence from morphometrics and phylogenetic comparative methods. *Evolution* **2020**, *74*, 1335–1355. [CrossRef] [PubMed]
19. Gokdogan, E.Y.; Burun, B. The studies on seed germination and in vitro cultures of *Salvia* L. species from turkish Flora. *Nat. Prod. Biotechnol.* **2022**, *2*, 60–73.
20. Grzegorczyk, I.; Bilichowski, I.; Mikiciuk-Olasik, E.; Wysokińska, H. In vitro cultures of *Salvia officinalis* L. as a source of antioxidant compounds. *Acta Soc. Bot. Pol. Pol. Tow. Bot.* **2005**, *74*, 17–21. [CrossRef]
21. Grzegorczyk-Karolak, I.; Matkowski, A.; Wysokińska, H. Antioxidant activity of extracts from in vitro cultures of *Salvia officinalis* L. *Food Chem.* **2007**, *104*, 536–541. [CrossRef]
22. Santos-Gomes, P.C.; Seabra, R.M.; Andrade, P.B.; Fernandes-Ferreira, M. Phenolic antioxidant compounds produced by in vitro shoots of sage (*Salvia officinalis* L.). *Plant Sci.* **2002**, *162*, 981–987. [CrossRef]
23. Santos-Gomes, P.C.; Fernandes-Ferreira, M. Essential oils produced by in vitro shoots of sage (*Salvia officinalis* L.). *J. Agric. Food Chem.* **2003**, *51*, 2260–2266. [CrossRef]
24. Wielgus, K.; Luwanska, A.; Szalata, M.; Mielcarek, S.; Gryszczynska, A.; Lipinski, D.; Slomski, R. Phytochemical estimation of Sage (*Salvia officinalis* L.) cultivated in vitro-flavonoids and phenolic acids. *Acta Fytotech. Zootech* **2011**, *14*, 8–11.
25. Bolta, I.; Evi, D.B.; Bohanec, B.; Andren, S. A preliminary investigation of ursolic acid in cell suspension culture of *Salvia officinalis*. *Plant Cell Tissue Organ Cult.* **2000**, *62*, 57–63. [CrossRef]
26. Mehalaine, S.; Chenchouni, H. New insights for the production of medicinal plant materials: Ex vitro and in vitro propagation of valuable Lamiaceae species from northern Africa. *Curr. Plant Biol.* **2021**, *27*, 100216. [CrossRef]
27. Zewdinesh, D.Z.; Beemnet, M.K. Effect of cutting size and position on propagation ability of Sage (*Salvia officinalis* L.). *Int. J. Adv. Biol. Biomed. Res.* **2016**, *4*, 68–76.
28. Murashige, T.; Skoog, F. A revised medium for rapid growth and bio assays with tobacco tissue cultures. *Physiol. Plant.* **1962**, *15*, 473–497. [CrossRef]
29. Esmaeili, G.; Fatemi, H.; Baghani avval, M.; Azizi, M.; Arouiee, H.; Vaezi, J.; Fujii, Y. Diversity of chemical composition and morphological traits of eight iranian wild *Salvia* species during the first step of domestication. *Agronomy* **2022**, *12*, 2455. [CrossRef]
30. Herraiz-Peñalver, D.; Asensio, S.; Manzanera, M.C.; Herrero, B.; Martin, H.; Santiago, Y.; Zalacaín, A.; Berruga, M.I.; Sánchez-Vioque, R. Evaluation of the morphological variability in Iberian *Salvia lavandulifolia* Vahl accessions. *Genet. Resour. Crop Evol.* **2017**, *64*, 1797–1806. [CrossRef]
31. Mossi, A.J.; Cansian, R.L.; Paroul, N.; Toniazzo, G.; Oliveira, J.V.; Pierozan, M.K.; Pauletti, G.; Rota, L.; Santos, A.C.A.; Serafini, L.A. Morphological characterisation and agronomical parameters of different species of *Salvia* sp. (Lamiaceae). *Braz. J. Biol.* **2011**, *71*, 121–129. [CrossRef]
32. Leontaritou, P.; Lamari, F.N.; Papasotiropoulos, V.; Iatrou, G. Exploration of genetic, morphological and essential oil variation reveals tools for the authentication and breeding of *Salvia pomifera* subsp. *calycina (Sm.) Hayek*. *Phytochemistry* **2021**, *191*, 112900. [CrossRef]
33. Leontaritou, P.; Lamari, F.N.; Papasotiropoulos, V.; Iatrou, G. Morphological, genetic and essential oil variation of greek sage (*Salvia fruticosa* Mill.) populations from Greece. *Ind. Crops Prod.* **2020**, *150*, 112346. [CrossRef]
34. Zigene, Z.D.; Asfaw, B.T.; Bitima, T.D. Phenotypic diversity of rosemary (*Salvia rosmarinus* Schleid.) accessions for qualitative characters. *Heliyon* **2022**, *8*, e11895. [CrossRef] [PubMed]
35. Araghi, A.M.; Nemati, H.; Azizi, M.; Moshtaghi, N.; Shoor, M.; Hadian, J. Assessment of phytochemical and agro-morphological variability among different wild accessions of *Mentha longifolia* L. cultivated in field condition. *Ind. Crops Prod.* **2019**, *140*, 111698. [CrossRef]
36. Devi, A.; Iqbal, T.; Wani, I.A.; Sharma, G.; Verma, S.; Noureldeen, A.; Darwish, H. Assessment of variability among morphological and molecular characters in wild populations of mint [*Mentha longifolia* (L.) L.] germplasm. *Saudi J. Biol. Sci.* **2022**, *29*, 3528–3538. [CrossRef]

37. Ayanoglu, F.; Özkan, C. Change in tissue mineral elemental concentration during root initiation and development of *Salvia officinalis* L. cuttings and effects. *Turk. J. Agric. For.* **2000**, *24*, 677–682.
38. Davies, F.; Geneve, R.; Wilson, S.; Hartmann, H.; Kester, D. *Hartmann and Kester's Plant Propagation: Principles and Practices*, 9th ed.; Pearson Prentice Hall: Upper Saddle River, NJ, USA, 2017; 915p.
39. Kacar, O.; Azkan, N.; Çöplü, N. Effects of different rooting media and indole butyric acid on rooting of stem cuttings in sage (*Salvia officinalis* L. and *Salvia triloba* L.). *J. Food Agric. Environ.* **2009**, *7*, 349–352.
40. Martini, A.N.; Bertsouklis, K.F.; Vlachou, G.; Dariotis, E.; Papafotiou, M. Comparative evaluation of rooting cuttings of five Mediterranean sage species (*Salvia* sp.) native to Greece. *Acta Hortic.* **2020**, *1298*, 587–592. [CrossRef]
41. Vârban, D.I.; Vârban, R.; Duda, M.M.; Muntean, S.; Moldovan, C. The influence of substrates on rooting of *Salvia officinalis* L. cuttings. *Hop Med. Plants* **2014**, *24*, 52–56.
42. Lemraski, M.G.; Eftekhari, M.; Faraji, M.; Zarrini, S.S. Study of callus induction in common sage (*Salvia officinalis* L.). *Int. J. Agric. Crop Sci. (IJACS)* **2014**, *7*, 386–389.
43. Ioja-Boldura, O.M.; Radu, F.; Popescu, S.; Borozan, A. Regeneration, micropropagation, callus cultures and somatic embryogenesis of common sage (*Salvia officinalis* L.). *Bull. Univ. Agric. Sci. Vet. Med. Cluj-Napoca. Hortic.* **2010**, *67*, 308–313.
44. Villegas-Sánchez, E.; Macías-Alonso, M.; Osegueda-Robles, S.; Herrera-Isidrón, L.; Nuñez-Palenius, H.; González-Marrero, J. In vitro culture of *Rosmarinus officinalis* L. in a temporary immersion system: Influence of two phytohormones on plant growth and carnosol production. *Pharmaceuticals* **2021**, *14*, 747. [CrossRef]
45. Tsoulpha, P.; Alexandri, S.; Tsaktsira, M. Critical factors affecting an efficient micropropagation protocol for *Pyrus spinosa* Forskk. *J. Appl. Hortic.* **2018**, *20*, 190–195. [CrossRef]
46. Tawfik, A.A.; Mohamed, M.F. Regeneration of salvia (*Salvia officinalis* L.) via induction of meristematic callus. *In Vitro Cell. Dev. Biol. Plant* **2007**, *43*, 21–27. [CrossRef]
47. Ngomuo, M.; Mneney, E.; Ndakidemi, P.A. The in vitro propagation techniques for producing banana using shoot tip cultures. *Am. J. Plant Sci.* **2014**, *5*, 1614–1622. [CrossRef]
48. Rout, G.R.; Samantaray, S.; Das, P. In vitro manipulation and propagation of medicinal plants. *Biotechnol. Adv.* **2000**, *18*, 91–120. [CrossRef]
49. Mohamed, M.A.H.; Aly, M.K.; Ahmed, E.T.; Abd El-latif, S.A.H. Effect of plant growth regulators on organogenesis of *Salvia officinalis* L. plants. *Minia J. Agric. Res. Dev.* **2019**, *3*, 401–414.
50. Cuenca, S.; Amo-Marco, J.B. In vitro propagation of two spanish endemic species of *salvia* through bud proliferation. *In Vitro Cell. Dev. Biol. Plant* **2000**, *36*, 225–229. [CrossRef]
51. Arikat, N.A.; Jawad, F.M.; Karam, N.S.; Shibli, R.A. Micropropagation and accumulation of essential oils in wild sage (*Salvia fruticosa* Mill.). *Sci. Hortic.* **2004**, *100*, 193–202. [CrossRef]
52. Mascarello, C.; Mantovani, E.; Ruffoni, B. In vitro culture of several ornamental and medicinal *Salvia* species. *Acta Hortic.* **2006**, *723*, 375–380. [CrossRef]
53. van Staden, A.B.; Lall, N. Medicinal plants as alternative treatments for progressive macular hypomelanosis. In *Medicinal Plants for Holistic Health and Well-Being*, 1st ed.; Lall, N., Ed.; Elsevier Academic Press: London, UK, 2018; pp. 145–182.
54. Zuzarte, M.R.; Dinis, A.M.; Cavaleiro, C.; Salgueiro, L.R.; Canhoto, J.M. Trichomes, essential oils and in vitro propagation of *Lavandula pedunculata* (Lamiaceae). *Ind. Crops Prod.* **2010**, *32*, 580–587. [CrossRef]
55. Ghanbar, T.; Hosseini, B.; Jabbarzadeh, Z.; Farokhzad, A.; Sharafi, A. High-frequency in vitro direct shoots regeneration from axillary nodal and shoot tip explants of clary sage (*Salvia sclarea* L.). *Bulg. J. Agric. Sci.* **2016**, *22*, 73–78.
56. Jafari, S.; Daneshvar, M.H.; Salmi, M.S.; Abdi, L.J. Indirect organogenesis and plant regeneration in common sage (*Salvia officinalis* L.): An important medicinal plant of Iran. *Mod Appl. Sci.* **2017**, *11*, 22–29. [CrossRef]

horticulturae

MDPI

Review

Seed Germination within Genus *Rosa*: The Complexity of the Process and Influencing Factors

Roxana L. Stoian-Dod [1], Catalina Dan [1,*], Irina M. Morar [2], Adriana F. Sestras [2], Alina M. Truta [2], Gabriela Roman [1,3] and Radu E. Sestras [1]

[1] Faculty of Horticulture and Business in Rural Development, University of Agricultural Sciences and Veterinary Medicine, 3-5 Mănăştur Street, 400372 Cluj-Napoca, Romania; roxana.stoian@usamvcluj.ro (R.L.S.-D.); gabrielaroman33@yahoo.com (G.R.); rsestras@usamvcluj.ro (R.E.S.)
[2] Faculty of Forestry and Cadastre, University of Agricultural Sciences and Veterinary Medicine, 3-5 Mănăştur Street, 400372 Cluj-Napoca, Romania; irina.todea@usamvcluj.ro (I.M.M.); adriana.sestras@usamvcluj.ro (A.F.S.); alina.truta@usamvcluj.ro (A.M.T.)
[3] Horticultural Research Station, 400457 Cluj-Napoca, Romania
* Correspondence: catalina.dan@usamvcluj.ro

Abstract: Seed germination is a crucial stage in the life cycle of plants, and understanding the factors influencing germination is essential for successful cultivation, plant breeding, and conservation efforts. The genus *Rosa*, commonly known as roses, encompasses a diverse group of flowering plants renowned for their beauty and fragrance. *Rosa* germination is influenced by a variety of factors, including seed dormancy, environmental conditions, and seed treatments. Many *Rosa* species exhibit different types of seed dormancy, such as physical dormancy caused by hard seed coats and physiological dormancy due to internal mechanisms. Overcoming seed dormancy often requires specific treatments, including cold stratification, scarification, or chemical treatments, to promote germination. Environmental factors, including temperature, moisture, light, and substrate, play vital roles in *Rosa* germination. Temperatures ranging from 15 to 25 °C, moisture, and exposure to light or darkness, depending on the species, constitute suitable conditions for seed germination. Many studies have been conducted to investigate the germination requirements of different *Rosa* species, thereby expanding our understanding of their propagation and conservation. Additionally, advancements in techniques such as in vitro germination and molecular approaches have further enhanced our understanding of *Rosa* germination biology.

Keywords: gibberellic acid; H_2SO_4; methods; roses; scarification; seed stimulation; stratification

Citation: Stoian-Dod, R.L.; Dan, C.; Morar, I.M.; Sestras, A.F.; Truta, A.M.; Roman, G.; Sestras, R.E. Seed Germination within Genus *Rosa*: The Complexity of the Process and Influencing Factors. *Horticulturae* **2023**, *9*, 914. https://doi.org/10.3390/horticulturae9080914

Academic Editors: Konstantinos Bertsouklis, Epaminondas Kartsonas, Angela Carra and Dimitrios Fanourakis

Received: 7 July 2023
Revised: 28 July 2023
Accepted: 8 August 2023
Published: 10 August 2023

1. Introduction

1.1. Distribution of the Genus Rosa

With more than 200 species spread across the northern Hemisphere, the *Rosa* genus within the Rosaceae family is one of the most significant ornamental plant genera regarding economic and cultural history [1,2]. The genus is divided into four subgenera according to conventional taxonomy [3], three of which are monotypic: *Platyrhodon* (Hurst) Rehder, *Hulthemia* (Dumort.) *Focke*, and *Hesperhodos* Cockerell. Approximatively 95% of all species are found in the subgenus *Rosa*, which is split into 10 sections, one of which is *Caninae*, the subject of this review. The genus is native to North America, East Asia, and Europe/West Asia (Figure 1).

Roses are among the most significant and versatile horticultural and industrial products. According to Guimares [4], they can be utilized as cut or garden flowers. Additionally, roses have long been employed in the cosmetics, food, and perfume industries [5]. Strlsjö and Larsen [6,7] stated that fruits (rose hips) are a good source of bioactive substances, such as vitamin C, carotenoids, tocopherol, phenolic acid, bioflavonoids, tannin, pectin, organic acids, amino acids, essential oils, and unsaturated fatty acids. Rose hips are composed of 29% seed and 71% pericarp, and the hues range from red to orange.

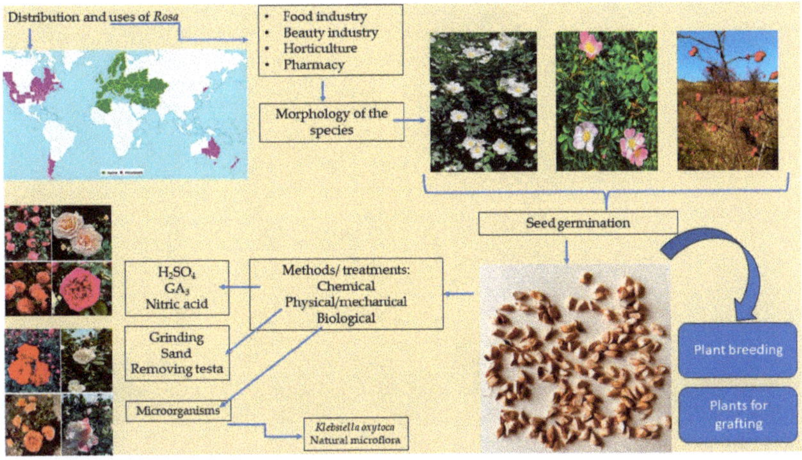

Figure 1. Distribution, uses, and seed germination in the *Rosa* genus.

The species reproduces through layering, cutting, tissue culture, basal shoots, and the use of its seeds [8]. Due to its resistance to drought stress, it serves as the most significant stock of ornamental roses [9].

Dog roses, which are part of the section *Caninae* (DC.) Ser. and dominate the territory of Europe and West Asia, are crucial to the development of rootstocks for the breeding of ornamental roses. Due to their distinct meiotic behavior and breeding system, dog roses occupy a special place among plants [9,10].

1.2. Botanical Aspect of the Species

The dog rose is a deciduous shrub that typically grows 1 to 5 m tall; however, it can, occasionally, climb higher into the crowns of taller trees by using the little, pointed, hooked prickles that cover its stems. The bark is light brown and very spiny, whereas the plant has spines, prickles, or thorns [5].

The leaves are compound, have 5–7 glabrous leaflets that are pinnate and have serrate margins, and are alternately inserted along the stem. Stipules are present at the base of the petiole. When injured, leaves release a pleasant scent [5].

The dog rose blooms between June and July, producing fragrant, sweet-smelling petals that are often pale pink, but they can also vary from deep pink to white (Figure 2). They have five petals and are 4–6 cm in diameter. It possesses a specific aestivation, typical of roses. There are three or more scales on the winter buds, and they overlap like shingles, with one edge covered and the other edge exposed. Unusually, though, two of its five sepals are whiskered on both sides, two are smooth, and one is hairy (or bearded) on just one side when viewed from beneath [1,2,5].

In general, an oval, red-orange, 1.5–2.0 cm hip develops after fertilization. Rose hip is a pseudo-fruit (botanically known as hypantium), developed from the inferior gynoecium and receptacle of the flower. It has a fleshy consistency, actually changing from green to orange or flaming red as it ripens (Figure 3). In terms of shape, hips may be globular, ellipsoid, obovate, pear-shaped, or bottle- or flask-like, and they vary greatly in size [2]. *Rosa* hips have smooth and polished-looking epicarps that might be glabrous, while others are dull and thorny. Many have resinous hairs, called trichomes. The hips have a sepal-like cap on top with variable aspect, depending mainly on length and margins. As the summer season progresses, a red, fleshy layer called the pericarp develops, and inside the aggregate pseudo-fruit, seeds mature. These seeds represent the real fruit of the species, being known as achenes, botanically classified as dry fruits, with one proper seed inside. They are 4.5–6.0 mm long, unevenly shaped, hairy, and yellowish in color [5].

(a) (b)

Figure 2. Flowers of *R. canina* of (**a**) white color and (**b**) pink hues, from the spontaneous flora of Transylvania, northwestern Romania (photo by R.L.S.-D.).

(a) (b)

Figure 3. (**a**) *Rosa canina* plant (shrub); (**b**) *Rosa canina* pseudo-fruit (hypantium). Bushes photographed in the spontaneous flora of Transylvania, northwestern Romania (photo by R.L.S.-D.).

1.3. Importance of the Species

1.3.1. Chemical Composition of the Hips

The fruits of the rose species are extremely beneficial to human health because of their remarkable quality and quantity of organic and inorganic components. The chemical composition of the hips can be found in Table 1 [11,12].

Table 1. Chemical composition of rose hip.

Nutrient	Value per 100 g	References
Water	58.66 g *	[13]
Energy	162 kcal *	[13]
Protein	1.6 g * 0.36 g	[13] [14]
Total lipid	0.34 g *	[13]
Ash	1.18 g *	[13]
Carbohydrate	38.22 g *	[13]
Fiber	24.1 g *	[13]
Sugars, total	2.58 g * 13.28 g	[13] [14]
	Minerals	
Calcium	169 mg * 944 mg	[13] [11]
Iron	1.06 mg * 1.1 mg	[13] [11]
Magnesium	69 mg * 122.6 mg	[13] [11]
Phosphorus	61 mg * 122.4 mg	[13] [11]
Sodium	4 mg * 15.8 mg	[13] [11]
Zinc	0.25 mg * 1.3 mg	[13] [11]
Copper	0.113 mg * 0.4 mg	[13] [11]
Manganese	1.02 mg * 5.9 mg	[13] [11]
Potassium	1025 mg	[11]
	Vitamins	
Vitamin C	426 mg * 411.0 mg 643.38 mg	[13] [11] [14]
Vitamin B-6	0.076 mg *	[13]
Vitamin A, IU	4345 IU *	[13]
Vitamin E (alpha tocopherol)	5.84 mg * 34.20 µg/g	[13] [11]
Tocopherol beta	0.05 mg *	[13]
Tocopherol gamma	1.34 mg *	[13]
Tocopherol delta	0.14 mg *	[13]
Vitamin K	25.9 µg *	[13]
Carotene alfa	31 µg *	[13]
Carotene beta	2350 µg * 2.60 µg/g	[13] [11]
Cryptoxanthin, beta	483 µg *	[13]
Thiamin	0.016 mg *	[13]

Table 1. *Cont.*

Nutrient	Value per 100 g	References
Riboflavin	0.166 mg *	[13]
Niacin	1.3 mg *	[13]
Lycopene	6800 μg * 390 mg/kg DW	[13] [14]
Lutein + Zeaxanthin	2001 μg *	[13]
Pantothenic acid	0.8 mg *	[13]

* Nutrient Data Laboratory, ARS, USDA National Food and Nutrient Analysis Program, Wave 9j, 2005, Beltsville MD USA.

1.3.2. Health Benefits of Rose Hips

The pseudo-fruit of the rose plant known as rose hip is well-known for being a beneficial source of vitamin C and polyphenols [5]. Studies conducted both in vitro and in vivo have shown fruits' anti-inflammatory, antioxidant, and antiobesogenic properties. The great variety of bioactive substances included in rose hip, particularly the anti-inflammatory galactolipid, have been linked to the health advantages of using the plant [14].

The fact that hips' chemical composition varies based on the genotype, growing area, climate, maturity, cultivation practice, and storage circumstances is an intriguing aspect. A wide range of biochemical activities, such as antioxidant, antimutagenic, and anti-carcinogenic capabilities, are present in phenols. *R. canina* contains 96 mg GAE/g DW of total phenols, according to one study [15]. The high ascorbic acid content of Rosaceae fruits may also play a role in their physiological actions. Numerous biochemical processes, such as antioxidant and anti-carcinogenic characteristics, are carried out by ascorbic acid. Necessary fatty acids, mainly long-chain polyunsaturated fatty acids that humans must consume since our metabolism cannot synthesize them, are another essential component of rose hip. Linoleic and -linolenic acids (ALA) are important fatty acids that control a variety of physiological reactions, including inflammation, immunological function, blood viscosity, and many others [15]. *R. canina* has 1.78% total fat [15]. It contains seven major fatty acids, according to fatty acid analysis: lauric acid (12:0), palmitic acid (16:0), linoleic acid (cis-C18:2 6), -linolenic acid (cis-C18:3 3), nonadecylic acid (19:0), cis-C19:1 6, and cis-C22:2 6 [15]. Another significant component of rose hip that deserves to be mentioned is the galactolipid GOPO. With no known toxicity, this bioactive molecule provides research evidence supporting both anticancer and anti-inflammatory effects [7].

R. canina hips appear to have many positive health benefits for humans, but more research is needed to properly understand how they could improve people's wellbeing before they can be regularly advised.

1.3.3. Horticultural and Economical Importance of the Species

Rosa is one of the most economically important genera in ornamental horticulture. Roses have the largest economic value of all the ornamental plants and have a long and well-documented tradition in selection and breeding [16]. High degrees of heterozygosity, documented issues with sexual reproduction from pollination through seed germination, and various levels of ploidy present challenges for rose breeding [17]. It includes many mountain species, natural hybrids, and cultivars of horticultural interest. It has ornamental value or is used as a soil improver species, and it is used as a rootstock for obtaining many types and cultivars of roses [18].

Roses are propagated both by seed and vegetatively. Seed propagation is practiced in rose breeding when different parental forms are crossed, and the hybrids obtained from seeds show a wide phenotypic variability due to the heterozygous structure of the parental forms and the genetic recombination that occurs [19]. Seed propagation is also widely used to obtain the rootstocks on which the various varieties of *Rosa* are grafted [20]. At the Horticultural Research Station in Cluj-Napoca (HRS), affiliated to the University of

Agricultural Sciences and Veterinary Medicine of Cluj-Napoca, more than 40 new varieties have been obtained through artificial hybridization [21], and *Rosa canina* is usually used as a rootstock to produce grafted roses in the nursery. In Figure 4, different varieties of roses are grafted on *R. canina* at the HRS, where a collection of over 400 rose cultivars has been established in a rosarium. Both in rose breeding and in the production of roses grafted on *Rosa canina* seedlings obtained from seeds or from other species used as rootstocks, seed germination is of great interest, both for obtaining a higher percentage of seedlings and for their proper quality [19].

Figure 4. Different rose cultivars grafted on *Rosa canina* from the HRS rosarium Cluj-Napoca, Romania (photo by R.L.S.-D.).

In the practice of growing roses, there are many techniques of vegetative propagation. The most commonly used method is budding, carried out in August-September [18]. The use of biotechnological techniques allows obtaining uniformed clonal material from rootstocks for ornamental roses with the best parameters of inoculation zone compared with plants propagated by green cutting [22]. The grafted roses demonstrated winter hardiness, longevity, and high flower productivity [23].

2. Seed Characteristics and Germination within the *Rosa* Genus

Regarding higher plants' ability to survive as species, the seed stage is crucial. It is the plant's dispersal unit that may endure the time between the maturity of the seed and the establishment of the following generation as a seedling once it has germinated. The seed, which is primarily in a dry state for its survival, is well-equipped to withstand prolonged durations under unfavorable conditions. The seed goes into dormancy to maximize germination over time [24]. Additionally, dormancy stops pre-harvest germination. To comprehend how different environmental conditions and applied chemicals regulate germination, numerous investigations have been carried out. However, very little is still known about the process by which the rose embryo emerges from the seed to complete germination and how embryo emergence is blocked in dormant seeds [25]. Seed dormancy has been defined as the incapacity of a viable seed to germinate under favorable conditions [25,26]. In roses, hips typically contain between one and thirty seeds, and hip set and seed germination are frequently less than 50% [27]. In the breeding of cut and garden roses, mature hips are collected in late September and early October, or 3–4 months following pollination. Hips are used to collect and count seeds. After that, seeds are stored for roughly 6–12 weeks at 2–5 °C and a moderate moisture level. Seeds can be sown directly in the germination substrate for vernalization, depending on the environment; otherwise, preservation in a cold chamber is required. A modern cut rose breeding program is thought to get off to a strong start with 120,000 seedlings after germination. Seeds are often sown on germination beds or benches at a density that can range from 150 to 400 seeds per square meter, depending on the available space and the number of seeds to start with. After two months, the vernalization period is deemed to be over [28]. Seeds of the *Rosa* genus are typically small, irregularly shaped (4.5–6 mm long), and vary in color depending on the species and variety (Figure 5). They are enclosed within a protective seed coat, which can range from smooth to slightly textured. *Rosa* seeds are generally characterized by their hard seed coat, which contributes to seed dormancy. The pericarp epi-, meso-, and endocarp layers make up a rose achene; the last structure is extremely impervious to water absorption (unpublished data). Therefore, the endocarpic layer may serve as a tegument-type physical barrier to achene germination [17].

Figure 5. *Rosa canina* seeds. The seeds were obtained from fruits collected from the spontaneous flora of Transylvania, northwestern Romania (photo by R.L.S.-D.).

The seed coat protects the embryo but can present problematic challenges for germination. However, *Rosa* seeds also possess the potential resources for germination and subsequent growth into mature plants when provided with the appropriate environmental conditions and treatments to overcome dormancy [29]. Seed germination within the *Rosa* genus is a crucial process that determines the successful establishment and propagation of further obtained plants. The germination of *Rosa* seeds is often influenced by factors such as seed dormancy and is strongly influenced by environmental conditions and seed treatments. Many *Rosa* species exhibit various types of seed dormancy, including physical dormancy caused by hard seed coats, as well as physiological dormancy due to internal mechanisms [30].

An achene is where rose seeds are created. This dry fruit has a solitary seed that almost completely fills the pericarp. Achenes and seeds are frequently used interchangeably in literature and breeding. Gudin [27] analyzed rose seed propagation from the perspective of generative multiplication, as is the case for the creation of perfume (*R. gallica* and *R.* × *damascena*), for the propagation of rootstocks (Caninae rose), and for landscaping. According to the winter environment in the rose's native regions, a specific, adapted seed dormancy was established. In species like *R. rugosa*, *R. gallica*, *R. canina*, and *R. soulieana*, it is known that dormancy can only be broken following intervals of chilling at 1–4 °C. Before germination, some roses require a second winter vernalization [31]. In contrast, the seeds of *R. persica*, which are adapted to the deserts of the Far East, do not respond to cold but can withstand dehydration. Depending on the rose species, there are differences in the period, kind, and cooling temperatures needed to break dormancy [27].

Overcoming seed dormancy often requires specific treatments, such as cold stratification or scarification, to break the inactivity process until favorable conditions are assured and promote germination. Nevertheless, abiotic factors like temperature, moisture, light, and substrate play important roles in the germination process. Understanding the germination requirements and dormancy mechanisms in the *Rosa* genus is essential for effective cultivation and conservation strategies for these valuable plants [17]. Roses are one of the most significant commercial crops in the world because of their significance as a decorative plant in landscape gardening, their high regard as a medicinal plant and for human nourishment, and the fact that the species is well adapted to a wide range of habitats [32,33]. Traditionally, stem cutting, layering, budding, grafting, and tissue culture are the primary vegetative methods used to propagate roses [34]. All of these techniques come with a number of issues, including a lack of rootstocks and a longer production period. The breeding of new cultivars, the restoration of native plants, the selection of rootstocks, and, in some varieties, the production of rose hips are all achieved through seed propagation; however, this process is challenging due to the low germination percentage that results from the prolonged seed dormancy [35–37].

According to Jackson and Blundell [38], Densmore and Zasada [39], and Bo et al. [37], inhibitors in the pericarp and testa, the hard pericarp, and physiological barriers in the embryo may be the main causes for the dormancy and delayed germination in rose achenes. Although it occasionally limits complete imbibition, the pericarp is permeable [36,37,40]. Some rose achenes experience dormancy as a result of a barrier in the shape of a hard pericarp [41], although this is not the only source of the condition [35,42].

The endocarp thickness in a rose achene pericarp can determine germination, as demonstrated by Gudin et al. [43]. Environmental factors, particularly temperature during achenes' maturation, and genetic factors, likely through their impact on the rate of embryo growth, regulate its thickness. The pericarpic tissues in their crosses, when only the male parent was different, were of the same maternal origin, whereas the embryos were of distinct hybrid origins; as a result, the endocarp (as the layer that is closest to the embryo), may be a crucial factor in determining an achene's ability to germinate [17].

High levels of abscisic acid (ABA) have also been found in the testa and pericarp of rose achenes, according to Bo et al. [37], which may prevent germination. The achenes' embryos have been shown to be fully formed and devoid of any morphological dormancy [37,38]. In addition, cold stratification has been used to get around physiological obstacles to embryo germination in a number of rose species [35,39,42].

Since only a few species of rose achenes have been studied, the process of dormancy in these plants is a complicated issue. Therefore, successful rose seed multiplication would benefit from better knowledge of dormancy in rose achenes. Only when the dormancy is broken can rose seeds have a better germination rate. Current efforts to break the dormancy have centered on two strategies: (a) removing the mechanical barrier known as the pericarp, which limits the embryo's growth and access to water and air; and (b) shortening the time the embryo must spend after ripening [44].

Achenes have not responded well to treatments that include soaking them in concentrated H_2SO_4 (sulphuric acid), exposing them to 100% oxygen-rich environments, dry storage, or cold stratification alone [44,45], or to various chemicals (such as GA_3). However, combining various therapies, such as H_2SO_4 scarification and cold stratification [39–41] or warm and cold stratification [39,42,45], can significantly enhance germination. To stimulate germination, several pretreatments have varying degrees of effectiveness depending on the species [41,45].

To successfully propagate horticultural plants, it is crucial to understand the specific kind of seed dormancy; however, at the moment, the majority of papers on seed dormancy do not specify the type of dormancy that was examined [46]. Researchers may not have looked at the many types of seed dormancy because there is no generally recognized system for defining dormancy. Morphological, physiological, morphophysiological, physical, and combinatorial dormancy are the five categories of dormancy that Baskin and Baskin [47] suggest as a new classification system for seed dormancy.

The definition of these various classes of dormancy is based on a number of characteristics, such as the embryo's morphology (underdeveloped or fully developed), the permeability of the seed coat to water (impermeable or permeable), and the physiological reactions of whole seeds to temperature or to a sequence of temperatures. With the use of such a classification system for seed dormancy, it is now possible to identify the type of dormancy by examining how different pretreatments affect germination [48].

3. Treatments Used to Stimulate Seed Germination in Species of the Genus *Rosa*

3.1. Inhibitory and Stimulatory Factors of Seed Germination in the Genus Rosa

In the generative propagation of *Rosa*, there are several factors that can impact germination negatively or positively. The low germination rate of *R. canina* seeds can be attributed to several factors, including:

- Seed dormancy: *R. canina* seeds often have seed dormancy mechanisms that prevent immediate germination. Seeds' dormancy can be caused by the hard seed coats or the presence of inhibitory substances that need to be broken down or leached out before germination can occur [21,23].
- Scarification requirements: Some *R. canina* seeds may require scarification, which is the mechanical treatment applied to the seed coat, to break dormancy. Scarification methods include nicking or sanding the seed coat or subjecting it to hot water or acid treatment within chemical treatments [21,49].
- Light requirements: Some *R. canina* seeds have light sensitivity and require exposure to light so that germination occurs. Seeds that require light will have low germination rates in darkness, so light can act as a germination stimulant for *Rosa* [49].
- Environmental factors: *R. canina* seeds may have specific environmental requirements for germination, such as moisture, temperature, and oxygen availability. If these conditions are not optimal, germination rates can be reduced [50].

Among the factors that can improve the germination of *R. canina* seeds, the most important are the following:

- Stratification: Many *R. canina* seeds require a period of cold stratification to break dormancy. Cold stratification involves subjecting the seeds to a period of cold, typically around 4 °C, for a certain duration. This mimics the natural conditions that the seeds would experience during the winter and helps overcome dormancy [51].
- Scarification: Some *R. canina* seeds have hard seed coats that can inhibit germination. Scarification techniques, such as mechanical scarification (e.g., nicking or scratching the seed coat) or chemical scarification (e.g., acid treatment), can help break the seed coat and enhance germination [52].
- Light exposure: While some *R. canina* seeds require light for germination, others may germinate better in darkness. Understanding the light requirements of specific seed lots can help optimize germination conditions accordingly [53].
- Moisture and temperature: Providing the seeds with adequate moisture and maintaining a suitable temperature range can promote germination. *R. canina* seeds generally require a moist environment, but excessive moisture can lead to fungal or bacterial issues. Optimal temperatures for germination typically range from 15 °C to 25 °C [54].
- Seed quality and age: Using high-quality, fresh seeds can improve germination rates. Older seeds may have reduced viability and lower germination rates [50].

Next, the main physical-mechanical, chemical, and biological treatments that can be applied to *Rosa* seeds to increase the germination rate will be presented, and a summary of the main methods, procedures, and techniques used to stimulate *Rosa* seed germination is shown in Table 2.

Table 2. Possibilities of increasing the germination percentage in the *Rosa* genus.

Method, Procedures and Technics	Details	Germination	Reference
Scarification	Grinding (1, 5, 10 min)	30%	[55]
Scarification	H_2SO_4 97% (1, 5, 10 min)	30%	[55]
Scarification	H_2SO_4 50% (30 and 60 s)	>30%	[52]
Scarification	H_2SO_4 (2, 4, 6 h)	No germination	[48]
Scarification	NaOCl	65.9%	[28]
Scarification	Fully removed testa	39%	[48]
Microorganism	Inoculation in *Klebsiella oxytoca* C1036	50%	[55]
Microorganism	Inoculation natural microflora	3%	[56]
Microorganism	Inoculation natural microflora + GarottaTM	95%	[56]
Stratification	8 weeks at 2.8 °C on moss	37.1%	[54]
Stratification	Dry storage 68 w + cold stratification 16–24 w	72–79%	[48]
Stratification	11 w warm stratification + cold stratification	13–18%	[51]
In vitro	GA_3 + manual scarification + agar medium + warm/chilling—cold/dark 21 days	80–85%	[57]
GA_3	2000 ppm GA_3 for 12 h (greenhouse)	74%	[30]
GA_3	300 ppm GA_3 for 24 h (Green house)	24.7%	[30]
GA_3	Chilled seeds + 200 ppm GA_3 for 6 h	11.7%	[30]
GA_3	Unchilled seeds + 400 GA_3 for 12 h	52%	[30]
GA_3	Stratification + pre-sowing GA_3	77.6	[28]

3.2. *Physical/Mechanical Treatments*

Various studies have been undertaken over the years on the species, many of them regarding the germination process. The scientific literature presents a low germination rate, often not more than 30%, which makes researchers all over the world have difficulties finding the proper methods to enhance seed germination within the *Rosa* genus [55].

- Harvesting time/period: For best seed quality, rose hips should be harvested when the fruits are mature, ripe, but firm [58]. Despite showing dormancy in the first year after fruition, the seeds typically germinate in the second year [59]. However, different harvesting periods have an impact on germination, with a previously documented 13–60% difference in germination frequency, with the best interval for harvesting rose hip being from late September until the beginning of October [60].
- Storage of seeds: In the study of Hoşafçı et al. [30], which aimed to investigate the effect of gibberellic acid (GA_3) treatments on the germination of dog rose seeds, it was discovered that even the seed storage type can influence the germination rate. The three pre-treatments used—(1) seeds kept at 4 °C; (2) seeds in fruit kept at 4 °C; and (3) seeds in fruit kept at room temperature (25 °C)—had considerable effects on the germination of dog rose seeds in the field experiment. The lowest (all-over mean) rate of germination (22.5%) was obtained for seeds kept in the refrigerator. The highest (all-over mean) rate of germination (39.1%) was obtained for seeds kept inside fruits at room temperature [30].
- Scarification: While most vegetable seeds germinate readily upon exposure to normally favorable environmental conditions, many seed plants that are vegetatively (asexually) propagated fail to germinate readily because of physical or physiologically imposed dormancy. Physical dormancy is due to structural limitations for germination, such as hard, impervious seed coats. Under natural conditions, weathering

for a number of years weakens the seed coat. Certain seeds have a tough husk that can be artificially worn or weakened to render the seed coat permeable to gases and water by a process known as scarification [61]. Scarification is a process that includes modifying, weakening, or opening a seed's covering in order to promote germination. Scarification frequently involves mechanical, thermal, and chemical processes. Many plant species have seeds that are frequently resistant to water and gases, which delay or prevent germination [48].

Scarification techniques, such as mechanical scarification (e.g., nicking or scratching the seed coat), as well as subjecting the seeds to hot water, can help break the seed coat and enhance germination [61]. According to Zhou et al. [48], scarified achenes have somewhat higher imbibition rates than intact achenes. However, the germination rate was only about 30% overall, which is a moderate improvement. The damage to the seeds during the grinding process may result in a slight reduction in germination. In a study by Lee et al. [55], several treatments were used in order to enhance germination in *Rosa* seeds. They used physical scarification. Grinding the seeds for 1, 5, and 10 min (+control) and stratifying them at 4 °C/12 weeks resulted in a germination percentage of not more than 30%; overall, there was no significant improvement. Like in the study of Zhou et al. [48], the fact that some of the seeds were damaged during the grinding contributed to the low germination rate. Pippino et al. [28] reported experiencing both chemical and microbiological seed scarification. Prior to stratification, a light scarification with sodium hypochlorite (NaOCl) encouraged more uniform germination and raised the percentage of germination from 49.2% to 65.9%.

UV-Irradiation: UV-B irradiation has some activity on the surface of the seed cells, stimulating rapid germination [62]. In a study by Lee et al. [63], UV-B radiation was applied to *R. rugosa* seedlings for 5, 10, and 30 min. Before 10 min of irradiation, there was no noticeable difference between the treated and untreated seedlings. Instead, the germination drastically decreased after 30 min of exposure. The inner cells of the seed appeared to be harmed by excessive UV-B exposure. In the germination study [62], such an effect was more or less expected. Finding the homeosis stage will be crucial to enhancing germination, but it will not be realistic due to the varied pericarp thickness of each seed caused by different cross-combinations. Living cells are susceptible to being harmed by UV-B. Future research should examine additional UV 185 sources [64]. When grown with the embryos facing the sun, half-cut achenes could germinate at a maximum rate of 100% in less than a week. Regardless of cold storage, half-cut achenes at 90 DAP germinated 100% of the time. To mimic the appearance of different light colors, various LED lights (red, blue, yellow, green, and white) were lit over the half-cut achenes. Ninety percent of *R. rugosa* seeds grew under blue LED lighting in a week of culture, and those seedlings afterwards had the best growth and development. Half-cut achenes were discovered to be a useful technique to enhance seed germination in *R. rugosa* without stratification or scarification. Using this method, rose cultivar breeding might be examined [63].

- Gamma ray irradiation: To remain competitive in the flower industry, breeders are constantly striving for fresh and innovative varieties. Low seed germination rates continue to be a significant issue in conventional rose breeding projects. To improve seed vigor and productivity and to improve the sprouting and emergence of buds that are carried out by seed coating, mutagenic agents, such as ionizing radiation, may be utilized. Ionizing radiation treatments must be tailored for each unique cultivar since the effects of radiation on seed vigor are typically genotype-dependent [64]. In a study by Giovannini et al. [64], hybrid tea rose seeds were exposed to gamma rays (0, 50, 100, and 200 Gy) in order to establish a radiation regimen for boosting seed germination. *Rosa* hybrid commercial cultivars' six different crossings' seeds were used to collect data on germination capacity and radio-tolerance. The final germination percentage and germination energy of the seeds, regardless of the cross, were not significantly affected by the range of gamma ray doses evaluated. These findings contrast with research done on seeds of various species, including *Vigna unguiculata* L. [65], *Citrus jambhiri* Lush. [66], and *Withania somnifera* L. [67], in which a gradual decline in seed germination and seedling vigor from lower doses to higher doses in given treatments of gamma rays was found. Since there are other findings that show that ionizing radiation has positive impacts on seed vigor, this is a contentious topic.

- Medium: Anderson and Byrne [68] tested the influence of stratification media and genotype over the germination process. They stratified fresh rose seeds from different hybrids for 10–12 weeks at 2.8 °C in milled sphagnum moss, sand, perlite, vermiculite, and moist filter paper. The achenes germinated the best on moss. When placed on moist milled sphagnum moss or agar and stratified for 8 weeks at 2.8 °C, once again, the best germination was on moss. The germination of the genotypes varied greatly, ranging from 0.7 to 37.1% [64]. According to Carter [69], rose seed germinated more effectively after stratification in damp peat moss or sand as opposed to moist vermiculite. The effectiveness of sphagnum moss as a stratification medium was demonstrated in their experiment. The outcomes provided more proof that the medium is essential for seed germination. The smallest rate of germination was encouraged by stratification on filter paper, while the best germination was obtained with sphagnum moss. In terms of stimulating germination, perlite fell somewhere between sphagnum moss and the other two media (sand and vermiculite). Sphagnum moss performed better or on par with the other stratification media across all genotypes examined. In contrast to Yambe et al.'s [70] findings for *R. multiflora*, leaching of WOB-28 rose seed for three or more days significantly reduced germination. Such a divergence may have been caused by changes in leaching technique, genotypic effects, or leaching water temperature.
- Stratification: Aches treated with warm and cold stratification were used to examine the effects of temperature and water stress. Freshly harvested achenes are devoid of any physical, morphological, or morphophysiological dormancy since the pericarps are permeable and the embryo is completely formed. Despite softening or even completely removing the pericarp, the germination percentage remained low (5%), while completely removing the testa greatly increased germination (39%), indicating the potential existence of germination inhibitors in the testa [52]. According to Zhou et al. [52], dry storage for 68 weeks followed by cold stratification for 16 or 24 weeks resulted in maximum germination (72–79%) in *R. multibracteata*. The most popular method for removing rose seed dormancy is chilling, as most species' achenes will eventually germinate if cooled for an extended length of time. For certain species, cold stratification periods equivalent to one field winter are sufficient. Only if the temperature of warm incubation was too high did the interruption of chilling with warm incubation result in subsequent dormancy induction. The dormancy of the seeds did not change when they were kept below this "compensating" temperature, and they may accrue the effects of chilling despite warm breaks [71].

The results of a study by Alp et al. [51] evaluated the effects of various warm stratification durations on the seed germination of some *Rosa* species, including *R. heckellana* ssp. *vanheurckiana*, *R. pulverelanta*, *R. dumalis*, and *R. canina* naturally grown. The seeds were stored for 10, 11, and 12 weeks at 25 °C before cold storage at 5 °C until the start of germination testing. In terms of germination, *R. heckellana* ssp. *vanheurckiana* seeds reacted to treatments differently than seeds of other species. When the seeds were stored in warm stratification followed by cold stratification, germination occurred promptly (between 1 and 3 weeks). The other species' seeds need 5 months of cold stratification after warm stratification in order to emerge from hibernation. The overall germination percentages were 13.80% in *R. pulverelanta*, 18.80% in *R. canina*, and 13.53% in *R. dumalis* at 25 °C of warm stratification and 5 °C of cold stratification. For these three taxa, an 11-week warm stratification period followed by a cold stratification period was determined to be the most successful stratification method [50].

- Temperature: Temperature is an important environmental component that inhibits seed germination. A hallmark of rose (*R. canina* L.) seedlings is the physical and physiological dormancy, which is often disrupted during warm weather, followed by cold stratification. When prepared seeds were subjected to a temperature of 20 °C, secondary dormancy was produced. A study by Pawowski [72] sought to discover and functionally describe the proteins linked to rose seed dormancy management. Using 2-D electrophoresis, proteins from primary dormant, following warm and cold stratification (nondormant), and secondary dormant seeds were examined. Mass spectrometry was used to determine which proteins were abundant in different ways.

The findings indicated that secondary dormancy was associated with more common spots than warm stratification, and that cold stratifications had the greatest impact on spot variability. The rise in actin and mitochondrial proteins at the end of dormancy suggests modifications to cellular processes and seed germination preparation. Low quantities of legumin, metabolic enzymes, and actin were found in secondary dormant seeds, indicating the use of storage resources, a decline in metabolic activity, and cell elongation [72]. When rose seeds were brought out of dormancy, more cellular and metabolic proteins that encourage germination were present in greater quantities. These proteins were reduced, and germination was arrested as a result of the induction of secondary dormancy. Both high temperature and water stress lowered germination in achenes treated with warm plus cold stratification [52].

The physical and physiological dormancy of rose (*R. canina* L.) seed is broken by warm, then cold stratification. Secondary dormancy is induced in pretreated seeds when they are exposed to a temperature of 20 °C. When rose seeds were brought out of dormancy, more cellular and metabolic proteins that encourage germination were present in greater quantities. These proteins are reduced, and germination is arrested as a result of the induction of secondary dormancy [72].

- Tetrazolium staining is typically used to assess the quality of rose seeds due to the vast variety of dormancy-breaking needs within each species [73]. The first step is to soak the achenes in water for 24 h. The pericarp is broken open by applying firm pressure with a knife. After the testa has been scratched or snipped at the cotyledon end, the seed is submerged for 6 h at room temperature in 1% tetrazolium chloride. The testa is slit along the side, and the embryo, which fills the seed cavity, is squeezed or teased out for evaluation [74]. The excised embryo method may also be used, although it has little advantage over tetrazolium staining [73]. For purposes of determining fill and chalcid infestation levels, x-radiography is suitable [74]. X-radiography is helpful for assessing the amount of fill and chalcid infestation [74].

3.3. Chemical Treatments

- Sulphuric acid: To boost ingestions and break physical dormancy, sulphuric treatment of the rose pericarp is advised rather than scarification. However, due to the uneven thickness of the rose pericarp, the application of sulphuric acid necessitates substantially more stabilization. This kind of scarification technique has been employed in certain research on roses, although it often only applies to a small number of cross-combinations or particular species [52,63,74]. Zhou et al. [48] treated seeds with sulphuric acid for 2, 4, and 6 h and observed no germination. The overall rate of germination was 30%. The treated and untreated samples did not differ significantly from one another. The longer the sulphuric acid treatment, the smaller the rate obtained. It appears that even during this brief course of treatment, sulfuric acid could harm the embryo [63].

Soaking the seeds in sulphuric acid (97%) for 1, 5, and 10 min (+stratification 12 weeks/4 °C) resulted in no more than a 30% germination rate [55]. Apparently, sulphuric acid can damage the embryo in a very short period of time. According to Younis et al. [75], applying H_2SO_4 50% for 30 and 60 s gives better germination. In a study by Zhou et al. [52] no germination was observed after the seeds were treated with H_2SO_4 for 2, 4, and 6 h. According to Roberts and Shardlow [76], a sulphuric acid treatment before warm + cold stratification can enhance germination. Cullum et al. [77] suggest that acid scarification can be eliminated all along, and the warm stratification time should shorten if the achenes warm stratification is made with compost activator. When paired with cold stratification, dry storage, scarification with H_2SO_4, and warm stratification increased germination and reduced the time needed for cold stratification to break dormancy [52]. Sulfuric acid treatment prior to warm and cold stratification promoted germination in nursery propagation of the rootstock rose *R. dumetorum* (*R. corymbifera*) 'Laxa' [76]. If the achenes are heated-stratified with compost activator, the acid scarification can be avoided, and the warm stratification duration decreased [77]. It appears that the purpose of these treatments, whether through acid or microbial digestion, is to weaken the pericarp at the sutures. Growth hormones like gibberellic acid or benzyladenine can be vacuum-injected into the achenes, as for *R. dumetorum* 'Laxa', to increase their response to warm and cold stratification, which suggests that something other than simple mechanical restriction may be at play [77]. Similar to this, it is possible to encourage the achenes of the relatively non-dormant multiflora rose to germinate without chilling by treating them with enzymes that weaken the pericarp sutures or by leaching the incubation solution with activated charcoal to remove any inhibitors [71].

Gibberelic acid: Neither gibberellic acid (GA_3) nor 'smoke water' (water through which smoke had been bubbled for 2 h) had any positive effect on germination, even on seeds that had been mechanically stratified or scarified [52]. According to Hoşafçı [30], there is a positive correlation between the concentration of GA_3 applied and the germination rate. He obtained 74% germination in greenhouse from seeds treated with 2000 ppm GA_3 for 12 h. A lower germination rate (24.7%) was observed when seeds were treated with 300 ppm GA_3 for 24 h. The same authors' field study found that germination varied from 11.7% for chilled seeds treated with 200 ppm GA_3 for 6 h to 52.0% for unchilled seeds treated with 400 ppm GA_3 for 12 h and stored inside fruits at room temperature [30]. Pre-sowing treatments with GA_3 were carried out following stratification. When compared to immersion in water (respectively 64.8%, 32.8 days), stratified seeds immersed in 1 g L^{-1} GA_3 had a significantly higher percentage of seed germination (77.6%), mean germination time (26.0 days), and uniformity of germination (2.5%) [28]. According to Meyer [71], in a non-dormant rose, the achenes might be stimulated into germination without chilling, either with macerating enzymes or by leaching with activated charcoal. Using macerating enzymes to break dormancy, Yambe et al. [70] demonstrated phytochrome-mediated light requirements. Acid scarification is reported to substitute for warm pretreatment in the cultivated rose, *R. gallica* L. [78]. Other closely related species, including *R. arvensis*, *R. floribunda*, *R. foetida*, *R. jundzillii*, *R. micrantha*, *R. obtusifolia*, and *R. rubiginosa*, have also been noted to enter a dormant state [60]. For *R. canina* seeds maintained in the refrigerator for 180 days, various treatments, including stratification and scarification with GA_3 and H_2SO_4, have been reported to remove dormancy and resulted in a 6.2% germination rate within the first year [60]. By employing solely GA_3 treatment, no germination was obtained in any of the examined species [60].

- Nitric acid: Younis et al. [75] tried to break achenes dormancy by using nitric acid. Nitric acid is a nitrogen oxoacid that has a role as a protic solvent and reagent. As is the case with sulphuric acid, it was used in order to break the seed's pericarp. Nitric acid (65%) with different time exposures (30, 60, and 90 s) was tested. Although it was expected, no significant effect of the acid was observed [78].

3.4. Biological Treatments

- Microorganism Treatment: Effective microorganisms (EM) are used to facilitate seed germination [79]. Microorganism: *Klebsiella oxytoca* C1036 is a strain that has been identified to act against the soft-rot pathogen in Tabacco [80]. Lee et al. [55] used this strain to enhance germination. The seeds were immersed in a suspension of the strain for 1 and 48 h. The germination rate rose to 50% with longer immersion. Treated seeds germinated twice as well as those of non-treated seeds. Other useful microorganisms should be investigated in further studies [55].

Rosa corymbifera 'Laxa' germination under typical conditions was no more than 2%. This was accomplished when the natural microflora on the seeds was present. The hips are where the microflora first appeared, and when the seeds are extracted, they are vaccinated. No germination occurred when microorganisms were not included in these pretreatments. Three percent of surface-sterilized seedlings that were then inoculated with natural microflora grew. GarottaTM, a commercial compost activator, was added to the industrial pretreatment to boost germination to 95% [56].

Remedier®, a combination of *Trichoderma harzianum* and *Trichoderma viridae*, and EmercalTM, bacteria and co-metabolites produced by bacterial fermentation, when added to the stratification sand, increased the proportion of germinated seeds but had no role in germination uniformity [28].

In a study by Kazaz [80], an investigation was undertaken to determine how microbial inoculation affected *Rosa damascena* Mill germination and the breaking of seed dormancy. The seeds underwent a 150-day cold stratification period at 4 ± 1 °C after 4 weeks of warm stratification at 25 °C. The seeds were injected with four different microbial fertilizers, including EM•1®, B:speelTM, BioplinTM, and PhosfertTM, prior to stratification. In the study, the treatments for microbial inoculation considerably ($p < 0.01$) increased the percentage of early germination during cold stratification. The EM1® yielded the highest rate of premature germination during stratification (69.3%). In terms of cumulative germination percentage, EM•1® (100.0%) had the highest germination rate, followed by PhosfertTM (84.0%) and B: seepelTM (84.0%), while the control treatment (69.3%) had the lowest germination rate. In comparison to the control, the EM•1® decreased the mean germination time by 1.7 days. In conclusion, it was found that dormancy was broken and germination was greatly enhanced with microbial inoculation (especially EM•1®) of oil rose seeds and a stratification time of 150 days [81].

- Macerating enzymes: *R. multiflora* Thunb. achenes were treated for 36 h with 1% Driselase, a macerating enzyme, which significantly enhanced the germination percentage. When the achenes were exposed to the enzyme for a longer time, the seeds germinated more quickly. In comparison to Driselase, treatment with Cellulase Onozuka increased seed germination at a lower dosage. Pure pectinase and cellulase preparations had outcomes that were comparable to those of the mentioned enzyme treatments. Pectinase treatment was more effective than cellulase treatment. These enzymes probably made the pericarp's suture less rigid, which caused the pericarp to split [82].
- Genetic make-up of the seed: One important factor contributing to poor germination is the genetic make-up of the seed. This is the reason many breeders maintain thorough records of seed set and germination in order to select the genitors that produce the greatest number of offspring. Techniques for treating seeds, such as harvesting, stratification, scarification, and leaching, are essential to maximizing germination [57].
- Break seed dormancy under in vitro conditions: In the study by Hajyzadeh [57], it was aimed to break the seed dormancy of rose hip under in vitro conditions by applying multiple strategies in an efficient manner. The seeds were given various doses of GA_3, mechanically scarified, stratified on agar-solidified MS medium containing GA_3 alone or in combinations with the suggested treatments, and then given a controlled physiological treatment that involved alternately giving the seeds warm/chilling and cold/dark treatments for 21 days, followed by 18 days of warm/light treatments.

It was found that the rosehip seeds might germinate in various ways depending on whether the scarified seeds were spread over an agar-solidified MS medium with or without GA_3. The best seed germination (80.00–85.00%) was observed when the three treatments were combined and the seeds were given controlled and alternately warm and cold treatments for 21 days, leaving them for 18 days in warm/light conditions. These crucial insights could be used in breeding and propagation operations to develop novel rosehip rootstock and fruit varieties [57].

4. Conclusions

Seed germination in the *Rosa* genus is a complex process influenced by various factors. Our understanding of these factors is crucial for successful cultivation, conservation, and propagation of *Rosa* species. Optimal conditions for germination vary among species, highlighting the importance of understanding the specific requirements for each *Rosa* species of interest. Advances in research techniques, such as in vitro germination and molecular approaches, have contributed to a deeper understanding of *Rosa* seed germination biology. Further research is needed to explore the germination ecology of specific *Rosa* species, uncover additional dormancy mechanisms, and develop efficient germination protocols for propagation purposes. The knowledge gained from studying *Rosa* seed germination can aid in the conservation of endangered species, the selection of suitable cultivars for horticulture, and the development of sustainable rose production systems.

Author Contributions: Conceptualization, R.E.S. and R.L.S.-D.; methodology, R.L.S.-D.; software, R.L.S.-D.; validation, R.E.S., A.F.S. and C.D.; formal analysis, A.M.T.; investigation, I.M.M.; resources, G.R.; data curation, R.L.S.-D.; writing—original draft preparation, R.L.S.-D.; writing—review and editing, R.E.S. and C.D.; visualization, R.E.S. and A.F.S.; supervision, R.E.S.; funding acquisition, R.L.S.-D. All authors have read and agreed to the published version of the manuscript.

Funding: This research was partially funded by the University of Agricultural Sciences and Veterinary Medicine of Cluj-Napoca (USAMVCN).

Data Availability Statement: Not applicable.

Acknowledgments: The research was carried out with the partial support of the Doctoral School of the University of Agricultural Sciences and Veterinary Medicine of Cluj-Napoca (USAMVCN) for R.L.S.-D.—student.

Conflicts of Interest: The authors declare no conflict of interest.

References

1. Rehder, A. *Bibliography of Cultivated Trees and Shrubs: Hardy in the Cooler Temperate Regions of the Northern Hemisphere*; Arnold Arboretum of Harvard University: Jamaica Plain, MA, USA, 1949. [CrossRef]
2. Wissemann, V. Conventional taxonomy of wild roses. In *Encyclopedia of Rosa Science*, 1st ed.; Roberts, A., Debener, T., Gudin, S., Eds.; Elsevier: Amsterdam, The Netherlands, 2003; pp. 111–117.
3. Wissemann, V. Hybridization and the evolution of the nrITS spacer region. Plant genome, biodiversity and evolution. *Part A Phanerogams* **2003**, *1*, 57–71.
4. Guimaraes, R.; Barros, L.; Carvalho, A.M.; Ferreira, I.C. Studies on chemical constituents and bioactivity of Rosa micrantha: An alternative antioxidants source for food, pharmaceutical, or cosmetic applications. *J. Agric. Food Chem.* **2010**, *58*, 6277–6284. [CrossRef] [PubMed]
5. Chrubasik, C.; Roufogalis, B.D.; Müller-Ladner, U.; Chrubasik, S.A. Systematic review on the *Rosa canina* effect and efficacy profiles. *Phytother. Res.* **2008**, *22*, 725–733. [CrossRef] [PubMed]
6. Strålsjö, L.; Alklint, C.; Olsson, M.E.; Sjöholm, I. Total folate content and retention in rosehips (*Rosa* ssp.) after drying. *J. Agric. Food Chem.* **2003**, *51*, 4291–4295. [CrossRef] [PubMed]
7. Larsen, E.; Kharazmi, A.; Christensen, L.P.; Christensen, S.B. An antiinflammatory galactolipid from rose hip (*Rosa canina*) that inhibits chemotaxis of human peripheral blood neutrophils in vitro. *J. Nat. Prod.* **2003**, *66*, 994–995. [CrossRef] [PubMed]
8. Rahnavard, A.; Asadian, G.; Pourshamsian, K.; Taghavi, M. Assessing genetic variation of dog rose (*Rosa canina* L.) in Caspian climate. *Biosci. Biotechnol. Res. Asia* **2014**, *10*, 119–125. [CrossRef]
9. Grant, V. Plant speciation. *New Phytol.* **1971**, *161*, 8–11. [CrossRef]
10. Wissemann, V. Molekulargenetische und morphologisch-anatomische Untersuchungen zur Evolution und Genomzusammensetzung von Wildrosen der Sektion Caninae (DC.) Ser. E. *Schweiz. Verlagsbuchhandlung* **2000**, *122*, 357–429.
11. Özkan, G.; Sagdiç, O.; Baydar, N.G.; Baydar, H. Antioxidant and antibacterial activities of *Rosa damascena* flower extracts. *Food Sci. Technol.* **2004**, *10*, 277–281. [CrossRef]
12. Szentmihályi, K.; Vinkler, P.; Lakatos, B.; Illés, V.; Then, M. Rose hip (*Rosa canina* L.) oil obtained from waste hip seeds by different extraction methods. *Bioresour. Technol.* **2002**, *82*, 195–201. [CrossRef]

13. Fan, C.; Pacier, C.; Martirosyan, D.M. Rose hip (*Rosa canina* L): A functional food perspective. *Funct. Foods Health Dis.* **2014**, *4*, 493–509. [CrossRef]
14. Rosu, C.M.; Manzu, C.; Olteanu, Z.; Oprica, L.; Oprea, A.; Ciornea, E.; Zamfirache, M.M. Several fruit characteristics of *Rosa* sp. genotypes from the Northeastern region of Romania. *Not. Bot. Horti Agrobot. Cluj-Napoca* **2011**, *39*, 203–208. [CrossRef]
15. Ercisli, S. Chemical composition of fruits in some rose (*Rosa* spp.) species. *Food Chem.* **2007**, *104*, 1379–1384. [CrossRef]
16. Leus, L.; Van Laere, K.; De Riek, J.; Van Huylenbroeck, J. Rose. In *Ornamental Crops*; Van Huylenbroeck, J., Ed.; Springer: Berlin/Heidelberg, Germany, 2018; pp. 736–738. [CrossRef]
17. Gudin, S.; Arene, L.; Chavagnat, A.; Bulart, C. Influence of endocarp thickness on rose achene germination: Genetic and environmental factors. *HortScience* **1990**, *25*, 786–788. [CrossRef]
18. Basu, S.; Zandi, P.; Cetzal-Ix, W.; Sengupta, R. The genus *Rosa*: An aristocrat from the plant family with class, color and fragrance. *Iran. Soc. Environ. Newsl.* **2014**, 1–9.
19. Wagner, S.; Ardelean, M.; Sestras, R.; Ghidra, V.; Pamfil, D.; Roman, G.; Florincescu, A.; Coredea, M.; Catana, C.; Budiu, V. Achievements in rose breeding at Cluj-Napoca, Romania, in the last thirty years. *Biotechnol. Biotechnol. Equip.* **2000**, *14*, 37–41. [CrossRef]
20. Izadi, Z.; Zarei, H.; Alizadeh, M. Effect of time, cultivar and rootstock on success of rose propagation through stenting technique. *Am. J. Plant Sci.* **2014**, *5*, 1644–1650. [CrossRef]
21. Wagner, Ş. *Trandafirul-de la Mit la Mileniul Trei: O Carte Pentru Toţi Iubitorii Rozelor*; Wagner, S., Ed.; Cluj-Napoca, Romania, 2002.
22. Ambros, E.V.; Vasilyeva, O.Y.; Novikova, T.I. Effects of in vitro propagation on ontogeny of *Rosa canina* L. micropropagated plants as a promising rootstock for ornamental roses. *Plant Cell Biotechnol. Mol. Biol.* **2016**, *17*, 72–78.
23. Vasilyeva, O.Y. Reproduction systems of representatives of the genus *Rosa* L. under conditions of continental climate. *Contemp. Probl. Ecol.* **2009**, *2*, 361–368. [CrossRef]
24. VonAbrams, G.J.; Hand, M.E. Seed dormancy in *Rosa* as a function of climate. *Am. J. Bot.* **1956**, *43*, 7–12. [CrossRef]
25. Bewley, J.D. Seed germination and dormancy. *Plant Cell* **1997**, *9*, 1055–1066. [CrossRef]
26. Finch-Savage, W.E.; Leubner-Metzger, G. Seed dormancy and the control of germination. *New Phytol.* **2006**, *1716*, 501–523. [CrossRef] [PubMed]
27. Gudin, S. Seed propagation. In *Ornamental Crops*; Reference Module in Life Sciences; Elsevier: Amsterdam, The Netherlands, 2017; pp. 719–767.
28. Pipino, L.; Scariot, V.; Gaggero, L.; Mansuino, A.; Van Labeke, M.C.; Giovannini, A. Enhancing seed germination in hybrid tea roses. *Propag. Ornam. Plants* **2011**, *11*, 111–118.
29. Khosh-Khui, M.; Teixeira da Silva, J.A. In vitro culture of the *Rosa* species. In *Floriculture, Ornamental and Plant Biotechnology; Advances and Topical Issues*, 1st ed.; Jaime, A., da Silva, T., Eds.; Global Science Books: East Sussex, UK, 2016; Volume 2, pp. 513–526.
30. Hoşafçı, H.; Arslan, N.; Sarıhan, E.O. Propagation of dog roses (*Rosa canina* L.) by seed. *Acta Hortic.* **2004**, *690*, 159–164. [CrossRef]
31. Werlemark, G.; Carlson-Nilsson, U.; Uggla, M.; Nybom, H. Effects of temperature treatments on seedling emergence in dogroses, *Rosa* sect. *Caninae* (L). *Acta Agric. Scand.* **1995**, *45*, 278–282. [CrossRef]
32. Hornero-Mendez, D.; Mınguez-Mosquera, M.I. Carotenoid pigments in *Rosa mosqueta* hips, an alternative carotenoid source for foods. *J. Agric. Chem.* **2000**, *48*, 825–828. [CrossRef]
33. Uggla, M. Domestication of Wild Roses for Fruit Production. Ph.D. Thesis, Swedish University of Agricultural Sciences, Alnarp, UK, 2004.
34. Pati, P.K.; Rath, S.P.; Sharma, M.; Sood, A.; Ahuja, P.S. In vitro propagation of rose—A review. *Biotechnol. Adv.* **2006**, *24*, 94–114. [CrossRef] [PubMed]
35. Tincker, M.A.; Wisley, M.A. Rose seeds: Their after-ripening and germination. *J. R. Hortic. Soc.* **1935**, *60*, 399–417.
36. Xu, B.M.; Zhang, Z.M.; Zhang, H.J. Germination and dormancy of *Rosa* seed. *Seed* **1993**, *63*, 5–9.
37. Bo, J.; Huiru, D.; Xiaohan, Y. Shortening hybridization breeding cycle of rose—A study on mechanisms controlling achene dormancy. *Acta Hortic.* **1995**, *404*, 40–47. [CrossRef]
38. Jackson, G.A.D.; Blundell, J.B. Germination in *Rosa*. *J. Hortic. Sci.* **1963**, *38*, 310–320. [CrossRef]
39. Densmore, R.; Zasada, J.C. Germination requirements of Alaskan *Rosa acicularis*. *Can. Field-Nat.* **1977**, *91*, 58–62.
40. Svejda, F. Water uptake of rose achenes. *Can. J. Plant Sci.* **1972**, *52*, 1043–1047. [CrossRef]
41. Bhanuprakash, K.; Tejaswini, Y.; Yogeesha, H.S.; Naik, L.B. Effect of scarification and gibberellic acid on breaking dormancy of rose seeds. *Seed Res.* **2004**, *32*, 105–107.
42. Svejda, F.J.; Poapst, P.A. Effects of different after-ripening treatments on germination and endogenous growth inhibitors in *Rosa rugosa*. *Can. J. Plant Sci* **1972**, *52*, 1049–1058. [CrossRef]
43. Gudin, S. Breeding/Overview. In *Encyclopedia of Rose Science*; Roberts, A., Debener, T., Gudin, S., Eds.; Elsevier: Amsterdam, The Netherlands; Academic Press: Oxford, UK, 2003.
44. Stewart, R.N.; Semeniuk, P. The effect of the interaction of temperature with after-ripening requirement and compensating temperature on germination of seed of five species of *Rosa*. *Am. J. Bot.* **1965**, *52*, 755–760. [CrossRef]
45. Semeniuk, P.; Stewart, R.N. Effect of temperature and duration of after-ripening period on germination of *Rosa nutkana* seeds. *Proc. Am. Soc. Hortic. Sci.* **1966**, *89*, 689.
46. Baskin, C.C.; Baskin, J.M. *Seeds: Ecology, Biogeography, and, Evolution of Dormancy and Germination*; Elsevier: Amsterdam, The Netherlands, 1998. [CrossRef]

47. Baskin, J.M.; Baskin, C.C. A classification system for seed dormancy. *Seed Sci. Res.* **2004**, *14*, 1–6. [CrossRef]
48. Zhou, Z.Q.; Bao, W.K.; Wu, N. Dormancy and germination in *Rosa multibracteata* Hemsl. EH Wilson. *Sci. Hortic.* **2009**, *119*, 434–441. [CrossRef]
49. Bonny, L. *Seed Germination Test Methods Used for Australian Tree Species at Coffs Harbour Research Centre*; Technical Paper; New South Wales Forestry Commission: Coffs Harbour, NSW, Australia, 1987.
50. McDonald, M.B. Seed quality assessment. *Seed Sci. Res.* **1998**, *8*, 265–276. [CrossRef]
51. Alp, S.; Çelik, F.; Türkoglu, N.; Karagöz, S. The effects of different warm stratification periods on the seed germination of some *Rosa* taxa. *AJB* **2009**, *8*, 5838–5841. [CrossRef]
52. Zhou, Z.; Bao, W.; Wu, N. Effects of scarification, stratification and chemical treatments on the germination of *Rosa soulieana* Crépin achenes. In *Floriculture and Ornamental Biotechnology*; da Silva, T., Ed.; Global Science Books: London, UK, 2009; pp. 75–80.
53. Woolley, J.T.; Stoller, E.W. Light penetration and light-induced seed germination in soil. *Plant Physiol.* **1978**, *61*, 597–600. [CrossRef] [PubMed]
54. Roberts, E.H. Temperature and seed germination. In *Symposia of the Society for Experimental Biology*; Company of Biologists Ltd.: Cambridge, MA, USA, 1988; Volume 42, pp. 109–132.
55. Lee, J.Y.; Kim, Y.C.; Han, T.H.; Kim, S.T.; Gi, G.Y. Study on increasing rose seed germination. *Acta Hortic.* **2010**, *855*, 183. [CrossRef]
56. Morpeth, D.R.; Hall, A.M. Microbial enhancement of seed germination in *Rosa corymbifera* 'Laxa'. *Seed Sci. Res.* **2000**, *10*, 489–494. [CrossRef]
57. Hajyzadeh, M. Application of multiple strategies to efficiently break seed dormancy of permanently odd pentaploid rose hip (*Rosa canina* L.) under in vitro conditions. *Int. J. Agric. Environ. Food Sci.* **2022**, *6*, 471–479. [CrossRef]
58. Tipi, E. Kusburnu fidan üretim teknikleri ve üretim hedefleri. In Proceedings of the Kusburnu Sempozyumu, Gümüshane, Bildiri Kitabı, Gümüşhane, Turkey, 5–6 September 1996; pp. 127–133.
59. Yılmaz, S. Kusburnu bitkisinin erozyon kontrolundeki yeri ve onemi. In Proceedings of the Kusburnu Sempozyumu, Gümüşhane, Bildiri Kitabı, Gümüşhane, Turkey, 5–6 September 1996; pp. 167–168.
60. Rowley, G.D. Germination in *Rosa canina*. *Am. Rose Annu.* **1956**, *41*, 70–73.
61. Ercişli, S. Farklı uygulamaların kusburnu tohumlarının cimlenmesi üzerine etkisi. II. In Proceedings of the Ulusal Fidancılık Sempozyumu, Bademli, Odemis, Turkey, 25–29 September 2000.
62. Noble, R.E. Effects of UV-irradiation on seed germination. *Sci. Total Environ.* **2022**, *299*, 173–176. [CrossRef] [PubMed]
63. Lee, J.; Lee, J.; Ki, G.; Kim, S.; Han, T. Improvement of seed germination in *Rosa rugosa*. *Korean J. Hortic. Sci. Technol.* **2011**, *29*, 352–357.
64. Giovannini, A.; Scariot, V.; Caser, M.; Buttafava, A.; Mansuino, A.; Ghione, G.G.; Balestrazzi, A. Mutation breeding using gamma rays to increase seed germination in *Rosa hybrida*. *Acta Hortic.* **2015**, *1087*, 373–378. [CrossRef]
65. Bind, D.; Dwivedi, V.K. Effect of mutagenesis on germination, plant survival and pollen sterility in M1 generation of in cowpea (*Vigna unguiculata* (L.) Walp). *Ind. J. Agric. Res.* **2014**, *48*, 398–401. [CrossRef]
66. Sharma, L.K.; Kaushal, M.; Gill, M.I.S.; Bali, S.K. Germination and survival of *Citrus jambhiri* seeds and epicotyls after treating with different mutagens under in vitro conditions. *Middle-East J. Sci. Res.* **2013**, *16*, 250–255.
67. Bhosale, R.S.; More, A.D. Effect of gamma radiation on seed germination, seedling height and seedling injury in *Withania somnifera* (L.) Dunal. *Int. J. Life Sci.* **2014**, *2*, 226–228.
68. Anderson, N.; Byrne, D.H. Methods for *Rosa* germination. *Acta Hortic.* **2007**, *751*, 503–507. [CrossRef]
69. Carter, A.R. Rose rootstocks-performance and propagation from seed. *Comb. Proc. Int. Plant Propagators Soc.* **1968**, *19*, 172–180.
70. Yambe, Y.; Hori, Y.; Takeno, K. Levels of endogenous abscisic acid in rose achenes and leaching with activated charcoal to improve seed germination. *J. Jpn. Soc. Hortic. Sci* **1992**, *61*, 383–387. [CrossRef]
71. Meyer, S.E. Rosa *L. The Woody Plant Seed Manual*; USDA Forest Service Agriculture Handbook; USDA Forest Service: Washington, DC, USA, 2008; Volume 727, pp. 974–980.
72. Pawłowski, T.A.; Bujarska-Borkowska, B.; Suszka, J.; Tylkowski, T.; Chmielarz, P.; Klupczyńska, E.A.; Staszak, A.M. Temperature regulation of primary and secondary seed dormancy in *Rosa canina* L.: Findings from proteomic analysis. *Int. J. Mol. Sci.* **2020**, *21*, 7008. [CrossRef]
73. Gill, J.D.; Pogge, F.L. *Rosa* L., rose. In *Seeds of Woody Plants in the United States*; Schopmeyer CS, Tech. Coordinator; USDA Agriculture Handbook 450; USDA Forest Service: Washington, DC, USA, 1974; pp. 732–737.
74. Belcher, E. *Handbook on Seeds of Browse-Shrubs and Forbs*; Technical Publication R8-8; USDA Forest Service, Southern Region: Atlanta, GA, USA, 1985; p. 246.
75. Younis, A.; Riaz, A.; Ahmed, R.; Raza, A. Effect of hot water, sulphuric acid and nitric acid on the germination of rose seeds. *Acta Hortic.* **2007**, *755*, 105–108. [CrossRef]
76. Roberts, L.; Shardlow, A.D. Practical aspects of the acid treatment of rose seed [*Rosa dumetorum* 'Laxa', dormancy breaking]. *Plant Propagator* **1979**, *25*, 13–14.
77. Cullum, F.J.; Bradley, S.J.; Williams, M.E. Improved germination of *Rosa corymbifera* 'Laxa' seed using a compost activator. *Comb. Proc.—Int. Plant Propagators Soc.* **1990**, *40*, 244–250.
78. Svejda, F. Hybrid *Rugosa* Rose' martin Frobisher'. *Can. J. Plant Sci.* **1969**, *49*, 100. [CrossRef]
79. Taylor, A.G.; Harman, G.E. Concepts and technologies of selected seed treatments. *Annu. Rev. Phytopathol.* **1990**, *28*, 321–339. [CrossRef]

80. Park, M.R.; Kim, Y.C.; Lee, S.W.; Kim, I.S. Identification of an ISR-related metabolite produced by rhizobacterium *Klebsiella oxytoca* C1036 active against soft-rot disease pathogen in tobacco. *Pest Manag. Sci.* **2009**, *65*, 1114–1117. [CrossRef] [PubMed]
81. Kazaz, S.; Erbas, S.; Baydar, H. Breaking seed dormancy in oil rose (*Rosa damascena* Mill.) by microbial inoculation. *AJB* **2010**, *9*, 6503–6508.
82. Yambe, Y.; Takeno, K. Improvement of rose achene germination by treatment with macerating enzymes. *HortScience* **1992**, *27*, 1018–1020. [CrossRef]

MDPI AG
Grosspeteranlage 5
4052 Basel
Switzerland
Tel.: +41 61 683 77 34

Horticulturae Editorial Office
E-mail: horticulturae@mdpi.com
www.mdpi.com/journal/horticulturae

www.ingramcontent.com/pod-product-compliance
Lightning Source LLC
Chambersburg PA
CBHW061516210926
51850CB00008B/1937
* 9 7 8 3 7 2 5 8 1 5 2 5 8 *